COLLINS BOTANICAL BIBLE

William Collins
An imprint of HarperCollins*Publishers*
1 London Bridge Street
London SE1 9GF
WilliamCollinsBooks.com

First published by William Collins in 2018

Text © HarperCollins*Publishers* 2018 except where otherwise specified
Images © individual copyright holders, see page 415 for details
Cover illustrations © Lynn Hatzius 2018

Text by Sonya Patel Ellis except where otherwise specified
Original illustrations © Lynn Hatzius, see page 415 for details
Design by Eleanor Ridsdale
With design assistance from Gareth Butterworth
Picture research by Sonya Patel Ellis and Jo Carlill

10 9 8 7 6 5 4 3 2 1

The author asserts her moral right to be identified as the author of this work.
All rights reserved. No parts of this publication may be reproduced, stored in a retrieval system or transmitted, in any form or by any means, electronic, mechanical, photocopying, recording or otherwise, without the prior permission of the publishers

A catalogue record for this book is available from the British Library

ISBN 978-0-00-826227-3

All reasonable efforts have been made by the author and publishers
to trace the copyright owners of the material quoted in this book and of any images reproduced in this book. In the event that the author or publishers are notified of any mistakes or omissions by copyright owners after publication, the author and publishers will endeavour to rectify the position accordingly for any subsequent printing.

Printed and bound in China

The health, beauty and foraging sections of this book are intended to be strictly informational; they are not and should not be deemed to be medical opinion or medical advice. Furthermore, the information included in this book should not be considered a substitute for the advice of a medical professional. Please consult a medical professional if you have any health-related questions. Moreover, it is not possible to predict every individual's reactions to a particular recipe or treatment included in this book. The author and publisher expressly disclaim any responsibility for any effects arising in any way from the advice, information, recipes, treatments or products referred to or described in this book.

MIX
Paper from
responsible sources
FSC™ C007454

This book is produced from independently certified FSC™ paper
to ensure responsible forest management.

For more information visit: www.harpercollins.co.uk/green

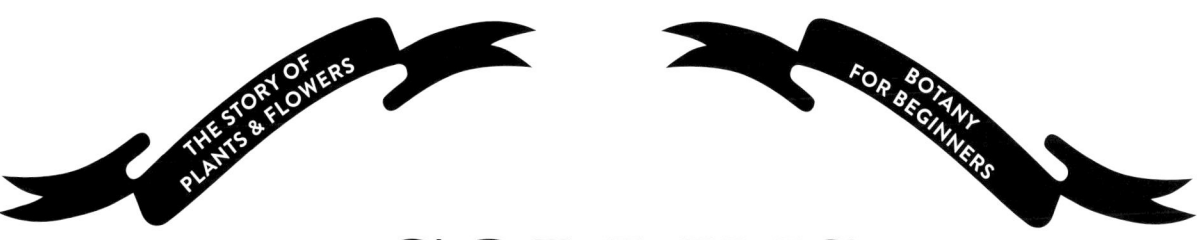

THE STORY OF PLANTS & FLOWERS

BOTANY FOR BEGINNERS

COLLINS BOTANICAL BIBLE

A PRACTICAL GUIDE TO
WILD AND GARDEN PLANTS

SONYA PATEL ELLIS

GROWING, GATHERING, RECIPES & REMEDIES

NATURE-INSPIRED ART, CRAFT & DESIGN

WILLIAM COLLINS

CONTENTS

Introduction A Life Botanical 7
How to Use This Book Make the Connection 13
A Brief Glossary 18

Chapter 1 The Story of Plants & Flowers 21
Pathways of the Plant Kingdom 24
Plants & People 58

Chapter 2 Botany for Beginners 85
Looking Closer 88
Botanically Speaking 120

Chapter 3 Growing & Gathering 145
Reasons to Nurture 148
Reasons to Gather 160

Chapter 4 Botanical Recipes 183
Eat the Seasons 186
Spring 188
Summer 202
Autumn 216
Winter 228
Botanical Drinks & Bowls 240

Chapter 5 Botanical Remedies 257
The Healing Power of Plants 260
Health 272
Wellbeing 280
Beauty 290
Home 300

Chapter 6 Inspired by Nature 307
Art forms in Nature 310
Botanical Art & Illustration 312
Photography & Film 332
Painting & Drawing 344
Sculpture & Installation 356
Design, Craft & Style 368

Acknowledgements 390
Further Reading 392
Index 400
Picture Credits 415

Opposite Just one of the exquisite plates from Dr Robert J. Thornton's *The Temple of Flora* (1807), 'The Nodding Renealmia' now answers to the name of *Alpinia zerumbet* or shell ginger.

Introduction

A LIFE BOTANICAL

Our lives are so ubiquitously intertwined with the Plant Kingdom, it's easy to sometimes take it for granted. *Collins Botanical Bible* tells the story of plants – from how they evolved and what they do, to the ways in which they feed, heal and inspire us – and celebrates our connections with the natural world.

What are plants? I've asked this question of many people during the course of writing this book and the combined answers are hugely revealing. My own children (aged four and six at the time) approached the query with a refreshing innocence, largely based on what they could see in their immediate environment or how plants made them feel: 'Plants are cacti, trees and bluebells', 'Plants grow', 'Plants have roots', 'Plants are in my garden', 'I love flowers', 'I love playing in the forest', and the suitably abstract, 'Sometimes I am a plant.'

The response from most adults was much more complex, revealing as much about the person themselves as it did about the Plant Kingdom. A significant number felt compelled to describe plants in a scientific manner, in many cases weaving together botanical facts and figures learned at school: 'Plants belong to the Plant Kingdom', 'Plants photosynthesise', 'Plants turn carbon dioxide into oxygen', 'Plants need the sun', 'Plants are all around us'. With just a small amount of further prompting – But what do plants do? How do plants make you feel? Do we need plants? Do you work with plants? – what many people felt was the 'right answer' (i.e. an educationally formulated one) evolved into a more personal narrative about their own botanical world. Conversations were initiated about plant-based food, materials and medicine, stories were shared about gardening, walks in the woods, favourite flowers, joyful plant-filled occasions, travels to different biomes, herbal medicines, nostalgic floral perfumes and working with plants, and thoughts were expressed about our role in the future of plants based on how much we rely upon and therefore need them.

The point of asking such a question was not to expose a lack of botanical understanding but rather to highlight just how much most people actually *do* know about plants from their everyday interactions with them – and by doing so, to encourage a more dynamic foregrounding of the wonders of the Plant Kingdom and thus greater enthusiasm for our understanding, care and pleasure of it. You don't need to have a Masters degree in botany to relate to plants (although a deeper knowledge of the Plant Kingdom can open as many doors of perception as an Aldous Huxley-inspired mescaline trip for some people), but the idea of looking closer at plants and elevating them to the same lofty status as other living organisms (animals and humans, for example) is ultimately inspiring – the Plant Kingdom really is a wonderful multifaceted prism through which to view the workings and abstractions of the world at large.

For my part, I've inhabited a 'botanical world' for as long as I can remember. As a young child, my mum would take my siblings and me for regular walks down the Nagger Lines (an abandoned railway between two former collieries) of the West Yorkshire village of Stanley, where I grew up. Between illicit jaunts into adjacent cabbage or rhubarb fields (this being the

Opposite Picking your own fruit is a lovely introduction to the botanical world, especially when the bounty is as tempting as these ruby red 'Discovery' apples.

heart of 'The Rhubarb Triangle'), we'd pick the wildflowers that grew along its scruffy banks, carefully placing ox-eye daisies, dandelions and mother die (cow parsley) into beloved flower presses bequeathed by a neighbour. At home, we'd turn them into pictures like the ones my mum had hanging on our lounge walls, in classic 1970s hippy-meets-Victoriana style, faded petals pinned upon deep black velvet or creamy handmade paper.

These flower presses would go everywhere with us: on plants-meet-art days out to the Yorkshire Sculpture Park; on holiday to Wales where we marvelled at our first sighting of a snake's-head fritillary (no one would believe that we had seen a chequered flower, as our intuitive feeling that it was rare saved it from the press); or further afield when we experienced our first foreign climes, courtesy of the Spanish isle of Ibiza. There we saw and tasted olives for the first time, inhaled the uplifting aroma of native rosemary and thyme clustered around boulders or patches of parched earth, and witnessed our first palm tree. I also fully realised at the eye-opening age of seven that my dad too, like these plants, was originally from another country, namely India; a place that came with its own culture, colours and natural scenery. This was my 'flower power' summer: July 1981 – the year that the narrative of my own personal 'botanical world' went global.

My sister, brother and I owe much to our mum and dad for our knowledge of plants. Although my dad studied botany while training to become a doctor, and thus has a good understanding of the science of plants, it was my combined parents' inquisitive minds and creative outlook that really laid the foundation for much of our learning and sustained interest in nature. We were also lucky enough to have a small garden while growing up, and we spent most of our time in it making mud pies, bashing succulents to extract the 'medicinal' juice, crafting with anything we could get our hands on – including foraged sticks, leaves, nuts, seeds, fruits and flowers – and helping grow spring bulbs, vegetables and berries, some of which we would then eat.

My botanical knowledge grew organically and osmotically – including grasping the very concept of osmosis and the idea of a xylem and phloem by way of a split-stemmed white carnation watered by two separate vases of food dye (infinitely more powerful than the words of any textbook) – and along the way I was shown the essential skills of how to marvel and inquire. A lifelong obsession with the natural history television presenter David Attenborough, who first appeared on my TV screen in 1979 with the documentary series *Life on Earth*, immeasurably expanded my botanical world. These horizons were then further personalised by an art degree specialising in fashion and textiles (mostly spent pivoting around aspects of nature), 20 years of editorial experience (for illustrated books, magazines and environmental organisations, exploring everything from alternative health to plant-inspired design) and a garden to call my own, not far from the ancient botanical wonders of Epping Forest (a new favourite day out).

I am not a botanist per se, but I can enthuse for hours about plants and all that they have taught me (for those with the inclination to listen or the wisdom to teach me more) and the importance of passing on the seeds of this knowledge has also become increasingly vital since having children of my own. Indeed, inspiring plant lust and conservancy in future generations may well be the main catalyst and *raison d'etre* for this book, a narrative that began to unfold in early 2013.

Around that time, through my work as a commissioning editor for an illustrated

'With each "big reveal" of my flower press, from the tiny spiral tendrils of some species of sweet pea to the elongated seedpods of a California poppy, I discovered new morphological details but also a little bit more about what makes that particular plant or species unique.'

Left Our magnolia tree blooms every annum, without fail, in the fourth week of March, a true harbinger of spring. It's impossible not to look up in wonder at this radiant plant. **Bottom** Spring is also the season to make Eggs Flora, the prettiest combination of pressed botanical stencils and natural dyes.

book publisher, I had begun noticing a trend for all things botanical – from naturecraft, edible flowers and healing herbs to the emergence of what is now a fully fledged houseplant revival. I began researching ideas around a book on pressed flowers and quickly became enamoured with the art, science and beauty of herbarium pressings. When maternity leave beckoned in July 2013, I was inspired enough to make my own traditional large-scale herbarium press (along the lines of those used by the Victorian plant hunters) and intuitively set about pressing and recording plants and flowers from my garden over the course of a year. The Herbarium Project was born (now part of A Botanical World, www.abotanicalworld.com).

Paying close attention to conservational herbarium methodologies, I was amazed by the results that ensued. Not only had I managed to preserve the seasonal beauty of possibly my most green-fingered year ever – where beds and borders burst with cosmos, love-in-a-mist, sweet peas, lavender, poppies and hellebores (some of the best specimens to press) – I had inadvertently stimulated a possibly obsessive appetite for the taxonomy and nomenclature of plants, the mythology and language of flowers and a deeper desire to know how plants worked. I had liberated the story within, as it were. With each 'big reveal' of my flower press, from the tiny spiral tendrils of some species of sweet pea (opposite) to the elongated seedpods of a California poppy (see page 174), I discovered new morphological details but also a little bit more about what makes that particular plant or species unique: how it grows, reproduces and ultimately survives, for example, or what plant family it belongs to.

My renewed foray into the world of pressed botanicals has brought numerous rewards, including exhibitions, events, workshops, writing and bespoke commissions, some of which you can read about on page 386 as just one of a collection of nature-inspired artist profiles.

In parallel with my pursuit of botanical artistry, I continued to develop an idea for a book that embraced all aspects of the botanical world, from how plants evolved to the many ways in which people, past and present, engage with them – from plant scientists, gardeners, nature writers, herbalists and plant-based chefs to nature-inspired artists, photographers, designers and makers. *Collins Botanical Bible* is the result (and is a sequel to its pollinating counterpart, *Collins Beekeeper's Bible*, 2010); a reminder of how much one already knows about the Plant Kingdom coupled with numerous expert-led ways of exploring it further.

What are plants? Hopefully you will find your own unique answer within the pages of this book. I hope you enjoy growing and developing your botanical world as much as I have and continue to do.

Sonya Patel Ellis, Author, *Collins Botanical Bible*

Opposite The delicate tendrils of this old-fashioned sweet pea 'Prima Donna' (*Lathyrus odoratus* 'Prima Donna'), pressed in August 2013 for the launch of The Herbarium Project, always remind me to look closer – you never know what you might find.

How to Use This Book

MAKE THE CONNECTION

Collins Botanical Bible is arranged to help you design, grow, develop and celebrate your own inspired version of the botanical world. For ideas on how to explore further, or illuminating examples of relevant plants, look for the Make the Connection guides throughout the book.

What do you love about plants? Perhaps you like cooking with plant-based ingredients such as fruits and vegetables. Or perhaps you are drawn to a particular style of botanical art, are inspired by an aspect of botany that might help you in the garden or have been thinking about making your own botanical remedies or greening up your home. *Collins Botanical Bible* has six chapters that explore all these elements – each one has been beautifully divided into handy sections by specially commissioned white-on-black, botanically inspired summary diagrams by illustrator and print-maker Lynn Hatzius (see page 373) – to help you further pinpoint areas that most interest *you*.

The book opens with Chapter 1: The Story of Plants & Flowers – a narrative that traverses billions of years, from the formation of the Earth to the botanical world of the present day. If you've ever stood in a green space – a garden, park or forest – or a room of houseplants, and wondered where all those amazingly diverse plant species came from, this could be the ideal entry point for you.

The story begins with a journey through the evolutionary Pathways of the Plant Kingdom (see page 24), from cyanobacteria, green algae and the first land plants to more complex seed-bearing and flowering plants (gymnosperms and angiosperms). Beside the description of each plant type you'll find a Make the Connection guide to plant species that you might already know.

Learning about cycads (see page 38) is so much more fun if you know that your Japanese sago palm is one of them, while the concept of monocots and eudicots is much clearer when backed up with tangible plant species or genera such as daffodils, pineapple, cacti or passionflowers (spot the difference on pages 42 and 45).

For those who love to travel, there's also a short introduction to the world's biomes (see page 50), with helpful links to some recognisable plants that grow in each one so that you can more easily make connections of your own or even recreate a mini biome at home. The plot then thickens as *Homo sapiens* (modern human) enters the scene in a section devoted to Plants & People (see page 58), exploring our quest to understand, harness and ideally help conserve the wonders of the Plant Kingdom through Gardening for Purpose (see page 64), Gardening for Pleasure (see page 68) and Flower Power Now (see page 82). The latter is a glimpse into what the future may hold for our botanical world, from flowers in space to the implications of issues such as population growth and climate change.

Chapter 2 covers Botany for Beginners, offering a visually rich invitation to study the basic science of plants. It begins at the beginning, with What is Botany? (see page 86), moves on to the all-important, Looking Closer (see page 88), then delves deeper into How Plants Appear (see page 90) – both out in the field and under the microscope – What Plants Do (see page 110) and why we are Compelled to Classify (see page 122), explaining the inspirations,

Opposite The story of plants and people is laced with lavender (*Lavandula* spp.), beloved for its sweet, therapeutic fragrance and that wonderful signature colour.

people and logic behind botanical plant names and terms, such as plant family, genus and species.

There really is a liberating beauty behind all that Latin – the universal language of the botanical world. Once embraced, it's also easier than you think to become obsessed with 'plant jargon', especially if you're drawn to patterns, communication, order or, indeed, connections. On that note, the Make the Connection lists continue here, a handy selection of known plant species provided for each textbook explanation to draw you back to the botany you already know . . . or think you know. (Hint: When is a nut not a nut? When it's a fruit, seed or a vegetable – find the answer on page 102).

On a serious note, plants are so important to our survival and that of planet Earth that it's vital to inspire the next generation of botanists/plant scientists to follow in the footsteps of past and present ones. It's my hope that if botany is viewed outside the usual textbook format and within the colourful context of history, food, wellbeing and art, as well as science, it will attract more fans to its worthy fold. Plants can be fun, plants can be sexy, plants feel good and plants certainly need future superheroes around the world to champion them.

Chapter 3 is designed to help you take more proactive steps to creating your own *personal* version of the botanical world, presenting Reasons to Nurture (see page 148) and Reasons to Gather (see page 160) as thought-provoking introductions to practical guides on Growing for You (see page 150), Growing for Nature (see page 156), Foraging for Food (see page 162), for Provisions (see page 170) and for Naturecraft (see page 174). For those who enjoy dipping in and out of books, a final contemplation on Living with Nature (see page 180) provides an interesting discourse on the idea of nature as 'home', rather than a 'place to visit', pivoting on the words of American poet and conservationist Gary Snyder. Look out for quotes by similarly inspired nature- and plant-lovers throughout the book, from Walt Whitman and Henry David Thoreau to Johann Wolfgang von Goethe and Gertrude Jekyll.

Such food for thought continues into chapter 4, with a collection of plant-based recipes from an international selection of chefs, cooks, foragers, wild-food educators and drinks experts. The contents of Nature's Larder (see page 184) are contextualised further by a look at why and how we Eat the Seasons (see page 186). Sweet and savoury recipes are then divided, by key ingredient, into Spring (see page 188), Summer (see page 202), Autumn (see page 216) and Winter (see page 228), as well as a chapter on Botanical Drinks & Bowls, covering smoothies, juices, cocktails and nourishing breakfast bowls (see page 240).

Orange, Fennel Seed and Almond Cake (see page 221) by British Michelin-starred chef Tom Harris of The Marksman, London, is just one of the highlights. Or try your hand at wild food classics such as Wild Green Pesto (see page 206), a master recipe à la American foraging and feasting expert Dina Falconi. Recipe contributors were chosen for their expertise but also in many cases as a way to present some of the wonderful plant-based or plant-championing recipe books that are currently out there. Find out more about each contributor's personal botanical world – how they became interested in plants – with inspiring quotes, and see the Make the Connection guides for links to websites, Instagram accounts and further reading.

The journey then continues with the similarly formatted chapter 5, a guide to understanding, appreciating and making your own Botanical Remedies. The Healing Power of Plants (see page 260) sets the scene for a section on natural healing from practical guides on Making Preparations

'Collins Botanical Bible *tells the story of the Plant Kingdom for its own merit but also our ongoing relationship with it – our combined past, present and future.*'

Left Turn to Botany for Beginners (see page 85) for how plants appear, what they do and why the study of plants is so vital. **Bottom** Growing & Gathering (see page 145) is all about getting and giving back to nature, from blackberry-picking to planting for pollinators.

Top Prepare plant-based feasts with Botanical Recipes (see page 183) such as Bernadett Vanek's Pea and Nettle Gnudi. **Right** Be Inspired by Nature (see page 307) by artworks such as Viviane Sassen's *Ultramarine* (2017). **Bottom** Find inner peace and wellbeing with Botanical Remedies (see page 257) such as Mama Medicine's Inner Radiance Ritual Bath.

(see page 262) and a collection of remedies geared towards Health (see page 272), Wellbeing (see page 280), Beauty (see page 290) and the Home (see page 300). Boost your immunity with the esteemed American herbalist Rosemary Gladstar's classic recipe for Fire Cider Vinegar (see page 277) – and be further inspired by a window into her botanical world plus tips on where to find out more. There's also a Warming Body Oil (see page 284) from leading British aromatherapist Julia Lawless, an Inner Radiance Ritual Bath (see page 287) from New York City's wellbeing guru Deborah Hanecamp, a.k.a Mama Medicine, and a non-chemical, naturally fragrant Herbal Cleaning Spray (see page 304) from the Herbal Academy.

Last but by no means least is Chapter 6: Inspired by Nature, which is devoted to the many and varied creative interpretations of the Plant Kingdom throughout its Blooming History (see page 308). Through an inspiring selection of artist profiles, it investigates how wonderfully varied our view of the botanical world can be, illustrated by easy-to-navigate sections on Botanical Art & Illustration (see page 312), Photography & Film (see page 332), Painting & Drawing (see page 344), Sculpture & Installation (see page 356) and Craft, Design & Style (see page 368).

From the vintage botanical art of Georg Dionysus Ehret (see page 313) to the sensual photographic portrayals of plants and people by leading fashion-art photographer Viviane Sassen (see page 340), the prismatic, interconnected potential of our botanical world is presented in all its multifaceted glory.

This is the story of the Plant Kingdom, celebrated for its own merit but also for our ongoing relationship with it – our combined past, present and future. Immense thanks to all those who have so generously and expertly contributed to the vision that is *Collins Botanical Bible*.

MAKE THE CONNECTION
www.abotanicalworld.com
#abotanicalworld

Talking about the botanical world
A BRIEF GLOSSARY

Before you journey through the pages of the *Collins Botanical Bible*, take a moment to consider some of the words and phrases commonly used within the pages of this book, describing and defining plants, people and the botanical world itself. These terms, presented thematically, highlight some key themes of the book.
For a more comprehensive botanical glossary, please look through the further reading section beginning on page 392 for a selection of illuminating reference titles.

Botanical
Of or relating to plants, or the study of plants; ingredients derived from plants (most usually medicinal)

The Botanical World
Synonym for the Plant Kingdom; more poetically, your personal vision of a plant-inspired life

Botany
The scientific study of plants; or relating to the plant life of a particular region, habitat or period

Botanist
An expert or student of the scientific study of plants; from the Greek *botanikos*, meaning 'of herbs'

Botanic Garden (see page 75)
An establishment where plants are grown for scientific study and display to the public

Botanical Illustration (see page 79)
A scientifically accurate depiction and representation of a plant or plant part, identifiable to species level

Botanical Art (see chapter 6)
Botanical illustration with aesthetic appeal; more widely, art that explores the botanical world

The Plant Kingdom
The collective plants of the world (living and extinct) estimated to include over 400,000 different species

Plants
Living organisms that typically grow in a permanent site and synthesise nutrients by photosynthesis

Flowers
The seed-bearing, usually petalled reproductive organs of the most diverse group of plants – the angiosperms

Herbs (in botany)
A seed-bearing plant that does not have a woody stem and dies to the ground after flowering

Herbs (medicinal or culinary)
A plant with leaves, seeds or flowers used for flavouring, food, medicine or perfume

Family (see page 129)
A taxonomic category that ranks above genus; in botany, families usually end in –aceae, e.g. Rosaceae – rose family

Genus (see page 130)
A taxonomic category that ranks above species; genera (plural) have capitalised Latin names e.g. *Rosa* – rose

Species (see page 133)
Principal taxonomic unit, denoted by a Latin binomial (two-part) name, e.g. *Rosa canina* – dog rose

The Earth
Our planet; where the Plant Kingdom, *Homo sapiens* (modern humans) and other organisms live side by side

Biome (see page 50)
A large, naturally occurring community of flora and fauna occupying a major habitat, e.g. forest, tundra or desert

Biodiversity
Generally referring to the variety and variability of life on Earth; the result of billions of years of evolution

The Environment
Relating to the natural world, as a whole or in a particular geographical area or biome

Evolution
How the diversity of the Earth's living organisms are thought to have developed from earlier forms

Nature
The phenomena of the physical universe, including plants, animals and landscape; can include humans

The Natural World
A coming together of nature; alluding to that which is not manmade; somewhere to get back to

Pollinator (see pages 98 and 157)
An animal that carries pollen cells from a male anther to a female stigma of a flower, so aiding reproduction

People
Human beings collectively, past and present; *Homo sapiens* (anatomically modern humans); us

' *There really is a liberating beauty behind all that Latin – the universal language of the botanical world. Once embraced, it's also easier than you think to become obsessed with "plant jargon", especially if you're drawn to patterns, communication, order or, indeed, connections.* '

Right *Amorphophallus titanium*, less lasciviously known as titan arum or corpse flower due to its titanic inflorescences and fly-attracting scent, is a potent reminder of the extraordinary diversity of the Plant Kingdom.

'In short, the animal and vegetable lines, diverging widely above, join below in a loop.'

Asa Gray, American botanist,
Natural Science and Religion (1880)

CHAPTER ONE

THE STORY OF PLANTS & FLOWERS

OUR BOTANICAL WORLD

The story of plants and flowers is an enduring tale of Earth-changing events, extraordinary evolution and ultimate survival. Although humans have only been part of this narrative for the slimmest chapter of time, we owe much to the wonders of the botanical world – not least the oxygen that gives us life.

Stop and look around you. Somewhere in your vicinity there will be a plant or flower: a prized bloom, a beautiful wildflower, a lowly patch of moss, a potentially troublesome weed, a treasured houseplant, a towering tree or perhaps a plant-inspired design. Now stop and take a deep breath. It's easy to take for granted the oxygen that we inhale, but don't forget that it comes from somewhere – or indeed, from nearly 400,000 different known species that make up the Kingdom Plantae.

The story of the plants on Earth spans Great Oxygenation Events, eras of extreme evolution, mass extinctions, divided kingdoms, powerful pollinators and the evolution of modern humans (*Homo sapiens*). But while the Animal Kingdom developed from the same eukaryotic, membrane-bound organisms as plants (at least 1.8 billion years ago – see page 24), it would be millions of years before humans were bipedal and brainy enough to quite literally stand up and 'smell the roses'.

By the advent of anatomically modern humans – estimated between 200,000 and 100,000 BCE – the Plant Kingdom had well and truly conquered the Earth. Thanks in part to a 'Big Bloom' of reproductively advanced flowering plants at least 140 million years ago (see page 41), biomes around the world were infiltrated by one promiscuous bloom or another, and thus were forever changed. The world of the first *Homo sapiens* had forests to roam in, fruits and seeds to eat, leaves to provide shelter, wood to carve and build with, herbs to heal and myriad plant species to inspire wonder, invention and creativity.

By at least 10,000 BCE, the story of plants and flowers was also one of plants and people (see page 58), beginning with humanity's first foray into forest gardening, the precursor of agriculture and cultivation. By 3000 BCE, we were sharing our botanical knowledge through writing. By 1500 BCE, evidence from Egypt shows that we had developed systems of ornamental horticulture (see page 60). By this point in history, humans were actively harnessing the pleasure-enhancing powers of the botanical world, either for their aesthetic appeal or to be used to benefit our health, wellbeing and happiness. We were Gardening for Purpose (see page 64) and Gardening for Pleasure (see page 68).

A thousand years later, around 340 BCE, the great Greek philosopher Aristotle established the first botanical gardens in Athens, in part prompting his student Theophrastus to compile his pioneering *Historia Plantarum* or 'Enquiry into Plants' (see page 62). This seminal text laid the foundations for a wealth of discovery to come – the amazing, unfurling, eternal discovery of our botanical world.

Just 3.8 billion years since Planet Earth's very first cellular life, the story of plants and flowers continues, and this chapter follows, exploring the realms of the Plant Kingdom, the biomes it helps create and the people who have interacted with it through millennia of plant-inspired existence.

Opposite The genome sequence of *Amborella trichopoda*, a rare tropical tree endemic to New Caledonia, has revealed it to be the sole survivor of the oldest line of flowering plants.

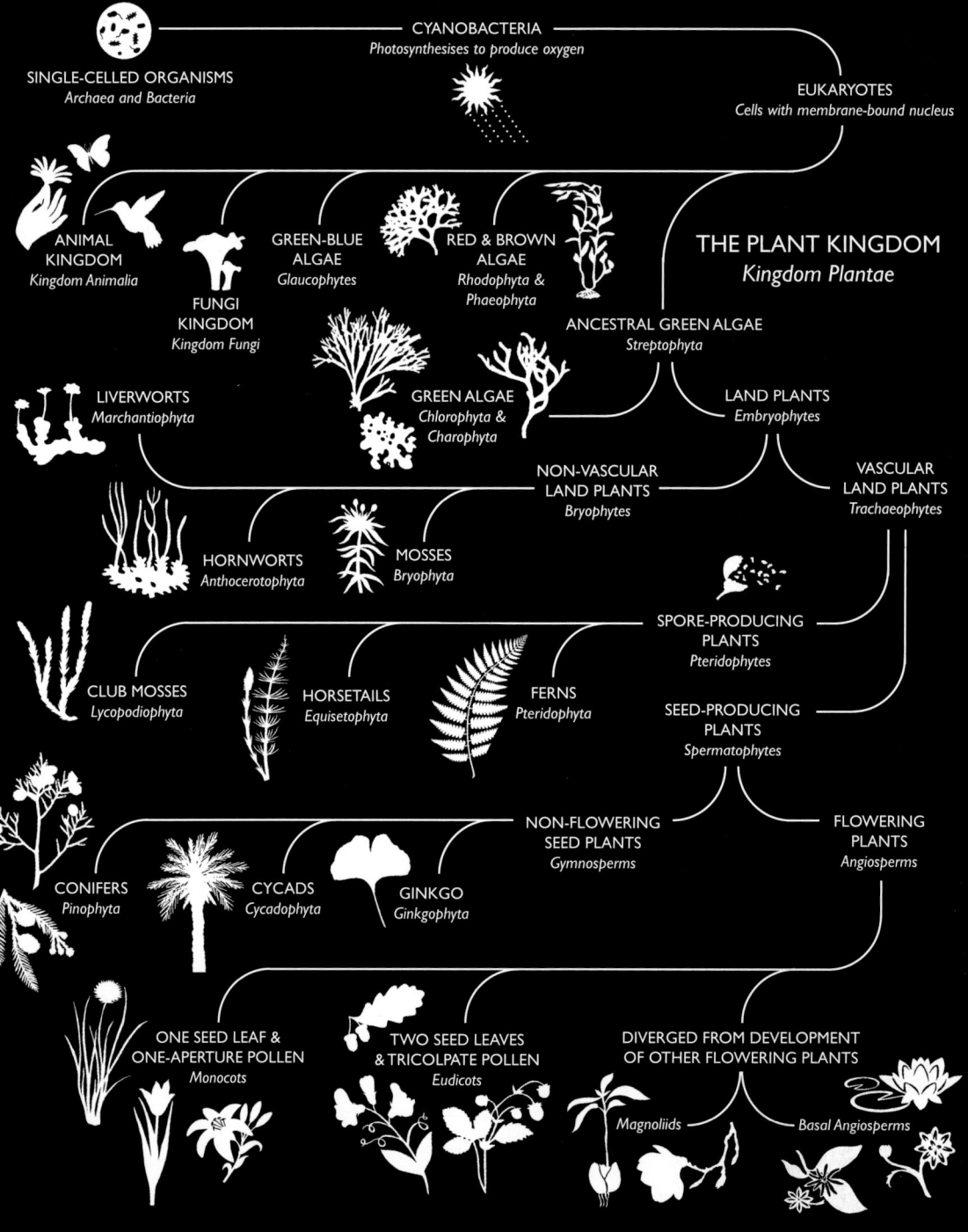

Where do pineapples come from? How are humans related to plants? Make the connection between the evolutionary wonders of the Plant Kingdom – within itself and in relation to other divergent life forms – and everything is illuminated.

Pick a plant, any plant. Your favourite houseplant or a plant derivative such as a coconut. Think about where that plant came from and how it got here. How is it connected to the lavender bush in your garden or the giant redwoods you long to visit one day, and how is it connected to you – a walking, talking, plant-curious *Homo sapiens*?

This chapter attempts to put these questions into context, via a journey from the first single-celled organisms that appeared some 3.8 billion years ago to the botanical world as we experience it today: currently the greatest biomass on Earth, comprising 400,000 known plant species and collectively providing the structural canvas and oxygen for a huge diversity of other life forms.

This monumental journey, illustrated opposite by Lynn Hatzius (see page 373), traces the route of the earliest oxygen-producing cyanobacteria (see page 27) to the first eukaryotes from which algae, plants, fungi and animals (including humans) would evolve. Then it's *Viridiplantae* (green plants) all the way, in all their verdant glory. Ancestral green algae evolved into new and more complex forms of aquatic green algae, some cells found a way to exist on dry land and developed into land plants (see page 28). Some land plants then diverged into liverworts, hornworts and mosses (see pages 30–31), while others developed vascular tissues to help transport food, water and products of photosynthesis (see page 111).

Vascular plants spawned enormous diversity, branching off into spore-producing plants – the club mosses, horsetails and ferns (see pages 34–35) – before evolving into seed producers (see page 36).

The first seed producers emerged around 380 million years ago, divided into groups of non-flowering plants or gymnosperms (see page 37) – conifers, cycads, gnetophytes and ginkgo (see pages 38–39) – and flowering plants or angiosperms (see page 40), which make up an estimated 90 per cent of the known Plant Kingdom today. Of these, 20 per cent are known as monocots (see page 42), 75 per cent are known as eudicots (see page 45), and a few thousand refuse to be typecast quite so neatly. These are currently split between the magnoliids (see page 46) and the basal angiosperms (see page 49).

Looking more closely at how the Plant Kingdom most likely evolved (even with the proviso that 30 per cent of all estimated plant species are yet to be discovered and any assumed evolutionary 'pathway' could reroute at any point) can be a wonderfully revealing experience, not least for the reminder of just how intrinsically connected plants and people really are.

Opposite Explore the amazing diversity of the Plant Kingdom further in the pages to follow, or turn to page 58 for more on the deep connection between plants and people.

THE FIRST PLANTS

Trace back any life form on Earth and at some point, several billion years back, you will arrive at the most primitive single-celled organism. What set the 'first plants' apart and allowed them to evolve and populate land was their ability to photosynthesise – all thanks to a little blue-green microbe called cyanobacteria.

Cyanobacteria and photosynthesis

Earth is thought to have formed around 4.6 billion years ago, with the first cellular life appearing about a billion years later. These were the single-celled prokaryotic organisms, the earliest bacteria and archaea. Around 2.5 billion years ago, some ocean-dwelling bacterial cells harnessed the power of the sun to create energy. In doing so, they could only absorb invisible light and produce sulphur, but nevertheless this is considered to be the first example of photosynthesis.

Blue-green cyanobacteria used chlorophyll a, one of the pigments that gave these aquatic cells their colour, to absorb visible light, and in the process made the very first oxygen, triggering the Great Oxygenation Event some 2.4 billion years ago.

Slowly, the Earth's atmosphere began to build up with life-giving O_2. Slowly, meaning another billion years or so – 'the boring billion', as some scientists refer to it – and as oxygen built up to levels that we now take for granted, at around 21 per cent of the Earth's atmosphere, the first eukaryotic cells began to evolve. These cells played host to a number of organelles, including a membrane-bound nucleus, within which lay the DNA of the cell. Eukaryotic cells then grouped together to form multicellular organisms and tissues before eventually evolving into three main kingdoms of similar characteristics – Kingdom Animalia, Kingdom Fungi and Kingdom Plantae – plus various non-green algaes (red, brown and green-blue).

So what set the green plants apart? The simplest answer lies in an organelle called a chloroplast, which contains pigments such as green-giving chlorophyll a and chlorophyll b. Chloroplasts convert solar power into energy-storing sugars and also produce oxygen. How these chloroplasts become part of plant cells is up for discussion, but one theory suggests that early eukaryotic cells 'ate' smaller cyanobacteria, thus beginning an endosymbiotic relationship. The engulfed cyanobacteria continued to photosynthesise and make energy and oxygen while the host eukaryote came to rely on its contribution. Over time, plant cells evolved to include a chloroplast and developed photosynthesising abilities of their own – a skill shared by each and every plant on Earth today.

Non-green algae, such as red and brown seaweed, including bladderwrack, kelp, Irish moss, Sargassum and Pyropia (the nori used to wrap sushi), also contain chloroplasts, but their cellular structure and potential evolutionary lineage is sufficiently different to set these 'plant pretenders' apart from the green stuff. They also lack the pigment called chlorophyll b, which is only found in green algae and plants. Cyanobacteria are also still very much part of the Earth's ecosystem today. Although single-celled, they often group together in huge oceanic swathes or blooms and are sometimes visible to the naked eye.

Various forms of single-celled prokaryotes are still around, too, in the form of archaea and bacteria, which along with eukaryotes make up the three main domains of life on Earth. Archaea are often found in the most inhospitable places in the world, such as salt lakes and thermal springs. Bacteria, on the other hand, are found almost everywhere on the planet: throughout the oceans, in the atmosphere and inside our own bodies. They also help break down plants. It's easy to give bacteria a bad name, but don't forget that this is where 'the first plants' all began.

Evidence of early photosynthesis

So how do we know the role that cyanobacteria played in oxygenating the Earth? Fossil hunters are continuously scouring the globe for rocks betraying signs of even more ancient life forms. These rocks include 'biosignatures' from the cyanobacteria, which create patterns and minerals that could only be formed by the presence of life. It is thought that the first free oxygen produced by cyanobacteria actually bonded with iron dissolved in the oceans and formed red-coloured iron oxide. Over time, sedimentary rocks called banded iron formations were created by these iron oxide deposits. Once the iron in the ocean was used up, the oxygen was free to start building up in the atmosphere, marking a change in the rocks. Other significant fossils that signify the presence of cyanobacteria are stromatolites, rock-like structures of calcium carbonate created by the bacteria's excretions.

Opposite Satellite image of what appears to be a large bloom of cyanobacteria in the Baltic Sea. Blooms like this flourish during summertime, when there is ample sunlight and high levels of nutrients in the water.

Green algae
Chlorophytes & Charophytes

There are several divisions of algae in the world but only one kind – green algae – makes it into the Plant Kingdom. This is due to a number of cellular features, including the presence of chlorophyll *a and* b, which gives the cells their bright green colour. Green algae are then split into chlorophytes and charophytes.

The majority of chlorophytes live in freshwater, while a large number also dwell in marine habitats and a few have adapted to living on land in summer alpine snowfields or on rocks or trees. Some chlorophytes have found a way to live symbiotically with sponges, jellyfish and molluscs, while others have learned to connect with fungi to form lichens.

Charophytes, on the other hand, are freshwater green algae, alongside which the land plants are thought to have emerged. One example, Charales – also known as stoneworts or rockweeds – are thought to be the closest known relatives of land plants. What look like their stems and branches, however, are actually colonies of cells. While land plants absorb their nutrients through a root system, stoneworts do it directly through these cells.

MAKE THE CONNECTION
Estimated known species: More than 7,000 Charales (stoneworts or rockweeds); *Ulva* (sea lettuce), pictured; *Codium* (dead man's fingers); lichen, symbiotically with fungi

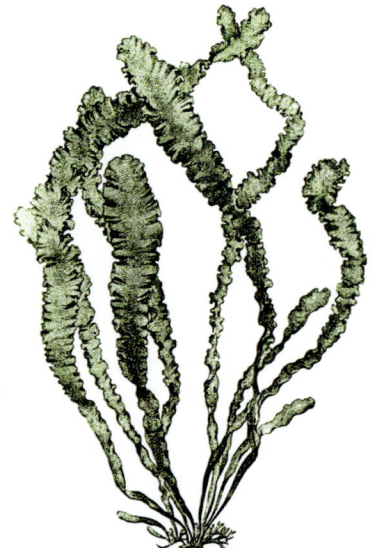

Land plants
Embryophytes

The first land plants, members of the group Embryophyta, are thought to have evolved around 470 million years ago, during the rapidly diversifying Ordovician period. As some green algae began to colonise shallow freshwater pools, they had to adapt to rain exposure, varied temperatures, high levels of UV light and seasonal dehydration. In the process, non-vascular liverworts and mosses began to form, possibly alongside symbiotic fungi. These newly formed non-vascular plants then evolved stem- and leaf-like components, as well as subterranean stems called rhizomes. Some plants went a step further and developed a waxy cuticle to stop them drying out. By the Silurian and Devonian periods (*c.* 444–359 million years ago), these plants had developed vascular tissues to transport water, carbon dioxide, minerals and oxygen.

Plants began to compete for sunlight and water using an innovative array of evolutionary tactics: leaves, roots, woody stems and seeds. In the Early Devonian period, the landscape reached no higher than waist-height, but by the Late Devonian some 'woody' plants and giant cousins of today's tiny club moss had begun to grow up to 50m (160 ft) high.

By the time of the Carboniferous or 'coal-bearing' period (*c.* 359–299 million years ago), swampy forests of spore-shedding ferns, horsetails and tree-like lycophytes dominated the landscape, and seed plants began to diversify to include now-extinct 'seed ferns' plus early cycads and conifers. Mountains formed and oxygen levels reached an all-time high, followed by a mass extinction of marine- and land-based life. Yet still some land plants survived – Earth's architects, continuously adapting and evolving to create and support the ecosystems that we are familiar with today.

Opposite From Ernst Haeckel's *Art Forms in Nature* series, circa 1904. This illustration was originally titled *Muscinae*, which is an obsolete term for mosses, liverworts and hornworts. This collective group of non-vascular land plants are now referred to as bryophytes (see pages 30–31).

NON-VASCULAR LAND PLANTS

Non-vascular quite literally means a plant without a vascular system of tissues – namely the xylem and phloem – that transport water and other substances. Species groups include the liverworts, hornworts and mosses, which together are known as the bryophytes. They share a peculiar life cycle and rely on a thin layer of water on the surface of the plant to enable fertilisation. You can find a bryophyte in almost every environment apart from the sea. Some occupy cold Arctic landscapes, others alpine, desert or wet rainforest conditions, while some find their way into terrariums, floristry displays and peace-enhancing moss gardens.

NON-VASCULAR LAND PLANTS

Liverworts
Marchantiophyta

In ancient times it was believed that non-vascular land plants from the group Marchantiophyta cured diseases of the liver, hence the common name. This belief probably stemmed from the way some types of liverwort resemble the human organ.

What liverworts do best, however, is reduce erosion along stream banks, help retain water in tropical rainforests and form soil crusts in deserts and polar regions. Most are a bright green colour and look leafy, lobed and possibly leathery, with spores produced in small capsules. Typically, individual plants aren't much bigger than 10-cm (4-in) long, but certain species can grow to cover large patches of ground, rocks or trees. Some of these you may recognise as the troublesome imposters you find in the damp and shady recesses of your greenhouse, in your lawn, between paving stones, or on the soil of plants bought from the garden centre. Others, such as the aquatic thallose liverworts (less leafy, with a flat plate of tissue, called a thallus), feature in aquariums to create fish-friendly habitats.

MAKE THE CONNECTION
Estimated known species: More than 7,000
Lunularia cruciata, thick, green, leathery and found in nurseries and greenhouses; *Riccia fluitans* (an aquatic liverwort); *Marchantia polymorpha* (umbrella liverwort), pictured

Hornworts
Anthocerotophyta

The 'horn' that gives these anthocerotophytes their common name refers to the tall, narrow sporophyte embedded in the top of the parent plant, a unique elongated feature that continues to grow throughout its life. This horn-like structure emits spores that germinate to produce flat, leaf-like, greasy-looking plants. These are the gametophytes, the more prominent, long-living sexual generation of plant. Their characteristic blue-green colour derives from the oft-found colonies of cyanobacteria in the lobes. (Under a microscope it is possible to see the mucilage where these cyanobacteria thrive and the discerning feature of one large chloroplast per hornwort cell.)

Hornworts grow worldwide but are mostly found in places that are damp or humid – in tropical forests, alongside streams and rivers, in cultivated fields, on rocks or trees and potentially in your garden. These plants are not to be confused with the common hornwort (*Ceratophyllum demersum*), often used in aquariums, which is actually a flowering plant.

MAKE THE CONNECTION
Estimated known species: More than 200
Phaeoceros laevis (smooth hornwort), pictured; *Phaeoceros carolinianus* (Carolina hornwort); *Anthoceros agrestis* (field hornwort)

NON-VASCULAR LAND PLANTS

Mosses
Bryophyta

Moss has been employed for myriad uses throughout history: as bandages, pillows, insulation and clothing, in pit ovens, as fuel and to extinguish fires, and for aesthetic purposes. Although most lawn-lovers consider moss to be a weed, in Japan it is thought to add serenity and antiquity to a garden scene and it is also used in smaller bonsai landscapes to cover the soil and suggest age. The natural properties of moss are also harnessed for use in floristry, for terrariums and for green walls and roofs.

Since moss has thread-like rhizoids rather than true roots, designed to anchor it to its chosen surface, it requires less planting, and with proper species selection it needs little or no irrigation. No wonder naturalistic gardeners across the world are now beginning to embrace moss, finding inspiration in the velvety undulations of iconic gardens such as the Bloedel Reserve in Washington State.

Individual moss plants are usually composed of simple 'leaves' that are generally one cell thick. These are attached to a 'stem' that may be branched or unbranched. The non-vascular structure, minus the water-transporting xylem and phloem, means this stem has a limited role in conducting water and nutrients. The taller unbranched stalks that you sometimes see on moss are the sporophytes, which develop after fertilisation. These are topped with single capsules that contain the spores. While most grow to be between 0.2–10cm ($^1/_{16}$–4 in) high, the tallest moss in the world, *Dawsonia*, can reach heights of 50cm (20 in).

Perhaps the most commonly known moss is *Sphagnum*, or peat moss. In decayed form this is the main constituent of peat, hence the name, but live *Sphagnum* is often used as a plant-growing medium and as a soft feature in gardens and floristry. Some species of this wonder moss can absorb up to 20 times their weight in water and shoot their spores out with remarkable force due to the build-up of air pressure inside the capsules. Not content with these superpowers, in total *Sphagnum* species are estimated to cover about 1 per cent of the Earth's land surface. A humble, insignificant 'weed' it is not.

True mosses make up the taxonomic division Bryophyta, though liverworts and hornworts are informally called bryophytes as well. Moss should also not be confused with lichen, which is actually algae or cyanobacteria living among fungi. This includes the lichen known as reindeer 'moss' (*Cladonia rangiferina*), which is also commonly used in floristry.

MAKE THE CONNECTION
Estimated known species: More than 10,000
Sphagnum (peat moss), used in floristry;
Brachythecium rutabulum (rough-stalked feather moss); *Polytrichum* (haircap moss), pictured;
Hypnum iponens (sheet moss), used for terrariums

VASCULAR PLANTS

Vascular plants comprise an estimated 391,000 known species, from tiny spore-producing club mosses to giant, flowering, seed-producing trees. What this diverse bunch all have in common is an inbuilt transportation system (the vascular bit), which carries water, minerals and products of photosynthesis (see page 111) around the plant by way of xylem and phloem (see page 107). Put simply, a plant with conducting tissues that can self-distribute resources through its roots, stems and leaves has a better chance of diversifying and thus surviving.

Vascular spore producers
Pteridophytes

Like their non-vascular, spore-producing counterparts the bryophytes (liverworts, hornworts and mosses), the vascular pteridophytes (comprising the ferns, horsetails, club mosses, spike mosses and quillworts) reproduce via an alternation of generations, meaning the plant has two distinct physical forms, and a plant's complete life cycle includes both. The sexual phase, known as the haploid generation or gametophyte, produces gametes or reproductive cells (*gamete*, from the Ancient Greek, meaning 'to marry'), and is followed by an asexual phase – the diploid generation or sporophyte – which produces spores.

What sets the broadly grouped pteridophytes *apart* from the bryophytes is that both generations of these plants are independent and free-living. The sporophyte is also much larger and more conspicuous.

How spores work

If you're lucky enough to live near a forest that's carpeted with ferns (probably the most recognised type of pteridophyte), or you have one growing in your garden, stop and take a look. What you're most likely seeing is the asexual or sporophyte stage of the plant. Turn over a fern frond and you'll find clusters (called sori) of spore cases that contain the spores. In some ferns, such as the *Platycerium* or staghorn fern, the arrangement is slightly different, with a combination of sterile and fertile fronds.

If the spores are ripe, the sporangia (spore-producing organs) that make up the sori appear golden or brown – you'll see them in small, bumpy, often symmetrical rows. Press a frond onto a piece of white paper or let it dry in a white envelope for two days at room temperature and the spores will reveal themselves as an almost microscopic brown or sometimes green dust. In nature, the spores are the crucial element; the bit that hopefully gets blown further away by the wind and finds a suitable spot to grow into a gametophyte.

If you want to try growing a fern you'll first need to separate the spore from the chaff. This is easily done by placing the dust on a piece of paper and gently tapping it. The spores will stay put while the chaff will bounce forward. Sprinkle your spores lightly onto watered soil (without covering, as they need light to germinate) and seal in a plastic bag to create a mini greenhouse. Place the bag in a modestly lit spot away from direct sunlight and after a few weeks to months you should hopefully have your own collection of small, green, heart-shaped growths not unlike liverworts. These are gametophytes, the sexual phase of the plant.

Each leaf-like prothallus, or gametophyte, contains half the number of chromosomes of an adult fern plant or sporophyte. It produces male and female components (sperm and egg) which fuse (fertilise) to reconstitute the full chromosome complement in the embryo. A bit of steamy humidity will be required here, as the male gametes need a thin layer of water to swim to the female ones. After a few weeks, a small diploid sporeling, or baby fern, should raise its first frond, and as bigger fronds appear, the prothallus should fade away.

You're now ready to acclimatise your fern to the world outside. You're also hopefully more able to recognise the sporophyte and gametophyte phases of spore-producing plants in the wild. What you first thought was a moss or liverwort on the forest floor, might actually be a baby fern about to make its way into the world. And what you thought was a fern – the frondy bit – was really only half of its story.

This, essentially, is how all pteridophytes work, although the horsetails and club mosses have their own unique back-up versions, too, including various vegetative methods of reproduction. As botanists throughout the ages have discovered, plants are wildly diverse, ensuring their continued survival.

A word about fungi

You may be wondering at this point how spore-producing fungi fit into all this? Head back to the Pathways of the Plant Kingdom diagram on page 24 and you'll see that the Fungi Kingdom exists in its own right. One school of thought suggests that a common ancestor of both fungi and animals diverged from plants some 1.1 billion years ago, but the fascinating debate goes on with each new discovery.

Opposite
The coiled fiddleheads of the ostrich fern (*Matteuccia struthiopteris*) unfurl into feathery fronds in late winter, followed by spore-bearing leaves in midsummer.

VASCULAR PLANTS

Club mosses, quillworts and spike mosses
Lycopodiophyta

The Lycopodiophyta – also referred to as the lycopods or lycophytes – are one of the oldest lineages of living vascular plants. Known species include club mosses, quillworts and spike mosses. Their division also includes the now-extinct genus *Baragwanathia*, fossils of which have been found dating back to the Late Silurian or Early Devonian era (*c.* 427–393 million years ago), and the long-gone zosterophylls, which was rated among the first vascular plants in the fossil record.

Species of Lycopodiophyta from the Carboniferous period included 10-m (33-ft) tall trees that contributed to the formation of coal. Today's Lycopodiophyta are a lot smaller, with distinctive microphylls (leaves that have only a small vascular trace or vein). Club mosses are thought to be structurally similar and therefore bear the most resemblance to early vascular plants, with their tiny scale-like leaves, spores borne in sporangia at the base of the leaves, branching stems and simple form.

MAKE THE CONNECTION
Estimated known species: Around 1,290
Lycopodium (ground pine), spores of which were used in Victorian theatre to produce flame effects; *Selaginella* (only genus of spike moss), pictured; *Isoetes lacustris* (lake quillwort or Merlin's grass)

Horsetails
Equisetophyta

Today's horsetails belong to the only living genus of the entire class of Equisetophyta, or Equisetopsida, as some botanists prefer to call it. This genus is called *Equisetum* (*equus* from the Latin meaning 'horse' and *seta* meaning 'bristle') and includes about 20 herbaceous species of horsetail-like plants. Horsetails comprise a single vertical stem, from which whorls of needle-like branches radiate at regular intervals. They can be found growing along streams and the edges of woodlands.

Horsetails reached maximum diversity during the Late Devonian and Carboniferous periods. Some now-extinct species grew to heights of 30m (100 ft) and their remains are abundant in coal deposits from this time. In more recent times, humans have used the rough, gritty, often silica-filled stems of horsetails as 'scouring rushes' to clean dishes and as tools to polish wood. Some species have also been used in various schools of herbal medicine to treat fluid retention and kidney stones or to cool a fever.

MAKE THE CONNECTION
Estimated known species: Around 20
Equisetum arvense (field horsetail), pictured, a common garden weed, worldwide; *Equisetum giganteum* (southern giant horsetail), used in South America as a diuretic

Ferns
Pteridophyta

Ferns appear in the fossil record some 360 million years ago, in the Late Devonian period. However, many of the known species living today did not appear until 145 million years ago in the Early Cretaceous – after flowering plants had become the dominant force. Some species, such as *Osmunda claytoniana*, have remained seemingly unchanged for at least 180 million years. Apparently *this* plant thought it got it right the first time – or settled itself down for the long run when it was confident that it had the best chance of survival.

The great Scottish-American naturalist John Muir wrote in *My First Summer in the Sierra* (1911):

'Only spread a fern-frond over a man's head and worldly cares are cast out, and freedom and beauty and peace come in.' In the same book he describes how 'broad-shouldered fronds' create 'a complete ceiling where one may walk erect over several acres without being seen'; how the light streams through this living ceiling and reveals 'the arching, branching ribs and veins of the fronds as the framework of countless panes of pale green and yellow plant-glass nicely fitted together – a fairyland created out of the commonest fern-stuff'; eloquently putting into words the peace that many people feel when they come across a fern-filled forest.

But this is far from the complete picture. Ferns grow in a wide variety of habitats – from mountain crevices to dry deserts, in bogs and swamps and in open fields – and often they succeed where many flowering plants have not. Some ferns form aggressively-spreading colonies and are considered weeds. A large number grow as epiphytes, living harmlessly upon another plant, such as a tree.

Uses for ferns are widespread, featuring in floristry, herbal medicine and – in fiddlehead form (the furled fronds of a young fern) and only with expert advice – as food. Be warned, however: fiddleheads can be toxic if not fully cooked, and some types, such as bracken (*Pteridium*) fiddleheads are thought to be carcinogenic.

MAKE THE CONNECTION
Estimated known species: Around 10,560
In the wild *Pteridium* (bracken); *Dryopteris filix-mas* (male fern), pictured; *Athyrium filix-femina* (lady fern); *Dicksonia/Cyathea* (larger tree fern species)
Kept as houseplants *Platycerium* (staghorn fern), an epiphyte; *Nephrolepis exaltata* (Boston fern); *Asplenium nidus* (bird's nest fern)
Used horticulturally *Adiantum* (maidenhair fern); *Polystichum munitum* (western sword fern)
Used in floristry *Asparagus setaceus* (asparagus fern); *Rumora adiantiformis* (leatherleaf fern)

Seed producers
Spermatophytes

The evolution of seeds was a key moment in the story of the botanical world. Not only did seeds make it possible for most of today's known species of plants and flowers to exist, they inadvertently helped to create significant changes in the Earth's climate, the structure of its terrain and the way in which plants and other life forms interact.

Today, around 1,100 species of gymnosperm (non-flowering seed plants) and around 369,000 species of angiosperm (flowering seed plants) are known to exist. The seeds they produce include pine nuts and bean seeds. Seed cases, nutrients and appendages are also ubiquitous in everyday human life, from the hard, protective outer shell of the coconut seed with its tasty inner flesh and multitasking oil, to the tiny hairs that propel cotton seeds through the air.

How seeds evolved

Where seeds come from is a fascinating tale of adaptation and survival. Their evolution began in the Devonian era (*c.* 419 to 359 million years ago) with the spore-producing club mosses, tree-like horsetails and progymnosperms (plants from which gymnosperms are thought to have emerged) such as *Archaeopteris*. In order to override features of vulnerability – namely tiny spores travelling relatively long distances without built-in nutrition or physical protection – these plants developed a new system of reproduction called heterospory. This involved producing a smaller male microspore and a larger female megaspore. (Some modern species of pteridophytes are heterosporous as well.)

The now-extinct Pteridospermatophyta, or 'seed ferns', were the first true seed plants. In appearance, these woody, tree-like plants with frondy foliage were not unlike the tree ferns of today. As relationships go, we now know they're not technically ferns, with fossils showing features more reminiscent of modern-day cycads. Still, seed ferns dominated the landscape during the Carboniferous (*c.* 359–299 million years ago) and Permian (*c.* 299–252 million years ago) periods – so how did they get there?

It's thought that some ancestral megaspores began to germinate into gametophytes *within* their sporangia (spore-producing cases) rather than *outside* it to potentially avoid exposure to the elements and therefore protect their developing spores. As a result of this adaptation, additional protective tissues and inbuilt nutrients were required to further aid the chance of reproduction and thus survival. These adaptive, all-in, megaspore environments would eventually evolve into the female ovules of seed plants.

In gymnosperms these ovules are naked – in female pine cones, for instance – while in angiosperms they are hidden in ovaries located at the base of the carpels. It's all there in the name: *gymno* translates from Ancient Greek as 'bare' or 'naked'; *angio* as 'enclosed'. In both cases the ovules contain megagametophytes (female gametophytes) that produce egg cells. When these cells are fertilised by the male gametes carried in pollen, they mature into seeds. Given the right conditions, some of these seeds will grow into seedlings, or, by another successful stroke of adaptation, they can lie dormant until the right conditions for germination prevail.

The advent of pollen

For their part, pollen grains are actually male gametophytes, also known as microgametophytes, and are thought to have first evolved from the free-sporing heterosporytes. Each grain of pollen will potentially germinate upon contact with the reproductive organs of a compatible female. In flowering plants this organ is the stigma, the uppermost receiving end of the carpel. When it germinates, the pollen grows a pollen tube in order to gain access to the ovule, where fertilisation can then take place to produce a seed. The pollen is *produced* in the male part of plants, such as the catkin-like cones of the pine tree or the anthers of stamens in flowers.

Fossil evidence suggests that the evolution of pollen coincided with the evolution of seeds. Microscopic evidence also shows that pollen grains come in a wide range of shapes and sizes (see page 108) – as do the methods of pollination and the pollinators that evolved with them.

Opposite The non-flowering bread tree (*Encephalartos altensteinii*), part of the palm-like Zamiaceae family of cycads, reproduces via cones and seeds. It takes its name from the 'bread' that can be made from the pith of its stem.

NON-FLOWERING SEED PLANTS

The gymnosperms (also known as non-flowering seed plants) are so-called for their 'naked seeds' – or rather, their unenclosed ovules – although many of these defining features are not visible until maturity. Their rank includes the conifers, cycads, ginkgo and gnetophytes such as *Welwitschia*. They have seeds often modified to form cones or found at the end of short stalks, and rely primarily on wind for pollination. Dominant during the Mesozoic era (*c.* 252–66 million years ago), gymnosperms still cover vast swathes of the Earth's surface. This group boasts some of the world's oldest, largest and most magnificent plant species.

Conifers
Pinophyta

Conifers make up over half the known species of gymnosperm on Earth. They comprise the majority of the cone-bearing seed plants and are mainly trees with some woody shrubs, lianas and epiphytes.

For most people the word 'cone' probably conjures up the woody kind that make wonderful winter decorations. These wind-pollinated, mature female cones also make wonderful natural barometers: open when dry and shut when wet, even when fallen. Female cones, however, come in a wide range of guises – juniper or yew 'berries', for example. Each mature juniper cone or aril contains a single seed within the fleshy, developed cone stalk.

The oldest living tree on Earth is a conifer: a lone Norway spruce in Sweden, whose root system has been cloning versions of itself for around 9,550 years. The biggest tree on the planet is 'General Sherman', a giant sequoia in California's Sequoia National Park.

MAKE THE CONNECTION
Estimated known species: More than 600 *Pinus* (hard pine); *Araucaria araucana* (monkey puzzle tree); *Juniperus* (juniper), used to flavour gin; *Taxus baccata* (yew); *Picea abies* (Norway spruce), the original Christmas tree; *Juniperus virginiana* (red cedar), repels moths; *larix decidua* (larch); *Sequoiadendron giganteum* (giant sequoia), branch and cone pictured

Cycads
Cycadophyta

The evergreen crown of feathery or pinnate leaves on a typically stout, cylindrical trunk means that cycads are sometimes mistaken for palm trees or ferns. But they are only distantly related: palms are in fact flowering seed plants and ferns reproduce via spores. Non-flowering cycads have seed cones, including some of the largest on Earth (see previous page).

Thought to have existed for around 300 million years, cycads were hugely diverse during the Jurassic period (*c.* 201–145 million years ago) and were possibly a main food source for herbivorous dinosaurs. One species of cycad, endemic to eastern Australia, is even known as 'dinosaur food'. In times of food shortages, humans have turned to starch-rich cycads such as *Cycas revoluta* (Japanese sago palm) for sustenance, although extensive and careful processing (leaching) is required to remove the highly poisonous neurotoxin cycasin.

The seeds of these plants are produced when sperm cells in pollen from a male plant fertilise the ovules of a female plant with the help of insects, which is unusual for gymnosperms (most are wind-pollinated). Individual plants can also change sex when damaged or in very cold conditions.

MAKE THE CONNECTION
Estimated known species: Around 348 *Cycas revoluta* (Japanese sago palm), most widely cultivated, most hardy, used in gardens and as houseplants; *Zamia pungens*, pictured; *Zamia pygmaea*, smallest cycad and gymnosperm at 25-cm (10-in) tall; *Lepidozamia peroffskyana*, nicknamed 'dinosaur food'

NON-FLOWERING SEED PLANTS

Ginkgo
Ginkgophyta

The term 'living fossil' is often applied to the ginkgo. That's because *Ginkgo biloba*, also known as the ginkgo or maidenhair tree, is the only living ginkgophyte. All other species are now extinct.

This is quite a fact, and one that is best enjoyed lying under the leafy, yellow-green canopy of a ginkgo tree, staring up at the filtered sky. These trees are commonly identified by their beautiful fan-like leaves, although the appearance of female *Ginkgo biloba* trees are notably enhanced by the small apricot-like structures they produce, which gave the genus its name – in Chinese ginkgo translates as 'silver apricot'. These 'apricots' are actually seeds that have developed a fleshy protective covering in the absence of an ovary wall.

Ginkgo was once widespread throughout the world but by 2 million years ago its decline had begun and it is now native only to China. It has been widely cultivated there and elsewhere for hundreds of years.

It's easy to see why this tree is so popular; *Ginkgo biloba*'s unique foliage and backstory combine to create an alluring addition to any garden or street. Ginkgo adapts well to the urban environment, too, tolerating pollution, drought and insect invasions, and rarely suffering from disease. As a measure of its resilience, ginkgo even survived the atomic bombing of Hiroshima in 1945.

Ginkgo trees are also easy to propagate from seed and have proven to be popular subjects for penjing and bonsai. While the seeds are known for their rancid smell, the nut-like gametophytes inside are considered a real delicacy in China. Ginkgo leaf extract and ginkgo seeds have also been used in Chinese medicine for at least 2,800 years, and although their medicinal properties have yet to be scientifically substantiated, their historic health-giving benefits do appear to warrant further research.

MAKE THE CONNECTION
Known species: 1
Ginkgo biloba, the sole living species of Ginkgophyta, pictured, bottom

Gnetophytes
Gnetophyta

Diverse and dominant in the Tertiary period (c. 66–2.6 million years ago), and with fossilised pollen found even earlier, only three genera of the division Gnetophyta survive today: *Gnetum*, *Welwitschia* and *Ephedra*. They differ from other gymnosperms by the presence of trachea or water-vessel elements (special barrel-like cells found in the xylem) within the plant. Apart from that, the highly adaptive gnetophytes hardly resemble each other at all.

MAKE THE CONNECTION
Estimated known species: Around 70
Welwitschia mirabilis, strap-like leaves, endemic to the Namib desert, pictured, top; *Ephedra* ('jointfirs'), *Gnetum* (mostly lianas)

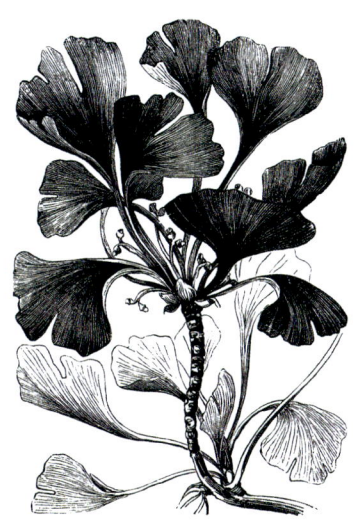

THE STORY OF PLANTS & FLOWERS 39

FLOWERING PLANTS

Flowering plants, or angiosperms, from obvious flowers to cacti to huge trees, make up around 90 per cent of the Plant Kingdom and include monocots, eudicots, magnoliids and basal angiosperms. The key to the huge diversity of flowering plant species known today – over 369,000 and counting – lies in their highly evolved reproductive systems. They have enclosed ovules, a wide array of pollination techniques, numerous ways to attract pollinators and an impressive ensemble of seed-dispersal methods, which in turn gives us much of our food.

The Big Bloom

Flowers provide much of the natural beauty in our botanical world but they are worth so much more than that. For most people, the word 'flower' conjures up images of brightly coloured petals or induces scent-related nostalgia: bright yellow sunflowers, red roses or jasmine on a warm night. In fact, flowering plants are wildly diverse and their number includes a range of perhaps unexpected members, too: the majority of deciduous trees; all cacti, succulents and grasses; many of our favourite houseplants; and a large proportion of our food. In these angiosperms, the 'flowering' part of the plant is not quite so obvious.

So what is a flower? Botanically speaking, it refers to the reproductive organs of a large number of the world's seed-bearing plants – more than 369,000 of all known plant species, to put things in context. The function of the flower is to effect reproduction via its various floral parts (see chapter 2 for more details of plant and flower anatomy and function). Stereotypically, this involves whorls of green sepals, pollinator-attracting petals, pollen-producing stamens and carpels, each with a style, a stigma and an egg-producing ovary.

The word stereotypical is key, as flowering plant species actually show a wide variation in floral structure. In some, petals are greatly reduced or lost altogether, in others the sepals are colourful and petal-like. Flower parts can be fused together or symmetrical, have a reduced stalk or cluster together in what's known as an inflorescence.

Some plants, such as sunflowers or daisies, have flower heads that are made up of many small flowers. Other plants have protracted life cycles and produce flowers at long intervals. *Tahina spectabilis*, or Madagascar palm, grows to huge proportions, dies after fruiting and flowers only once – after 100 years. *Puya rainmondii*, or Queen of the Andes, takes 80 to 150 years to bloom. Sometimes the most distinctive part of a flowering plant is its fruit, for example, pineapple, acorn or peanut. All these features are extremely helpful to botanists in terms of sorting flowering plants into their family, genus and species.

Opposite The dramatic *Cattleya labiata* (ruby-lipped cattleya), of the family Orchidaceae, dominates a lush junglescape in Martin Johnson Heade's *Orchid and Hummingbird near a Waterfall* (1902).

March of the flowers

So when did flowers first appear? The great naturalist Charles Darwin called it an 'abominable mystery', but a patchwork of fossil and genetic evidence provides an increasing number of clues. A recent study into the DNA of known plant species suggests that a small, woody 'living fossil' called *Amborella trichopoda* (see page 23), found only on the South Pacific island of New Caledonia, has a lineage that goes back 130 million years. The oldest fossilised flower specimen also dates back to this time.

Many early flower fossils, however, appear unrelated to any modern flowering plant family. A recent find in China, a genus known as *Archaefructus,* supports the theory that the first flowers were not showy, but simple and inconspicuous. While a 2017 study using fossil and genetic data of past and present flowering plants resulted in a 3D model of 'the first flower', a 140-million-year-old magnolia-like species.

Short, rapid life cycles may have also helped early bloomers to seed new ground and thus evolve faster than their competitors, the spore producers and non-flowering seed plants. What definitely *did* create an advantage for flowering plants was their approach to pollination. Wind was too hit-and-miss; insects could do a much more efficient job. The question was: how to attract them? Research suggests that it took another 30 to 40 million years before flowers had evolved sufficiently to flaunt anything like a flashy petal at a passing insect, but once they did, they sparked a 'great radiation' of diversity.

Thus flowers evolved to include bright colours and irresistible scent in their quest for pollination. The sweet, nutritious liquid that they produced lured the ancestors of bees, butterflies and wasps onto their pollen-topped stamens and waiting stigmas. If a match was made, the ovule would be fertilised and mature into a seed complete with an embryo – the part that would grow into a new plant upon germination – and, in most cases, a nutritious endosperm to help feed it. When flowers developed the ability to cross-pollinate between the flowers of different plants of the same species, rather than flowers on the same plant, their gene pool became stronger and their dominance in the Plant Kingdom was secured.

FLOWERING PLANTS

Monocots
One seed leaf and one-aperture pollen

The monocots comprise around 20 per cent of all flowering plants, and around a third of these are orchids. A great many monocots are true grasses, including all of the world's major cereal crops, such as rice, wheat and maize. Add the edible components of fellow monocots such as bananas, coconuts, ginger, turmeric, onions, garlic and sugar cane, and you've got a delicious meal. Ornamentally, monocots range from horticultural bulbs (lilies, daffodils, tulips, irises, bluebells) to houseplant favourites (Swiss cheese plant, spider plant, peace lily), and fast-growing foliage (bamboo) to tropical icons (the palm tree). Monocots can also lay claim to the biggest and possibly sexiest seed courtesy of the coco de mer (archaically named *L. callipyge*, Greek for 'beautiful buttocks'), pictured on page 100; as well as the smallest, the orchid, which is unusual for its lack of endosperm.

MAKE THE CONNECTION
Estimated known species: Around 70,000
Houseplants *Monstera deliciosa* (Swiss cheese plant); *Chlorophytum comosum* (spider plant); *Spathiphyllum* (peace lily); *Tillandsia* (air plant); *Sansevieria trifasciata* (snake plant); *Aloe* (aloe)
Ornamental *Bambusa vulgaris* (common bamboo); *Cocus nucifera* (coconut-producing palm tree); Orchidaceae (orchid); *Lodoicea* (coco de mer)
Food *Saccharum.* (sugar cane); *Allium schoenoprasum* (chives); *Ananas comosus* (pineapple), pictured; *Musa acuminata* (cultivated banana); *Zea mays* (maize); *Allium cepa* (onion); *Allium sativum* (garlic); *Triticum aestivum* (most common wheat); *Oryza sativa* (Asian rice)
Bulbs *Lilium* (lily); *Tulipa* (tulip); *Iris* (iris); *Narcissus* (daffodil)

What makes a monocot?

* An embryo with a single cotyledon or seed leaf – look inside a kernel of maize or sweetcorn for an example of this

* Long and narrow leaves with parallel veins

* Unbranching fleshy stems with vascular bundles scattered throughout rather than arranged in a ring around the outside

* Short and stringy roots with no central main root

* Underground storage organs such as a bulb, corn or rhizome (an underground stem)

* Flower parts – including petals, sepals, stamens and later seed pods – in multiples of three

* Pollen grains with a single aperture (a softer furrow or pore, where the pollen tube can break through for germination)

* No secondary growth and therefore no wood, although palms and agaves can produce a wood-like substitute

FLOWERING PLANTS

Top Many well-known houseplants are monocots, including the holey-leaved Swiss cheese plant (*Monstera deliciosa*).
Bottom Spathe and spadix of the calla lily (*Zantedeschia aethiopica*) from the plant family Araceae. **Left** Aloe is a member of the lily family, and can be identified as a monocot in part by its leaf-by-leaf rosette of parallel-veined, strap-like leaves.

Eudicots

Two seed leaves and tricolpate pollen

The eudicots comprise more than 75 per cent of all flowering plants, including most trees, garden plants, herbs and food crops. The word 'eudicot' was introduced in 1991 to differentiate dicots that had tricolpate pollen – pollen with three softer areas through which the pollen tube could break through – from those that didn't: the magnoliids and the basal angiosperms. Eudicots either stem from a small group of early diverging basal eudicots that tend to show some ancestral characteristics: the Ranunculales (barberry, poppy and buttercup families) and the Proteales (water lotus and plane tree families), or they originate from a large clade of core eudicots. These are highly diverse and include families with five sepals, five petals, one or two whorls of five stamens each and a whorl of three or five carpels.

What makes a eudicot?

* An embryo with two cotyledons or seed leaves – look inside a bean seed for an example of this

* Leaves with reticulated venation – veins that are interconnected and form a web-like network

* Stems with vascular bundles in a ring around the edge

* Roots that develop from the lower end of the embryo, known as the radicle

* Underground storage organs such as a corm or rhizome

* Flower parts – including petals, sepals, stamens and later seed pods – in multiples of four or five

* Pollen grains with three apertures where the pollen tube can break through for germination

* Secondary growth that produces wood and bark that helps to increase the plant's diameter

Opposite *Passiflora laurifolia* is a remarkable 'paper mosaick' by Mary Delany (see page 372), illustrating one of around 400 species of passion flower (*Passiflora*), some of the most intricately-flowered eudicots on Earth.

MAKE THE CONNECTION
Estimated known species: Around 175,000
Trees *Quercus* (oak); *Acer* (maple); *Eucalyptus* (eucalyptus)
Ornamental *Lathyrus odoratus* (sweet pea); *Papaver rhoeas* (corn poppy); *Dianthus caryophyllus* (carnation); *Paeonia* (peony); *Rosa* (rose); *Jasminum* (jasmine); *Passiflora* (passion flower)
Houseplants *Sempervivum* (hen and chicks/houseleek); *Ficus lyrata* (fiddle-leaf fig); *Opuntia ficus-indica* (prickly pear cactus), pictured below
Food *Solanum* (potato); *Cucurbita* (courgette); *Fragaria* x *ananassa* (strawberry); *Prunus persica* (peach); *Helianthus* (sunflower); *Viola tricolor* (viola heartsease); *Tropaeolum* (nasturtium)
Herbs/medicinal *Foeniculum vulgare* (fennel); *Lavandula* (lavender); *Borago officinalis* (borage); *Digitalis* (foxglove)
Trade *Camellia sinensis* (tea); *Gossypium* (cotton)

FLOWERING PLANTS

Magnoliids
Two seed leaves and one-aperture pollen

The magnoliids are an informally named group of dicots that includes four orders: Laurales, Magnoliales, Canellales and Piperales. They are more closely related to monocots and eudicots than to the basal angiosperms. The magnolia is a conspicuous member of this group, with its showy creamy-white or pink flowers, while the avocado is the most common fruit. Other members provide valued spices, perfume and wood: cinnamon, nutmeg, black pepper, bay, camphor and ylang-ylang. Altogether the magnoliids are quite a feast for the senses. The plants in this group are also believed to retain the characteristics of more primitive flowers and thus are central to discussions about the earliest flowering plants – were they woody trees or shrubs (not unlike living magnolias) or tropical paleoherbs? The answer could help to decide whether these flowers really did catalyse the 'great radiation' of biodiversity or whether other factors, such as climate, were complicit.

MAKE THE CONNECTION
Estimated known species: Around 8,500
Laurales *Laurus nobilis* (bay laurel); *Persea americana* (avocado); *Cinnamomum verum* (true cinnamon), pictured below; *Cinnamomum camphora* (camphor laurel); *Sassafras* (sassafras);
Magnoliales *Magnolia* (magnolia); *Liriodendron* (tulip tree); *Myristica fragrans* (nutmeg/mace); *Cananga odorata* (ylang-ylang); *Asimina triloba* (common pawpaw); *Rollinia deliciosa* (biriba/wild-sugar apple); *Annona reticulata* (custard apple); *Annona muricata* (soursop)
Canellales *Canella winterana* (wild cinnamon); *Drimys winteri* (winter's bark); *Tasmannia lanceolata* (Tasmanian pepper)
Piperales *Piper nigrum* (black pepper); *Piper methysticum* (kava-kava); *Asarum caudatum* (western wild ginger)

What makes a magnoliid?

* An embryo with two cotyledons or seed leaves – look inside an avocado seed for an example of this
* Leaves with branching veins
* Large flowers with numerous, spirally-arranged tepals (indistinct petals and sepals), stamens and carpels
* Pollen grains with one aperture where the pollen tube can break through
* Woody or herbaceous
* Presence of aromatic oil

FLOWERING PLANTS

'O the pleasure with trees!
The orchard—the forest—the
oak, cedar, pine, pecan-tree,
The honey locust, black-walnut,
cottonwood, and magnolia'

Walt Whitman, 'Poem of Joys',
Leaves of Grass (1860)

Top *Magnolia Grandiflora* (2003), a watercolour on paper by contemporary artist Jenny Barron, depicts a strikingly flowered species from one of the most obvious and ancient magnoliid plant families, Magnoliaceae. **Left** *Foliage, Flowers, and Fruit of the Pepper Plant* by Marianne North illustrates *Magnolia grandiflora*'s relative *Piper nigrum* (the source of black and white pepper), one of a number of aromatic spice-bearing magnoliids, including cinnamon and nutmeg.

THE STORY OF PLANTS & FLOWERS 47

FLOWERING PLANTS

Basal angiosperms
Diverged from the lineage that leads to most other flowering plants

Like the magnoliids, the basal angiosperms don't fit so neatly into a flowering plant group. For many years, they were placed alongside the eudicots in a well-known collective division called the dicots, due to the presence of two seed leaves or cotyledons and similarities in leaf venation. However, they also share features with other angiosperms, such as one-aperture pollen and a lack of differentiation between petals and sepals. It is thus thought that they diverged from the ancestral angiosperm lineage before all the others diverged from each other. Their group includes the orders Amborellales, Nymphaeales and Austrobaileyales, referred to collectively as ANITA or ANA. The water lily, star anise and *Amborella trichopoda* (see page 23) are the standout species in terms of beauty, usefulness and plant science.

What makes a basal angiosperm?

* An embryo with two cotyledons or seed leaves – look inside a water lily for an example

* Alternate leaves with ovate to triangular, smooth-edged blades

* Solitary flowers or many small, spirally-arranged ones

* Large flowers or many small ones with numerous, spirally-arranged tepals (indistinct petals and sepals), stamens and carpels

* Pollen grains with a single aperture where the pollen tube can break through

* Woody shrubs or vines or aquatic herbs

* Presence of aromatic oils

MAKE THE CONNECTION
Estimated known species: More than 100
Amborellales *Amborella trichopoda,* endemic to New Caledonia, the only known species of *Amborellales* is currently placed by scientists at the most basal lineage of angiosperms following complete genome sequencing
Nymphaeales *Nymphaea alba* (white water lily); *Nymphaea lotus* (white Egyptian lotus); *Nymphaea gigantea* (giant water lily); *Nymphaea caerulea* (blue Egyptian water lily); *Nymphaea mexicana* (yellow water lily); *Caboma aquatica* (fanwort), frequently planted in aquariums
Austrobaileyales *Illicium verum* (star anise), pictured below; *Austrobaileya scandens* (a woody liana), the oldest species of flowering plant in Australia; *Illicium anisatum* (Japanese star anise), highly toxic

Opposite The relatively rare red *Nympaea alba* var. *Rosea*, discovered in Sweden in the 19th century, displayed in *Paxton's Flower Garden* (Cassell, 1882) by Professor Lindley and Sir Joseph Paxton, revised by Thomas Baines.

OUR BOTANICAL BIOMES

Humans relate to plants in numerous ways throughout the course of each day – we eat them, we grow them and we are visually inspired by their extraordinary structures, features and displays. However, the most powerful way to explore and understand the Plant Kingdom is in the wild, where plants grow in their native habitats and climates or have adapted to become an integral part of the scene, living alongside and playing host to a vast array of other life forms, and actively altering the landscape through their evolutionary tactics, growth patterns and diversity. These spaces are Earth's biomes.

Deciduous forest

Deciduous forests are temperate forests that go through four main seasons: spring, summer, autumn and winter; summer is warm, winter is relatively cold. Rain generally falls throughout the year. In autumn the leaves of these deciduous trees change colour and fall off and in spring they grow back again. This is because leaves lose water quickly, and the trees can't take in enough moisture to sustain them when the ground is potentially frozen, so dropping the leaves is a way of staying dormant through the winter.

The east of the United States and Canada, most of Europe and parts of China and Japan all contribute to this biome. Within this, the temperate forest levels include a tree stratum with 18–30-m (60–100-ft) tall hardwoods, such as maple, oak, birch, magnolia and beech and potentially some conifers; small trees and saplings; shrubs such as holly or laurel; herbs and ferns; and the forest floor where lichen and moss grow. Anything that lives in this biome has to adapt to the changing seasons, including plants, animals, birds and insects, as well as the many human colonies that have settled there.

MAKE THE CONNECTION
Trees *Fagus* (beech); *Quercus robur* (English oak); *Quercus alba* (American oak) *Acer* (maple); *Magnolia grandiflora* (magnolia bull bay); *Liquidambar* (sweetgum tree); *Carya* (hickory); *Tilia x europaea* (common lime); *Carya illinoensis* (pecan tree); *Betula papyrifera* (paper birch); *Juglans regia* (walnut)
Shrub/vine *Ilex aquifolium* (holly); *Hedera* (ivy); *Viburnum opulus* (guelder rose); *Rhododendron* (rhododendron); *Gaylussacia baccata* (huckleberry); *Viscum album* (European mistletoe)
Understorey *Mnium hornum* (carpet moss), *Athyrium filix-femina* (lady fern); *Allium ursinum* (wild garlic/ramsons); *Hyacinthoides non-scripta* (British bluebell); *Anemone nemorosa* (wood anemone); *Crocus vernus* (spring crocus)

Opposite Limahuli Garden and Preserve, Kauai, Hawaii. Hawaii boasts a unique biome called a cloud rainforest that obtains some of its moisture from 'cloud drip' in addition to precipitation.

Rainforest

Rainforests make up 6 per cent of the Earth's land surface, produce 40 per cent of the world's oxygen and contain the most diverse range and highest volume of plants in the world – up to 80,000 different species in the Amazon rainforest alone, including those that give us most medicines, such as quinine. This biome is mostly commonly found near the Equator, in a hot and humid climate where it rains virtually every day. Some call it the jungle.

There are four distinct layers in the forest: the widely spaced 40–60-m (140–200-ft) tall emergent trees with straight, smooth trunks, few branches, umbrella canopies and buttress roots; an upper canopy of 18–40-m (60–130-ft) tall trees providing homes and food for most animals; an understorey of 18-m (60-ft) tall trees, made up of tree trunks, small trees, shrubs and smaller plants; and an almost plantless forest floor – unless a gap in the canopy opens above.

Much of the 'rain' in the forest is in fact caused by plant transpiration (see page 112); many plants also have drip tips, grooved leaves and oily coatings that help the water run off onto the plants or ground below.

MAKE THE CONNECTION
Ornamental Orchidaceae (orchids); Bromeliaceae (bromeliads); *Bougainvillea* (bougainvillea); *Socratea exorrhiza* (walking palm); *Victoria amazonica* (Amazon water lily); *Ficus* (fig trees, including stranglers and banyan trees); *Rafflesia arnoldii* (corpse flower), world's biggest flower; Nepenthaceae (tropical pitcher plants); *Monstera deliciosa* (Swiss cheese plant)
Medicinal *Catharanthus roseus* (rosy periwinkle), used in anti-cancer drugs; *Cinchona* (cinchona tree), produces quinine
Food *Euterpe precatoria* (acai palm); *Vanilla planifolia* (vanilla orchid); *Musa* (wild banana and plantain); *Coffea* (coffee); *Bertholletia* (Brazil nut); *Theobroma cacao* (cocoa bean tree); *Curcuma longa* (turmeric); *Zingiber officinale* (ginger)
Materials rattan genera (rattan palm); *Hevea brasiliensis* (rubber tree); *Bambusa tulda* (Indian timber bamboo); *Swietenia macrophylla* (mahogany)

Tundra

The absence of trees is one of the defining features of the tundra. It's too cold for them to survive, so what takes their place are primarily dwarf shrubs, sedges, grasses, mosses and lichens, with maybe the odd tree in between.

There are three types of tundra: Arctic, Antarctic and alpine. The Arctic tundra is found in the far Northern Hemisphere, such as Alaska and Russia; here the subsoil is permanently frozen and it is basically impossible for trees to grow; however, moss, lichen and heath are happy to step in. Winters are very cold and dark but summers are warm enough for the permafrost to melt, so marshes, lakes, bogs and streams abound, as do around 1,700 known plant species. Global warming is an immediate threat in these regions.

Antarctic tundra occurs in Antarctica and the sub-Antarctic islands. Most of it is covered in ice fields but some parts of the Antarctic Peninsula have areas of rocky soil where mosses, lichens, liverworts and aquatic algae can be found. Meanwhile, alpine tundra does not have a permafrost and is found in mountainous regions worldwide. It supports dwarf shrubs growing close to the ground, and often transitions to subalpine forest, a conifer-dominated intermediate zone that may precede dense forest.

MAKE THE CONNECTION
Calliergon giganteum (Arctic moss); *Saxifraga caespitosa* (tufted saxifrage); *Lupinus arcticus* (Arctic lupine); *Salix pulchra* (diamond-leaf willow); *Ledum groenlandicum* (Labrador tea)

Taiga

This is the world's largest biome. Snow and coniferous forests generally equals taiga (other names for this biome include snow forest and boreal forest, after the Greek god of the north wind, Boreas) and much of this type of landscape stretches across Eurasia and North America, between the tundra and the temperate forests.

Despite its size, there are fewer plant and animal species in the taiga than in other forest biomes. Insects love the summer months and birds migrate here to nest and feed on them when it is warmer, but temperatures are below freezing for six months of the year – too cold to take up permanent residence for many creatures.

Some lichens and mosses do snuggle up to the pine, spruce and fir trees. These evergreen conifer trees have long, thin needles covered in wax to help protect them from the cold. But this is not the only way in which the trees have adapted to help them to thrive in their environment. Keeping their needles all year helps initiate photosynthesis as soon as the weather gets warm to maximise the growing season, and the trees also grow thin and close together to provide protection for each other from the wind. And the classic triangular, tiered Christmas tree shape so perfect for the layering on of tinsel and baubles? It's actually designed to stop branches breaking under the weight of all the snow.

MAKE THE CONNECTION
Abies (fir); *Pinus* (pine); *Picea* (spruce); *Vaccinium vitis-idaea* (lingonberry); *Vaccinium oxycoccos* (cranberry); *Ptilium ceista-castrensis* (peat moss); *Cladonia rangiferina* (reindeer lichen); *Rubus chamaemorus* (cloudberry)

Opposite Most of Central Siberia's pristine Putorana Plateau, a complex of high, flat-topped mountains and deep and wide canyons, gorges and cold-water lakes is covered by Siberian larch (*Larix sibirica*) taiga forests.

'The clearest way into the Universe is through a forest wilderness.'

John Muir, *John of the Mountains: The Unpublished Journals of John Muir* (1938)

OUR BOTANICAL BIOMES

Top Wild prairie flowers in Minnesota include coneflower species of *Rudbeckia*, *Ratibia* and *Echinacea* from the plant family Asteraceae. **Right** Spring rains in the Pawnee National Grassland, Colorado can give way to blankets of summer wildflowers around the geologically prominent Pawnee Buttes. **Bottom** Kenya's Maasai Mara is so called for the circles of trees, scrub, savannah and cloud shadows that mark the area – 'Mara' in Maa (the language of the Maasai peoples) means 'spotted'.

Temperate grassland

Temperate grasslands are landscapes defined by large, flat, rolling plains of grasses, some flowers, herbs and a few trees. Here the seasons alternate between hot summers and relatively cold winters. The climate does not produce enough rainfall to support the growth of a forest, but not so little as to form a desert. This biome includes the plains and prairies of central North America, the steppes of the former Soviet Union, the pampas of Argentina and Uruguay, the veldts of South Africa and the puszta of Hungary.

The majority of temperate grasslands are semi-natural and co-dependent on low-intensity farming, which maintains these grasslands through grazing and cutting regimes. Truly natural temperate grasslands are rare and their associated wild flora often threatened. The species and often the height of the primary grass that grows in a temperate grassland also varies due to temperature, rainfall and soil conditions.

When the rainy season arrives, many grasslands are blanketed in wildflowers. Colour also comes in the form of wildfires, which are important as a means of natural selection. Many grasses easily bounce back, while species of fire-intolerant trees and shrubs are prevented from taking over.

MAKE THE CONNECTION
Grasses *Andropogon gerardi* (big bluestem grass); *Bouteloua gracilis* (blue grama grass); *Elymus canadensis* (Canada wild rye); *Nasella pulchra* (purple needlegrass); *Bouteloua dactyloides* (buffalo grass); *Schizachyrium scoparium* (little bluestem); *Sporobolus heterolepis* (prairie dropseed); *Bouteloua curtipendula* (sideoats grama); *Koeleria macrantha* (junegrass); *Sorghastrum nutans* (Indian grass); *Cortaderia selloana* (pampas grass)
Flowers *Astragalus canadensis* (Canada milkvetch); *Asclepias* (milkweed); *Liatris* (blazing star); *Echinacea* (coneflower); *Solidago* (goldenrods); *Helianthus petiolaris* (prairie sunflower); *Dalea purpurea* (prairie-clover); *Baptisia australis* (blue wild indigo); *Aster* (aster); *Machaeranthera tanacetifolia* (Tahoka daisy); *Leucanthemum vulgare* (oxeye daisy)

Savannah

The savannah – where tropical grassland meets open woodland – is a landscape characterised by trees that grow sufficiently wide apart or regularly spaced to allow enough open canopy for grasses to flourish beneath them.

Found on either side of the Equator, on the edges of tropical rainforests, savannahs have a warm temperature all year round; winters are very long and dry, while the summers have a distinct 'monsoon season'. Most familiar of these are the acacia-studded East African savannahs, such as the Serengeti Plains of Tanzania. Many large grass-eating mammals such as the wildebeest and zebra live here, moving around and munching on the plentiful grasses, but also some hungry carnivores, such as lions and cheetahs, who will happily cash in on the roaming prey.

South America also has savannah in Brazil, Colombia and Venezuela. The vast Los Llanos plain, east of the Andes, is regularly flooded by the Orinoco River, and the plants here have adapted to grow for long periods in standing water. Australia also boasts savannah, complete with eucalyptus and kangaroos.

MAKE THE CONNECTION
Adansonia (baobab), pictured below; *Pennisetum purpureum* (elephant grass); *Acacia senegal* (Senegal gum acacia); *Anigozanthos manglesii* (kangaroo paw); *Vachellia tortilis* (umbrella thorn); *Euphorbia ingens* (candelabra tree); *Cynodon dactylon* (Bermuda grass)

Desert

Deserts cover around one-fifth of the Earth's land surface. They occur anywhere rainfall is less than 50cm (20 in) per year, meaning they can be hot and dry as well as cold, semi-arid and coastal.

Hot and dry deserts are found near the Tropic of Cancer and the Tropic of Capricorn. What little vegetation there is in these areas comes in the form of ground-hugging shrubs and short, woody trees. Many plants, such as cacti, have adapted to endure intense summer heat and drought, adopting long-term water-storage solutions: packing leaves with nutrition; developing waxy coatings to stop moisture from escaping and spines to keep animals at bay. The Sahara is the most well-known of the world's deserts, as well as the largest of the hot and dry kind. Others are found in North America (the Chihuahuan, Sonoran, Mojave and Great Basin), South and Central America, Southern Asia, Ethiopia and Australia.

Cold deserts largely occur in temperate regions at higher latitudes with relatively lower temperatures that can bring large amounts of snowfall in winter – as found in Antarctica and Greenland, for example. Vegetation is sparse and includes deciduous plants with spiny leaves. Coastal deserts such as Africa's Namib and Chile's Atacama deserts are found where the land meets the ocean and often boast a wider variety of drought-adaptive native plants.

MAKE THE CONNECTION
Cacti/succulents *Echinocactus/Ferocactus* (barrel cactus); *Opuntia ficus-indica* (prickly pear cactus); *Carnegiea gigantea* (saguaro cactus), popularly depicted in westerns; *Stenocereus thurberi* (organ pipe cactus); *Sempervivum* (hen and chick succulent/houseleek)
Trees/shrubs/flowers *Yucca elata* (soaptree yucca); *Bursera microphylla* (elephant tree); *Salvia eremostachya* (desert sage); *Baileya* (desert marigolds); *Hesperocallis* (desert lily); *Chilopsis* (desert willow); *Kali tragus* (prickly Russian thistle, common tumbleweed); *Agave tequilana* (blue agave), produces tequila; *Encelia farinosa* (brittlebush); *Olneya tesota* (desert ironwood); *Yucca brevifolia* (Joshua tree)

Wetlands

Wetlands can refer to marshes, swamps, bogs and fens and generally involve a lot of standing water. They are found in almost every terrestrial biome in the world, often around lakes, rivers, streams or near coasts, from the cypress swamps of Louisiana, to the salt marshes along the east coast of the United States, the huge peaty bogs of the West Siberian Lowlands and dense mangrove forests of the tropics.

Plants in these environments are mainly aquatic or hydrophytes (adapted to thrive in very moist and humid conditions), such as water lilies, cattails, sedges and rushes. Some are emergents (living partially in air), some float and some are submerged. Blooms of algae can also persist.

Wetlands often get a bad name, but they are hugely important for the environment for numerous reasons. They can take on excess water from other sources and therefore help prevent flooding. They can also supply drinking water and actively help clean it due to the ability of some plants to naturally filter out sediments. Plant matter from wetlands is also released into freshwater biomes where fish and other aquatic creatures can feed on it. And then there's rice, the staple diet of more than half the people on the planet, and a verdant feature of many wetlands around the world.

MAKE THE CONNECTION
Juncaceae (rushes); Cyperaceae (sedges); *Oryza sativa* (rice); *Acer saccharinum* (silver maple); *Nymphaea odorata* (white water lily), pictured below; *Rhizophora* (mangrove tree)

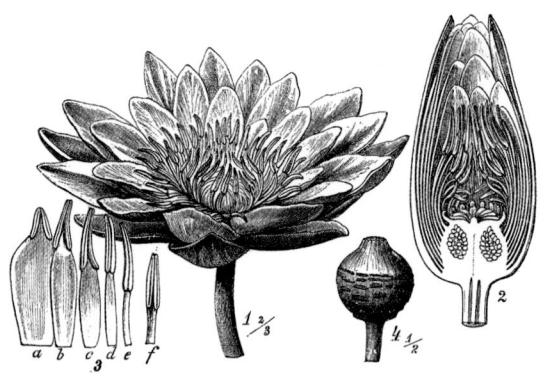

OUR BOTANICAL BIOMES

Left Santa Rita prickly pear cactus (*Opuntia santa-rita*) is native to desert regions of Arizona, California and Mexico – beware the hairy spines. **Top** Tree houseleek species, many native to the Canary Islands, are naturally dormant in summer to help conserve water. **Bottom** The mysterious swamps of Louisiana are forested with bald cypress trees (*Taxodium distichum*), dripping in Spanish moss (*Tillandsia usneoides*).

PLANTS & PEOPLE

4.6 billion years ago	Formation of the Earth
3.8 billion years ago	First cellular life
1.5 billion years ago	Animal and Plant Kingdoms diverge
First land plants	450 million years ago
55 million years ago	Primitive primates
First trees and forests	385 million years ago
5.8 million years ago	First bipedal humans
First seeds	380 million years ago
200,000 years ago	First modern humans, *Homo sapiens*
First flowers	250–140 million years ago
50,000 BCE	The 'Great Leap Forward' for mankind
Flowers evolve into a 'Big Bloom'	140 million years ago
10,000 BCE	Neolithic Revolution
First written evidence of medicinal herbs	3000 BCE
4000–3500 BCE	Dawn of civilisation begins in Mesopotamia
First vineyards	2800–1900 BCE
2000–1500 BCE	First evidence of ornamental gardens, Egypt
Alexander the Great conquers the ancient world	336–323 BCE
1500 BCE	First evidence of foreign plant exchange and trade, Egypt
The 'Silk Road' brings trade between East and West	130 BCE–1453 CE
350–287 BCE	Theophrastus publishes *Historia Plantarum* (Enquiry into Plants)
First monastic kitchen and herb gardens	529 CE
1440 CE	The printing press enables widespread reproduction of botanical texts and herbals
The Crusades bring war but also cultural exchange	1095–1291
1453–1750	Age of Exploration and colonisation of the Americas
Golden Age of Botanical Art	1600–1800
1544	Orto Botanico di Pisa opens as first university botanical garden in Europe
Tea introduced from China to the West	1607
1685–1820	Age of Enlightenment
Height of 'Tulip mania' and the Dutch Golden Age	1637
	Plant cells first illustrated in Robert Hooke's *Micrographia*	1665
1753	Carl Linnaeus publishes *Species Plantarum*, heralding binomial nomenclature
Royal Botanic Gardens open in Kew	1759
	Discovery of photosynthesis	1779
1859	Charles Darwin publishes *On the Origin of Species*
Birkenhead Park (world's first public park) inspires design of Central Park, New York	1859
	Mendel's Laws of Inheritance	1866
1914–1926	Claude Monet's water lilies
Photosynthetic equation	1931
1967–1970	Flower Power Generation
First plant genome sequenced	2000
2003	Human genome sequenced
First flower grown in space	2016

The complete story of plants and flowers spans billions of years, the story of plants and people, only 200,000 or so. Yet in that time, plants have provided sustenance, shelter, medicine and inspiration, defining the human experience.

Plants and people share the same backstory up until around 1.5 billion years ago, when membrane-bound eukaryotic organisms began to diverge into various kingdoms: plants, animals, fungi and non-green algae.

In the Late Cretaceous period (*c*. 145–66 million years ago), around the same time that flowers were really coming into bloom, primates diverged from other mammals. Several incarnations of ape later, the first hominins began to get a real bipedal, big-thinking, precision grip on the world. *Homo habilis* began using tools; *Homo erectus* swapped tree-sleeping for ground-dwelling, and began to use complex tools, fire-making and cooking techniques – or so evidence appears to suggest. *Homo sapiens* (anatomically modern humans) evolved around 100,000–200,000 years ago (a figure that morphs as new fossil and genetic evidence appears) and is now the only living species of the genus *Homo*. The question is, how did we get here, how did we set up camp in so many of the world's biomes and how did we adapt to survive there? Many of the answers lie within the Plant Kingdom.

For millions of years ancestral and modern humans had to find their own food, so they spent a large part of each day gathering plants or hunting for animals. Then around 10,000 years ago, we made the transition to producing food and changing our surroundings in the process. We created forest gardens, bred livestock and grew crops to feed them and ourselves.

We put our animals out to pasture, collected wood from trees and other natural materials to make fire, shelter and tools. We began to gather and socialise, wear clothing and create adornment and art. We drew on nature and in turn all these activities helped to transform Earth's landscape.

By 1500 BCE humans were gardening for purpose (see page 64) and gardening for pleasure (see page 68), setting a precedent for gardens and parks of the future (see page 75). Then between 350–287 BCE, the first botanical text was published in the form of Theophrastus's *Historia Plantarum*, laying the foundation for nearly 2,500 years' worth of books, magazines and papers (see page 76) that collectively shine a light on the Plant Kingdom. With our big brains and tool-wielding ways, we humans were not content to simply 'wander among the tulips'. We wanted to own them, paint them, cultivate them, trade them (see page 66) and display them as floral arrangements (see page 72), rooms full of houseplants, flower paintings (see page 79) or plant-inspired craft and landscape designs (see page 80).

For its part, modern plant science has given us extraordinary insight into the minutiae of the Plant Kingdom. We've discovered some amazing things along the way, but advanced plant science, along with increased agricultural knowledge, has also shifted the paradigm. Very real issues such as deforestation, global warming, genetic modification, increased monoculture and subsequent species and biome loss mean the next few chapters of the story of plants and flowers are very much in our hands.

Opposite Plants and people have a deep-rooted, cellular connection that goes back way further than the advent of Homo sapiens. Our paths, it appears, are forever intertwined.

Forest gardens

The first forest gardens were simply a way for early humans of around 10,000 years ago to secure food in tropical areas. They originated along rainforest river banks and in the wet foothills of monsoon regions. Plant species that provided certain foods or materials were promoted but those that didn't were weeded out. Eventually species that didn't naturally occur in an area were incorporated, too. Thus, hunter-gatherers became gardeners for the first time.

These 'forest gardens' mimicked the natural order of forests and included the various plant layers that would traditionally be found within them: tree canopies, smaller trees and shrubs, herbaceous plants, ground cover, roots and vines. These gardens were an integral part of a dwelling place and vital to human survival.

This kind of polyculture – using multiple crops in the same place – is still widely practised around the world today, although it is often referred to as an 'orchard garden' or 'home garden'. In Kerala, India, it is still the most common form of land use; in Africa it is practised on the slopes of Mount Kilimanjaro; while in Nepal most families have a home garden of some kind.

Not only do such initiatives help provide essential provisions – food, fodder, fuel, medicines, spices, herbs, construction materials and potential income – this kind of agriculture can be significantly more sustainable, too. Growing plants in successive layers makes maximum use of space, encourages biodiversity and thus has a lower impact on the natural habitat and ecosystems; ground cover eliminates the need for weed killers (and laborious weeding) and can fix nitrogen (transforming into a form useable by organic life); while companion planting can help naturally control pests, such as insects.

For those involved in today's permaculture or edible forest movements, it's the win-win benefits of working *with* nature that prove so attractive, both in the moment and – faced with very real issues such as biodiversity loss and climate change – in the long term.

First ornamental gardens

The first enclosure of outdoor space is thought to have naturally occurred around 10,000 BCE, as a way to keep unwanted visitors out. In fact, the word 'garden' stems from the Old English *geard,* for 'fence'.

By 3000 BCE Mesopotamia, the 'land between the rivers' of Tigris and Euphrates, was a bustling, literate conurbation. In Western tradition, the infamous Garden of Eden and the Hanging Gardens of Babylon allegedly once flourished in this region. Archaeological evidence has yet to be found to support this claim, but many factors point to the Assyrian city of Nineveh, the remains of which lie just outside modern-day Mosul in northern Iraq.

But an extract from one of the most popular pieces of Babylonian literature, written in cuneiform on clay tablets as early as 2000 BCE, tells of one of the earliest forms of garden. It describes a courtyard or temple space where trees with fragrant and edible fruits such as the date palm (*Phoenix dactylifera*) and tamarisk or salt cedar (*Tamarix*) were part and parcel of the reconstruction of Paradise. Various historic accounts of Babylon's 'hanging gardens' also describe plants such as cypress, myrtle, juniper, olive, willow, pomegranate, plum, fig, quince and grapevine. One thing they all needed was water.

Thanks to the Nile and its network of irrigation canals, water was abundant enough in civilised Ancient Egypt to promote the idea of the pleasure garden further. Large-scale palace gardens first appeared just before the Middle Empire (2035–1668 BCE). They were laid out geometrically with enormous ponds and borders of new species of plants discovered upon the conquest of neighbouring lands. Historical evidence from sources such as tomb paintings also suggests stylistic features including rows of shade- and fruit-giving trees, flowerbeds, water-loving lotus blossoms and papyrus plants, plus the addition of such green spaces to luxury residences and temples. Humans, it appeared, had now made their first documented steps towards gardening for purpose *and* pleasure.

Opposite Marten van Heemskerck's 16th-century engraving of the Hanging Gardens of Babylon (with the Tower of Babel in the background) was most probably based on descriptions from Ancient Greek sources.

Plants that shaped society

The first ornamental gardens such as those evidenced in Ancient Egypt set trends for layout and planting design but also for a wide range of plant species that played a significant role in the shaping of civilised society, among them the grapevine and papyrus sedge.

From the tall, grassy papyrus (*Cyperus papyrus*), which was widely cultivated in the wetlands of the Nile Delta, came a primitive form of 'paper' produced by soaking, overlapping, hammering, drying and polishing its fibrous pith. Papyrus was also made into baskets, sandals and even food. From the grapevine came vital shade for the garden created by twining the plant over pergolas, but also wine made from the luscious, sugar-rich fruit as evidenced by written records and organic materials found in ancient tombs, some dating back 3,000 years around the time when grape culture was first thought to have been introduced from the Levant. The contents of these *papyri* (papyrus texts) and *amphorae* (jars) show that wine was not simply an indulgence but a way to honour the gods, and also dispense medicinal remedies via the addition of beneficial herbs and tree resins.

Over in Sumer, located in the lower regions of Mesopotamia between 4500–1900 BCE, wine-making from grapes and dates was similarly prolific and drunkenness culturally embraced. Evidence of wine-making has also been found in Ancient China and Persia, while the first vineyards in Greece are thought to have been cultivated as early as 2800–2200 BCE. In Ancient Greece, wine also had a lot to answer for, used by philosophers such as Socrates, Plato and Aristotle as a means to the truth. By the time the Ancient Romans 'arrived', they had thousands of years of viticulture to draw on. Grapes were widely cultivated across their ever-expanding empire and wine was available to all. Consuming it was a democratic, everyday occurrence.

There are now 79 accepted species in the genus *Vitis*, or grapevine, with between 5,000 and 10,000 varieties of the common grapevine (*Vitis vinifera*) alone. It's come a long way since foragers and farmers first harvested its berries, and it has had a *lot* of influence on society, but it's by no means the only star of the show.

In his pioneering text, *Enquiry into Plants* (350–287 BCE), the Greek philosopher Theophrastus described numerous species that continue to shape society today: wood for building houses and ships; herbs for medicine and flavour; and legumes and cereals for food including species of wheat, tree resin and gum. Some game-changers weren't mentioned in writing until much later – for example, the first reference to coffee as a beverage only crops up in the sixteenth century (although cultivation of *Coffea arabica* is documented back to twelfth-century Yemen where Sufi priests used it to stay awake for religious rituals). Thriving global trade of coffee subsequently led to the first recorded European coffee house opening in Venice in 1645, and nearly 400 years on, the caffeine-rich drink is a vital start to the day for many people as well as a profitable cash crop.

A more ancient drinking tradition, though, is tea, the invention of which Chinese legend attributes to the mythical sage ruler Shennong. He is credited with identifying hundreds of medical and poisonous herbs by personally testing their properties via his transparent body. Tea was supposedly an antidote against the poisonous effects of some 70 herbs, first tasted when the leaves of the *Camellia sinensis* plant allegedly fell onto burning twigs around 2437 BCE. *Camellia sinensis* made its way to the Western world from China in 1607 via the Dutch East India Company. By the eighteenth century, England was firmly established as a great tea-drinking nation and, in a bid to break the Chinese monopoly on the thirst-quenching leaf, introduced it to British-ruled India. Tea would also have a significant bearing on global politics, with the American Revolution (1765–1783) and the Opium Wars (1839–1842 and 1856–1860) both attributed, in part, to the trade of this seemingly harmless plant.

MAKE THE CONNECTION
Society-shaping plants *Vitis vinifera* (European grapevine); *Camellia sinensis* (tea); *Coffea* (coffee); *Morus alba* (white mulberry), widely cultivated to feed silkworms; *Papaver somniferum* (opium poppy); *Saccharum* (sugar cane); *Triticum* (wheat); *Elaeis* (oil palm); *Hevea brasiliensis* (Pará rubber tree); *Quercus* (oak); *Gossypium* (cotton); *Nicotiana tabacum* (tobacco plant); *Chincona* (quinine); *Oryza sativa* (Asian rice); *Tulipa* (tulip)

PLANTS & PEOPLE

Top By the time this early 19th-century painting was produced, China had been artfully cultivating, processing and preparing tea for over two millennia. **Left** The leaves and leaf-buds of *Camellia sinensis*, the evergreen shrub from which tea is produced, contain significant levels of the stimulant caffeine, also found in coffee and cacao plants.

'*Plants underpin all aspects of our everyday life – from the food that we eat, to the clothes that we wear, the materials we use, the air we breathe, the medicines we take and much more. These essential services provided by plants are far too often taken for granted.*'

Professor Kathy Willis, Director of Science at the Royal Botanic Gardens, Kew, in *State of the World's Plants* report, 2017

THE STORY OF PLANTS & FLOWERS 63

GARDENING FOR PURPOSE

The first forest enclosures set the scene for the rise of purposeful gardening. Subsistence farms and orchards emerged and large-scale agriculture paved the way for new and larger civilisations. In turn these civilisations developed an interest in the art, science and technology of growing plants – the practice of horticulture. Today we use thousands of years of horticultural practice and knowledge to purposefully manage the landscape and the Plant Kingdom into providing for us – from the smallest kitchen gardens to huge cash crops.

Food

The early human diet was dictated by necessity, but likely was largely based around nuts, berries and other wild vegetation supplemented with meat when the opportunity arose. Even then plants were ever-present in the food chain.

The Neolithic or Agricultural Revolution brought about the domestication of plants and increased knowledge of the properties of these specimens. Around 9000 BCE, humans began to selectively breed cereals. Neolithic founder crops of the Southwest Asian 'Fertile Crescent' included flax, three cereals (emmer wheat, einkorn wheat and barley) and four pulses (lentils, peas, chickpeas and bitter vetch). Millet, rice, oranges and peaches were some of the first species cultivated in China. In parts of Africa, Asian yams, taro, plantains and bananas were popular crops, while in the Americas, maize, beans, squash, manioc and potatoes were commonly grown.

Today, more than 30,000 known plants have documented edible uses, and of those some 7,000 have been cultivated or gathered by humans at some point in our collective global history. Of those, however, fewer than 30 estimated plant species now provide 95 per cent of the world's combined calorie intake.

What plant-based food we do eat is generally divided into seeds (cereals, legumes, nuts and oils, such as wheat, rice, beans, peas, lentils and rapeseed), fruits (technically the ripened ovaries of plants such as apples, oranges, mangoes, avocados, tomatoes, pumpkins and aubergines) and vegetables (leaves, roots, stems, bulbs, inflorescences, seed cases and flowers, including potatoes, carrots, onions, spinach, lettuce, bamboo shoots, broccoli and cauliflower).

Some of these plant foods grow wild, much as they did in Paleolithic times, but most exist in their current incarnations via the horticultural hands and minds of humans. We grow them on a massive scale for commercial food production and we grow them in our gardens. What we are cultivating and eating are the results of around 10,000 years of agricultural history.

Opposite The inter-fertility of fruits from the genus *Citrus*, including citron (*Citrus medica*), bitter orange (*Citrus aurantium*) and sweet orange (*Citrus sinensis*) as depicted by Basilius Besler, allow for numerous hybrids and cultivars.

Medicine

Medicinal plants have also been identified and used by humans for thousands of years. Details about the use of herbal medicine during Prehistoric times are vague but, based on archaeological evidence and the practices of certain present-day indigenous populations, point to the divination of herbal remedies from plant spirits to shamans (with non-divined medicinal plant knowledge naturally picked up along the way). Women are also thought to have played a role in gathering medicinal herbs.

With written evidence, however, it becomes easier to observe specific plant use and prescription, from pictorial notes made about myrrh and opium by the Ancient Sumerians, to the prized Ancient Egyptian medical papyri, the *Ebers papyrus* (written around 1500 BCE), which describes more than 850 plant medicines, including aloe, cannabis, garlic and juniper. Similarly, there is much to be learned from Indian Ayurvedic medicine, the principal texts of which were transcribed between the first and sixth centuries BCE but date back at least as far as the Vedas (1500–500 BCE) – or, indeed, the first Chinese book on medicinal plants, written around 250–200 BCE but originated circa 2800 BCE.

In Ancient Greece and Rome, 'Father of Medicine' Hippocrates (*c.* 460–370 BCE), pharmacologist and botanist Pedanius Disocorides (author of *De materia medica*, 50–70 CE) and Pliny the Elder (author of *Naturalis Historia,* 77–79 CE) compiled their own studious lists of herbs, which would significantly influence the botanists and physicians of the medieval Byzantine and Islamic worlds. Such texts were also made available to western Europe following the Crusades (1095–1291) and later the advent of the printing press (*c.* 1440), thus inspiring a new wave of 'herbals', which also gleaned from the herbal lore of medieval monasteries, medicine men and women, and the garden patch and hedgerow.

Several hundred years on, such herbal lore remains relevant, despite the advances of modern medicine. More and more people are looking to plant-powered healing, wellbeing and nutrition, as delivered by botanical remedies (see chapter 5) and recipes (see chapter 4) – perhaps longing to get back to the natural in the face of so much that is manmade.

Science, exploration and trade

Plants also play a huge role in the story of science, exploration and trade. Early human explorations and trade happened naturally as the first Paleolithic hunter-gatherer peoples moved from place to place, or simply left the homestead to find provisions, and later barter for items of perceived similar value. The driving force behind both these activities? Survival.

As Neolithic civilisations evolved (10,200–2500 BCE), a need for food and tools expanded to include luxuries such as spices, textiles, wood and precious metals. Trading networks and routes were established, explorers and navigators employed to find new lands and resources, and nations plundered for their riches including curious or useful species of plant.

Organised plant exploration took longer to establish. Ancient Egyptian frescoes dating back to 1500 BCE show Queen Hatshepsut sending ships to bring back trees from their trading partner, the Land of Punt. While carvings from the Temple of Karnak from the reign of Thutmose III (1479–1426 BCE) show depictions of exotic plants and seeds.

Early writings from around the world also produce evidence of the use and incorporation of non-native global plant species. Records from the Western Zhou Dynasty in China (1100–771 BCE), for example, reveal the introduction of crops including wheat and barley. As trading routes opened up between East and West, so too, it appears, did the variety of foreign crops moving between nations. Over in Greece, grafting techniques that would allow the introduction of new fruit species are attributed to the Ancient Greek school of Hippocrates (460–370 BCE), while fellow Ancient Greek Theophrastus (371–287 BCE) also name-checked species from distant locations in his *Historia Plantarum* and *Causes of Plants*, including the banana.

Theophrastus's books show an awareness of plant anatomy, the process of germination and the roles of soil and climate in growing. Much of his knowledge would have been garnered from the plants he observed on his travels through Greece or in his botanical garden at the Academy of Athens. But he also profited from his association with the conqueror Alexander the Great (356–323 BCE), who spent most of his ruling years as the King of Macedonia on an unprecedented military campaign through Asia and northeast Africa. From Alexander's followers, Theophrastus heard of cotton, banyan, pepper, cinnamon, myrrh and frankincense.

The ever-expanding Roman Empire also brought new and exciting things to its rulers and citizens, as reflected in Pliny's *Naturalis Historia* (77–79 CE), which told of papyrus, grapes, wood and spices. Many goods arrived via a network of trade routes between China and Europe, now known as the Silk Road, with spices particularly prized for their ability to enhance flavour and mask the taste of spoiled meat. When the Ottoman Empire took Constantinople in 1453, however, it blocked many of these key land-based trade routes. As a result, nations were forced to turn to the sea to import these goods. The Age of Exploration was born.

This period of intensive, largely sea-based European exploration that took place between the fifteenth and eighteenth centuries was also driven by a curiosity to discover what else lay out there. Italian-born, Spanish-based navigator Christopher Columbus found one answer when he encountered the Americas in 1492. The great Portuguese explorer Vasco da Gama found another with a new maritime route around Africa to India in 1498. Voyages across the Atlantic and Pacific oceans and entire circumnavigations of the globe followed suit, as did explorations by the Dutch, French and English. Latecomer Captain James Cook landed upon Tahiti, Australia, New Zealand and Hawaii.

As a result of all these travels across the seas, new settlements and colonies were established, and this triggered the widespread transfer of plants, animals, culture, human populations, technology, ideas and, unfortunately, diseases, between Old and New Worlds. Scientists and artists travelled with the explorers on these voyages to collect evidence of 'new' plants and flowers and record them. By the end of the Age of Exploration, the study of our botanical world in terms of how plants work and evolved, as well as their uses, was very much underway – as was the visual depiction of plants and flowers in botanical illustration and art.

Opposite Sea Island cotton plant (*Gossypium barbadense*), as illustrated in Köhler's *Medicinal Plants* (1887), is coveted by clothmakers for its long, silky cellulose 'fibres'; for the cotton plant, it's all about dispersing seeds.

GARDENING FOR PLEASURE

Gardening for pleasure arose from the first ornamental gardens of Mesopotamia and Egypt. These ancient gardens were designed and built for shade, spirituality and relaxation but also for beauty, which was often enhanced by the inclusion of new and exciting plant species from faraway lands. The pleasure principle also began to extend to the act of gardening or being around plants, flowers and nature, and to the promotion of health, wellbeing and happiness – as well as the numerous ways in which a garden can inspire.

Private gardens

The concept of the private pleasure garden – for all, not just the culturally elite – was largely popularised by the Romans. By the first century CE, many basic 'kitchen gardens', or *horti*, had evolved into colonnaded 'peristyle' gardens decorated with statues, water features, pathways and painted frescoes that paid homage to the divine or symbolic. The upper classes also had huge showpiece villa gardens out in the countryside, where pleasure was derived from spending time in the garden and showing off its features – including an abundance of native and exotic plants.

Private pleasure gardens across the globe would continue in this vein for another 1,800 years. From the gardens of medieval castles and country houses to the formal and elaborate arrangements of Renaissance-era Italy, France and England, to the palace and scholarly gardens of Asia and the Islamic world, horticultural pleasure was for a long time very much about status, display or devotion, despite marked differences in style between various regions, cultures or philosophies.

However, the nineteenth century changed this dynamic once again. A period of great social and economic change was stimulated by the Age of Exploration, the 'discovery' of the New World, the rise of colonialisation and industrialisation. Plants became symbols of lands conquered and explored, bringing new riches and providing insight and evidence for academics eager to decipher the mysteries of creation.

Plants also brought pleasure to an increasing number of private gardens and gardeners worldwide, as evidenced through enhanced seed catalogues and gardening guides, which began to include ornamental specimens and notions of design as well as edibles and herbs. The horticulturally-inclined also had increasing access to larger and more awe-inspiring botanical gardens, botanical magazines and societies, new tools and greenhouses, architecturally-incorporated green spaces, and plant-inspired art and interiors. Gardening for pleasure and in private was going mainstream.

Public gardens

The orchards, tree-lined promenades, town squares, philosophical gardens, common pastures and communal areas of countryside evidenced throughout civilisation could all be considered 'public spaces'. The concept of 'the public park', however, as somewhere that people of all walks of life could enjoy, was first popularised in the seventeenth century when some of Europe's biggest houses and palaces opened their green spaces to the lower echelons of society.

Hyde Park in London was opened to the public as a 'Royal Park' in 1637 and was particularly popular during May Day parades. The decadent splendour of Versailles in France was also opened up to the people, although visitors were subject to a highly censored, politically motivated code of conduct. The seventeenth century also saw the rise of the botanic garden, often associated with university faculties of medicine.

The Orto Botanico di Pisa started the trend in 1544 and similar establishments quickly spread across Italy. The tradition also passed into Spain and northern Europe, with notable examples including the University of Oxford Botanic Garden (1621), the Hortus Botanicus in Amsterdam (1638), the Jardin des Plantes in Paris (1640), Sweden's Uppsala University (1655) and the Chelsea Physic Garden (1673). In 1759 came the Royal Gardens at Kew, developed as a public attraction as well as a centre for learning, followed by the Missouri Botanical Garden a century later in 1859.

Although many nineteenth-century 'public parks' were intrinsically plant-focused, most were primarily concerned with health and wellbeing, from New York's Central Park (1859) co-designed by Frederick Law Olmsted (see page 381) to Victoria Park in East London (1845), which was actively petitioned for by over 30,000 local residents.

Today there are thankfully numerous public parks to enjoy worldwide, from Beihai Park in Beijing (which has been around in some form or another since the eleventh century, complete with beautiful imperial palaces) to the conserved wild nature of Yellowstone National Park – the United States' and the world's first official national park.

Opposite Katsushika Hokusai's (see page 355) *Big Flowers* series (1832) of woodblock prints, including *Chrysanthemum and Bee,* rose to popularity during the Impressionist era – indeed Claude Monet owned a whole set.

The garden as muse

The botanical world has served as a source of inspiration for millennia, as have the earthly paradises that men and women have continually sought to create. 'I perhaps owe having become a painter to flowers,' said the French Impressionist Claude Monet (1840–1926) who created his own slice of Eden at Giverny, renowned for inspiring monumental works such as his Water Lilies series (see page 345).

Indeed, Monet was just one of an array of garden-loving artists of the late nineteenth-century Impressionist era, including the Spanish artist Joaquín Sorolla, who created his horticultural muse in Madrid, inspired by Islamic gardens such as the Alhambra in Granada, and fellow Frenchman Pierre Bonnard, who was heavily influenced by leading English garden designers of the period, William Robinson and Gertrude Jekyll. Both Monet and Bonnard's gardens appeared to run wild and free, as did the principles of Robinson's seminal text *The Wild Garden* (1870), which favoured loose drifts of colour over notions of formality. Bonnard's *'jardin sauvage'* at Vernonnet is very much captured in his landscape-esque artwork *Resting in the Garden* (1914), and such horticultural ideals were soon assimilated into the gardens and public spaces of a newly industrialised America.

Frida Kahlo's garden (see page 348) at the Casa Azul in Mexico City was also an undeniable muse displaying the same native species and colours in pots and flowerbeds as found in her paintings and hair. While the artist and architect César Manrique pretty much took a whole island as his plot, working his sympathetic building designs, murals and planting schemes into the volcanic terrain of the Canary island of Lanzarote – his terraced Jardín de Cactus is particularly inspiring, with more than 10,000 different plants selected by eminent botanist Estanislao González Ferrer. Over in Marrakech, the Jardin Majorelle is a similar riot of vibrant colour and exotica that took the painter Jacques Majorelle (1886–1962) 40 years to create, a lifelong devotion to plants and art echoed by Derek Jarman's garden at Dungeness, Barbara Hepworth's fern-strewn studio in Cornwall and William Morris's garden at Red House.

Richard Mabey's wonderful book, *The Cabaret of Plants* (2015), also reminds us that plants themselves are the principal source of inspiration when creating a garden – the garden being a space in which we best present plants and potentially manifest our creative urges. It opens with illustrations of what may be the first human representations of plants in cave art some 35,000 years ago, before meandering through various encounters between plants and people, including whole chapters riffing on the favoured flowers of Romantic poets, such as William Wordsworth and John Keats.

Tucked within the pages of Mabey's 'cabaret' also lies an image of a gigantic *Victoria amazonica* water lily leaf, the extraordinary natural engineering of which inspired the metal framework of the Crystal Palace that housed Britain's Great Exhibition of 1851. As Mabey attests: 'No large building had ever been built like this chimera of steel and cellulose before, but every large glass building since ultimately derives from it.'

A roll call of inspired gardeners throughout history – among them artists, designers, scientists, politicians and, of course, gardeners – displays similarly epic proportions, from the great naturalist Charles Darwin who wrote *On the Origin of Species* (1859) in a study overlooking his garden at Down House, to American Founding Father Thomas Jefferson who wandered through his extensively experimental gardens at Monticello, Virginia, before drafting America's Declaration of Independence in 1776.

The garden was also an obvious muse for many of history's most famous botanists, from Carl Linnaeus who worked out his idea for binomial nomenclature (see page 123) during his directorship of Uppsala's botanic gardens in Sweden, to English naturalist John Ray, who conducted his taxonomic and botanical experiments in the university gardens of Cambridge.

Possibly the most a-musing garden happening of all, however, may be the day that a young scientist by the name of Isaac Newton was allegedly hit on the head by an apple. The fruit in question fell from a Flower of Kent tree (*Malus pumila*) in his family's garden at Woolsthorpe Manor, Lincolnshire, thus taking things to another dimension – it inspired the theory of gravity.

Opposite The gardens at Claude Monet's Giverny spill over with the colours, fragrance, glorious nature and light that so obviously inspired his paintings, and kept him rooted even during the artillery fire of World War I.

The march of the flower

Flowers are designed to attract. Pollinators are their desired targets, of course, but humans can't seem to help falling in love with the Plant Kingdom's most lascivious features either – we're transfixed by their colours, shapes, textures and scents, and we feel compelled to display them wherever we can.

Perhaps the most potent example of such floral lust is illustrated by the tulip mania that gripped the seventeenth-century Netherlands. Tulips were expensive as well as beautiful, inspiring great desire and trading competition among the Dutch elite and leading to the first documented economic bubble. Painters such as Jan Brueghel the Elder and Ambrosius Bosschaert immortalised the prized exotic cultivars, heightening their value. Exorbitant bulb prices eventually crashed the market but the enchantment of these lustrous flowers endured, as a cut flower and as artist's muse.

The use of flowers for display grew to even greater heights in the Western world during the eighteenth and nineteenth centuries, when increasing numbers of people had access to land. The desire to garden was also piqued by the first horticultural publications and societies, by new and exciting botanical discoveries, and by numerous nurseries that sprang up to provide seeds and plants for budding gardeners. Among these were seductive new species, hybrids and varieties (see page 130) of flowers, such as roses and pelargoniums.

Floral trends were also set by the horticultural stars of the day: William Kent and Lancelot 'Capability' Brown championed a rolling pastoral style; Humphry Repton was all about the harmony of garden and landscape; J. C. Loudon proffered the Gardenesque style, displaying the character of each plant to its full potential; William Robinson, Gertrude Jekyll and Vita Sackville-West (see page 381) went 'wild' and mixed shrubs with perennials, annuals and bulbs in deep beds; while Helena Rutherfurd Ely, a founding member of the Garden Club of America, was in favour of 'an informal and sensual style'. By the mid-twentieth century, flowers were inextricably linked with ideals of beauty and peace, and were taken up by political protestors to represent the opposite of militarisation.

The flower garden displayed

The cultivation and trade of cut flowers as an industry began in earnest in the Victorian era, helped along by the development of greenhouses. Roses, chrysanthemums, dahlias, carnations and zinnia were all the rage and flower breeders introduced hundreds of new varieties of plant.

In response to all this fervent floriculture, the Royal Horticultural Society in England launched the Great Spring Show in Kensington in 1862, a precedent for the Chelsea Flower Show that followed in 1913. Flower shows and flower arranging on both sides of the Atlantic were big news, with men and women competing to win the title of prize vegetable, bloom or display, an especially popular pastime after World War II.

The nineteenth century also established flower arranging as an art form of its own. Where Victorians were often strict, formal or prudish in their daily lives and public manners, their rounded, opulent floral displays were anything but. Roses and the 'language of flowers' were paramount, with nosegay bouquets of aromatic herbs or sweet-scented flowers also favoured.

Gertrude Jekyll's book *Flower Decoration in the House* (1907) largely set Western floristry trends for the twentieth century, giving month-by-month advice on the best use of flowers in a domestic setting, including vases and holders for plants. Jekyll followed this up with her most famous book *Colour in the Flower Garden* (1908), where intelligent combination and harmony were key – the more you knew about plants the better.

Fast-forward 20 years and the doors of artistic expression were thrown wide open to embrace the anything-goes style of British society florist Constance Spry. She put hedgerow cuttings in old tins, daisies in jam jars, pussy willows in vases, arranged scarlet roses with kale leaves and insisted that – with a little imagination – everyone's life could be enriched by flowers. Spry didn't live to see the wilder-still embodiment of flower power of the late 1960s, or the naturalistic or wild displays popularised by floral designers like London's Flora Starkey (see page 376) and Simone Gooch of Fjura, New York's Sarah Ryhanen of Saipua, or the literally out-of-this-world Japanese artist, Azuma Makoto (see page 359), but the nature-loving eccentric in her would undoubtedly have approved.

GARDENING FOR PLEASURE

'*Follow your own star . . .
Just be natural and gay and
light-hearted and pretty and simple
and overflowing and general and
baroque and bare and austere and
stylised and wild and daring and
conservative, and learn and learn
and learn. Open your mind to
every form of beauty.*'

Constance Spry, British floral designer
(1886–1960)

Top *Still-Life with Flowers* (1665) by Dutch painter Rachel Ruysch (see page 348) pays homage to a profuse arrangement of much-coveted tulips, carnations and roses. **Bottom** Pulitzer prize-nominated *Flower Power* by Bernie Boston perfectly captured the mood and impetus of flower-powered protestors at the 1967 Pentagon March, actively helping to turn the tide of public opinion against the Vietnam War.

THE STORY OF PLANTS & FLOWERS   73

SPREADING THE WORD

The invention of the printing press in 1440 was a significant milestone in the story of plants and flowers. It spread the knowledge gained during the Age of Discovery faster and wider than previously possible, and promoted the subsequent flow of scientific and cultural exchange. It also facilitated the 'Golden Age of Botanical Art'. The masses were enthralled. Scientists and artists were further inspired. Botanical gardens were heralded for their on-the-ground presentations of exotic species and research. Letters and ideas were avidly exchanged; societies were founded that still thrive today and the celebration of botanical diversity continues.

Botanic gardens

Botanic Gardens Conservation International (BGCI) is the world's largest network for plant conservation and environmental education, currently listing more than 500 botanic gardens around the world from around 100 countries across every continent. Based at Kew Gardens in London, and with offices in the United States and China, BGCI is committed to spreading the message of conservation.

A look at current population statistics illustrates why it is so important for these institutions to relay the message about protecting and conserving the environment: around one in five of the world's 391,000 known plant species are already threatened with extinction while millions of people depend on wild plant resources for at least part of their livelihoods. Plants are vital to the existence of the world *and* to humans. One-third of the world's known plant species are also grown in botanic gardens, making these spaces vital to the ongoing conservation of plants, especially through seed-banking.

Modern botanical gardens are descendants of the medieval monasteries of the eighth century, where monks promoted the beauty of plants and flowers as a celebration of God, nurtured knowledgeable collections of medicinal herbs and sowed the seed for the sixteenth-century physic gardens of Italy. The first of these was created by the Italian physician and botanist Luca Ghini at the University of Pisa in 1543. Physic gardens at the Italian universities of Padova (1545), Firenze (1545) and Bologna (1547) were then swiftly established, set up purely for the academic study of medicinal plants.

These were followed by the Leipzig Botanical Garden (1580), Hortus Botanicus in Leiden (1587), Jardin des Plantes de Montpellier (1593), University of Oxford Botanic Garden (1621), University of Uppsala Botanical Garden (1655), Royal Botanic Garden Edinburgh (1670), Chelsea Physic Garden (1673) and Saint Petersburg Botanical Garden (1714), all still open today.

Extensive overseas exploration produced something of a paradigm shift. The chief concern of seventeenth-century botanic gardens was not medicinal research but an interest in exotic 'plant trophies' from outside Europe – the beautiful, the strange and the new. Plants poured in from around the world and as the Age of Enlightenment dawned in the late eighteenth and early nineteenth centuries, botanic enquiry took on a scientific dimension, inspiring publications such as Carl Linnaeus's *Systema Naturae* (1735) and *Species Plantarum* (1753), or Philip Miller's *Gardener's Dictionary* (1731).

Botanic gardens became more educational as a result, with 'order beds' demonstrating the latest plant classification systems devised by botanists working in the associated herbaria. Many of these gardens featured heated conservatories or orangeries that would keep cold-sensitive plants warm over winter, including the iconic Palm House at Kew. Founded in 1759, Kew was originally a royal 'physick garden' that, under the directorship of naturalist Sir Joseph Banks and botanists Sir William Hooker and thence his son Sir Joseph Hooker, grew in size and reputation to become the world-respected botanic garden it is today.

Official plant hunters and horticultural collectors were sent out to South Africa, Australia, Chile, China, Ceylon (now Sri Lanka) and Brazil to furnish the 'great botanical exchange house of the British Empire'. Joseph Hooker helped Charles Darwin classify the plants that he collected in the Galápagos – the trip that inspired Darwin's theory of evolution. He also corresponded regularly with the leading American botanist of the day, Dr Asa Gray, and helped establish tropical gardens in the colonies for identifying plants of commercial value.

By the end of the nineteenth century botanic gardens had gone truly global, including among their number Bartram's Garden, Philadelphia (1728), the Royal Botanic Garden Sydney (1816), the US Botanic Garden (1820), Singapore Botanic Gardens (1822) and the New York Botanical Garden (1891).

The BGCI now lists more than 3,000 botanical institutions in its 'Garden Search' worldwide but you don't need to be a botanist to visit them – from orchid festivals to nature-inspired art and photography exhibitions to glimpses of rare openings of the gigantic titan arum or corpse flower (*Amorphophallus titanum*) (see page 19) – which generally blooms once every seven years – there's inspiration for everyone.

Opposite *The Common Sunflower (Flos Solis Maior)*, a hand-coloured copper engraving by Basilius Besler (see page 319), first published as part of *Hortus Eystettensis* (1613), helped put the art into botanical illustration.

Correspondence

The eighteenth-century fascination for natural history as both science and pastime produced a plethora of correspondence. Letters of note include a series penned by the Swiss philosopher Jean-Jacques Rousseau to a Madame Delessert in Lyon with the purpose of helping her daughters learn botany. On Rousseau's death, they were translated and published to widespread acclaim, helping to cultivate Carl Linnaeus's system of plant taxonomy (see page 123).

But letters were as vital to the discussion of ideas as they were to their dissemination as illustrated by the 1,400 or so letters exchanged between Charles Darwin and Joseph Hooker, some of which include batches of the manuscript of Darwin's *On the Origin of Species* (1859) for Hooker's comment. Sir Joseph Banks was also keen on sharing and sounding out botanically inspired ideas, corresponding with almost 600 people around the world, his burning passion for natural history and exploration obviously matched by his talent for networking. Banks was also the recipient of letters containing botanical illustrations produced in the field or onboard ships by resident artists, such as the HMS *Endeavour*'s Sydney Parkinson (see page 323) whose works traversed jungles and oceans to make it into Banks's *Florilegium*.

But perhaps the most important aspect of correspondence was that it bridged the gender gap. Darwin is thought to have written to around 2,000 correspondents, of whom around 100 were women, a good portion of which were female botanists. Botany, like horticulture and floristry, was something that nineteenth-century women could do at home without judgement by society. As such, many women not only eagerly embraced it, they became highly skilled practitioners, going on to publish papers and books. Among their number were the American botanist and suffragist Mary Agnes Chase, the British plant morphologist Agnes Arber and the celebrated children's book author and illustrator Beatrix Potter.

While today's botanically inspired correspondence is now just as likely to come in the form of an email, blog, tweet or Instagram post, a quick search through such social media platforms shows that sharing notes on plants is very much alive and well, and may even be experiencing something of a renaissance.

Publications

Herbals were among the first literature produced in Ancient Egypt, China, India and Europe and were some of the first books to be printed. Most notable of these was John Gerard's *Herball or Generall Historie of Plantes*, first published in 1597, which arranged plants according to their physical similarities rather than their medicinal ones, heralding the beginnings of scientific classification. Gerard's book also helped bring herbalism to the layperson through subsequent publications, such as Nicholas Culpeper's astrologically themed *Complete Herbal* (1653), which had an extensive impact on medicine in early North American colonies, and Elizabeth Blackwell's *A Curious Herbal* (1737–1739), depicting exotic species from the New World.

The visual presentation of botanical knowledge came even further to the fore through the publication of *florilegia* and *flora* – a florilegium being a collection of flower illustrations and a flora a botanical record of plants associated with a specific place – the most iconic of which is surely the seasonally arranged, supremely illustrated *Hortus Eystettensis* (1613) produced by the German apothecary and botanist Basilius Besler (see page 319), which took 16 years to complete.

Over the following centuries, florilegia and flora would develop in response to studies in botany and ecology, avid exploration and colonial audiences eager for information about their new land and how to grow plants there, resulting in publications such as the *Codex Liechtenstein* (1776), William Curtis's *Flora Londinensis* (1777–1798), *Flora Graeca* (1806–1840) and Mark Catesby's *Natural History of Carolina, Florida and the Bahama Islands* (1729–1747).

British apothecary and botanist Curtis went on to launch the highly influential *Curtis's Botanical Magazine* (also known as *The Botanical Magazine; or Flower Garden Displayed*) in 1787 to an eager international audience, packed with colour illustrations of plants from leading botanical artists around the world. It is still published today (by Royal Botanic Gardens Kew), a true pioneer among a growing raft of plant-inspired magazines from *Gardens Illustrated*, *Garden Design*, *Horticulture* and *The Garden*, to a more avant-garde mix of plants and contemporary culture in *The Plant*, *Pleasure Garden*, *Rake's Progress*, *Garden Collage* and *The Planthunter* – find details for further reading on page 392.

Left The winter–spring flowering springfire shrub (*Metrosideros collina*), as observed by Sydney Parkinson while visiting Tahiti aboard the *Endeavour*. **Top** Detail of a tulip by Sydenham Edwards who produced thousands of botanical plates for *Curtis's Botanical Magazine* and his own *Botanical Register*. **Bottom** Botanist, expert on grasses and suffragette Mary Agnes Chase (1869–1963) often funded her own research trips when women were prohibited from attending.

Botanical illustration

The first herbals went to great lengths to describe the appearance and properties of plants and flowers, but when it came to truly capturing them, a picture could speak a thousand words. When herbals began to include detailed botanical illustrations, their scientific value and wide appeal considerably increased.

The first printed illustrated herbal was the *Pseudo-Apuleius Herbarius* in 1481. It undoubtedly influenced the dissemination of knowledge about medicinal plants, but unfortunately the illustrations were crude and inaccurate because they had been repeatedly copied from early manuscripts, errors and all. The German physician and botanist Leonhart Fuchs changed all that, spending some 30 years gazing intently at plants and flowers and then sharing his knowledge with fellow physicians through his pioneering work, *De Historia Stirpium* (1542).

Every illustration in this book was observed and drawn from real life, Fuchs advising the craftsmen not to 'indulge their whims as to cause the drawing not to correspond accurately to the truth'. Indeed, the drawings are amazingly accurate, especially given that they were created in the days pre-microscope. Unfortunately, the text was littered with errors, derived as it was from 'the best work of the ancients'. Fuchs also misidentified some plants by mistakenly pairing those he found near home with Mediterranean species described by classical-era scientists, such as Dioscorides.

For the main part, however, botanists became more knowledgeable about their subject, and thus botanical illustrators also upped their game in terms of accuracy and exploded detail. The Austrian artist Ferdinand Bauer covered his preliminary sketches with colour numbers so he could accurately replicate tone and shade of specimens found in the field, going on to publish the highly detailed *Illustrationes florae Novae Hollandiae* (1813), following his role as scientific illustrator on board HMS *Investigator* to Australia. Meanwhile, his similarly talented brother Franz (see page 316 for both Bauer brothers) participated in paintings and drawings of flower dissections at Kew.

The scientific world was significantly impressed and collaborations abounded.

The German artist Georg Dionysius Ehret (see page 313) worked with Carl Linnaeus and a wealthy Dutch banker, George Clifford, to produce the *Hortus Cliffortianus* in 1737. He also provided illustrations for later editions of Philip Miller's *The Gardener's Dictionary*. English naturalist Mark Catesby liaised with American botanists, such as John Bartram, to produce plant-filled backgrounds for his *Natural History of Carolina, Florida and the Bahama Islands* (1729–1747). While the English artist James Sowerby was even more prolific, providing illustrations for *Flora Graeca*, *Flora Londinensis*, *Curtis's Botanical Magazine* and his major works *English Botany* (1791–1814) and *A Specimen of the Botany of New Holland* (1793–1795), the first monograph of the flora of Australia. The extraordinary Victorian British painter and explorer Marianne North (see page 322) also visited Australia – as well as Syria, Brazil, Tenerife, California, Japan, Borneo, Java, Ceylon, New Zealand, India and South Africa – eventually founding a gallery of her life's work at Kew in 1882, which is still open today.

Perhaps the most revered botanical illustrator of all, however, was Pierre-Joseph Redouté (see page 313), nicknamed 'the Raphael of flowers' for his watercolours of roses and lilies among other fine plant species. Redouté's precise renderings of plants and flowers (over 2,100 plates of 1,800 different species in total) remain as fresh today as they did when first painted and are often reproduced as decorative art, along with fellow luminaries of the 'Golden Age of Botanical Art' (generally given as the years between 1750 and 1850).

While the primary aim of botanical illustration may have been to accurately depict plant specimens for scientific study – a closely guarded definition – there's no denying that works by some of its most visionary and poetic practitioners past and present are also more than worthy of the title of botanical art. Find just a few of them featured in chapter 6, which profiles traditional artists from Ehret and Redouté, Margaret Mee, Rory McEwen and Rosie Sanders, to those who play with the very concept of botanical art forms, such as Katie Scott and Macoto Murayama.

Opposite Under the patronage of Empress Josephine (Napoleon's first wife), Pierre Joseph Redouté collaborated with botanist Claude Antoine Thory to produce *Les Roses* (1817–1824), which soon became a collector's item.

Photography and film

The invention of photography is largely credited to the British scientist and inventor William Henry Fox Talbot and the chemist and astronomer John Herschel. The former invented the calotype 'photogenic drawing' process in 1840 (the first photographic process capable of producing negative images on paper using silver-chloride coated paper) and the latter introduced the art and science of the cyanotype (the developing process used to make blueprints) in 1842. Such technology was eagerly harnessed by family friend Anna Atkins (see page 333), who began producing her own cyanotype prints of botanical specimens. Her first publication, *Photographs of British Algae: Cyanotype Impressions* (1843), was widely considered to be the first book illustrated by photographic images.

By the turn of the twentieth century, photography had become an indispensable tool in botanical and horticultural laboratories, although the quality was blurry and certainly not good enough to rival the accuracy or reliability of botanical illustrations (the use of which still prevails today as the ideal visual accompaniment for scientific research).

Photography, however, certainly has its place, from the innovative flat lays of Roger Phillips's *Wild Flowers of Britain* (1977) and *Wild Food* (1983) to the visual feast that is *Plants of the World: An Illustrated Encyclopedia of Vascular Plants* (Kew Royal Botanic Gardens, 2017) – which uses more than 2,500 full-colour photographs to systematically explore every vascular plant on Earth – to the otherworldly microscopic photography of seeds and pollen by Rob Kesseler (see page 335). The same is true for the offshoot medium of film, making possible awe-inspiring nature documentaries, such as *The Private Life of Plants* (1995), presented by David Attenborough or the botanical timelapses of Louie Schwartzberg (see page 343).

More than 170 years since Talbot's first photogenic drawings of leaves and lace, it's hard to imagine a world without photography and film, from the plant-filled galleries of Instagram to awe-inspiring shots by artists such as Karl Blossfeldt (see page 333), Irving Penn (see page 336) Robert Mapplethorpe (see page 336), Ansel Adams (see page 339) or Viviane Sassen (see page 340) – surely some of the most persuasive arguments for looking closer at our botanical world.

Botanical style

Travel pretty much anywhere in the world, or indeed through the echelons of art history, and you'll find plants and flowers woven into the fabric of everyday life and art. The Plant Kingdom is an obvious muse, with its sculptural forms, alluring flowers, expressive foliage and abundance of naturally-occurring patterns (see Art Forms in Nature on page 310) – neutral enough to cross cultural or religious boundaries yet ripe for abstraction.

What's interesting, however, is the story behind a beautiful plant motif or an object adorned with one. Why was that particular plant or flower chosen, how do different peoples or cultures develop such distinctive decorative styles and which plant motifs have prevailed through time and why?

Fan-shaped palmettes (inspired by palm trees but also plants such as the lotus and papyrus) are thought to stem back to Ancient Egypt but prevail around the world. While leaf-inspired, scrolling arabesques – a prominent feature of Islamic art – can also be found in Ancient Roman decorations, in the ornament of Renaissance-era Europe and within the Arts and Crafts movement. Other popular plant-inspired patterns or floral fashions include vigorous vines, unfurling peonies, blooming chrysanthemums (traditional floral symbols of China and Japan), rambling roses and the humble daisy.

In ancient times, the portrayal of many of these plant species, or renditions of them, was often rooted in symbolism, from the 'tree of life' to plants representing longevity, power or fertility. Today, the language of flowers is more often side-stepped in favour of a stylish design. Still the language of plants lives on via the preferred plants and flowers that find their way into our wardrobes or into our homes – and the basic botany that we subliminally pick up along the way.

It's quite possible that you have never seen a chrysanthemum flower in real life but you may well be familiar with one from the designs of Arts and Crafts pioneer William Morris (see page 369). The same goes for flower power daisy motifs made iconic in the 1960s, the 'Jungalow-style' desert cacti or rainforest leaves popularised by Justina Blakeney (see page 379), or indeed any of the plants or flowers that have inspired millennia of art, craft, design or botanical style.

'*Art is the Flower – Life is the Green Leaf. Let every artist strive to make his flower a beautiful living thing, something that will convince the world that there may be, there are, things more precious more beautiful – more lasting than life itself . . .*'

Charles Rennie Mackintosh, Scottish designer (1868–1928)

Top *Strawberry Thief* (1883) is one of numerous plant-inspired prints by William Morris, surreptitiously furnishing wider audiences with a realm of botanical knowledge. **Bottom** This extraordinary hand-coloured micrograph of a pincushion flower seed (*Scabiosa cretica*, 2013) by artist and professor Rob Kesseler (see page 335), powerfully illustrates the potential rewards and allure of looking closer.

FLOWER POWER NOW

We hold a vast amount of knowledge about our botanical world, garnered from thousands of years of explorations and investigations. We've even grown flowers in space. But that same pioneering, inventive human spirit has also led to worrying modern-day issues such as climate change and biodiversity loss. What then for the future of plants and flowers, and indeed plants and people? The answer would appear to lie in the art and science of collaboration. The more we work with our botanical world, championing its interests as well as our own, the better it will repay us, and the longer we can continue to coexist.

The Earth laughs in flowers

On 16 January 2016, NASA astronaut and Expedition 46 Commander Scott Kelly shared photographs of a blooming zinnia flower. But this wasn't any old zinnia – this was the first flower to be grown in space. The aim of this project was to help scientists back on Earth to better understand how plants grow in microgravity and for astronauts to practise the kind of autonomous gardening they might be expected to undertake on a deep-space mission.

At one point Commander Kelly was worried that the plants weren't looking great, but rather than ask for rigid, scientific instructions, he asked for the green light to handle them based on what he *saw*, as he would do when gardening at home. The 'Veggie team' on Earth agreed and sent a very sparse guide dubbed 'The Zinnia Care Guide for the On-Orbit Gardener'. Soon the flowers were on the rebound.

This is a very extreme example of venturing out 'into the field' to learn firsthand what plants and flowers are all about. But it can also be taken as a touching, anecdotal story about the relationship between plants and people. If we choose to listen to the Plant Kingdom, it will repay us in flowers.

It's a sentiment explored in Ralph Waldo Emerson's terrestrial epic *Hamatreya* (1846). 'The earth laughs in flowers' is the poem's most famous line, a supposed celebration of the world's resplendent floral loveliness. But this only paints half the picture. Within context, *Hamatreya* is a cautionary tale about the 'boastful boys' – the people – who believe themselves to be possessors of the Earth and the ultimate power of nature to reclaim the land so proudly toiled or civilised for human gain.

The Earth does indeed laugh in flowers; each and every one of our blooming beauties is a wonder to behold. But 'Hear what the Earth say' – the line with which the poem introduces its intrinsic 'Earth Song' – might be a more fitting quote for our modern times. Had Emerson written the poem today, he may well have included direct references to climate change, pollution and biodiversity loss, some of the very real fall-outs of 'possessing' the land. What might his solutions be? Renewable energy instead of fossil fuels, a halt to deforestation, the end of genetically modified megacrops, the active conservation of endangered ecosystems and species, perhaps? What happens if we don't counteract the damage already done? Will the Earth ultimately reclaim . . . ?

There is certainly evidence of nature actively taking over where humans are long gone, or have been forced to abandon their plots: the huge fig-tree buttresses of Angkor Wat in Cambodia; the creeping greenery of the nuclear Exclusion Zones of Chernobyl and Fukushima; entire ancient cities 'lost' in the jungles of South America. The Plant Kingdom, in one form or another, has survived the many devastations of ice ages and mass extinctions over billions of years, so what then for plants and people?

The answer is overwhelmingly simple. Acknowledge the power of the botanical world and collaborate with it rather than work against it. One way is to support conservation initiatives or organisations such as the Nature Conservancy, the Sierra Club, Botanic Gardens Conservation International or the Millennium Seed Bank at Kew. Another is to lobby governments to act on or sign up to promises to meet agreements on climate change and biodiversity.

Creative collaborations and representations can also be hugely powerful in changing opinions. Rob Kesseler's microscopic photographs of pollen surely make the study of the Plant Kingdom all the more lush and fascinating, as do Katie Scott's wonderful illustrations for *Botanicum* (2016). See both artists in chapter 6. On the political front, Bernie Boston's iconic photograph *Flower Power* (1967) helped inspire thousands to join the movement against war in Vietnam, as well as an entire flower power generation. It portrayed the flower as a symbol of peace but also as an activist, uniting people and plants in the only conceivable cause – a sustainable, *shared* future.

Opposite Wat Mahathat in Ayutthaya Historical Park, Thailand is known for its iconic Buddha's head, cradled by the tangled roots of a sacred fig or bodhi tree (*Ficus religiosa*), a symbolic illustration of the power of nature to reclaim.

'I shall endeavour to find out how Nature's forces act upon one another . . . I must find out about the harmony in nature.'

Baron Alexander von Humboldt,
Prussian naturalist and explorer (1769–1859)

CHAPTER TWO

BOTANY FOR BEGINNERS

WHAT IS BOTANY?

Botany – the scientific study of plants and flowers – leads to greater understanding of our botanical world and all of the complex ecosystems it supports. Historically, this field was regarded as key to decoding the very purpose of life. Today more than ever, a deeper understanding of nature is vital to Earth's shared future.

In 1648, the results of a five-year experiment by the Flemish physician Jan Baptist van Helmont (1580–1644) were published by his son Franciscus Mercurius van Helmont. These results would help answer a mystery that had baffled philosophers and scientists for centuries – how exactly do plants grow?

Van Helmont's experiment, conducted while under house arrest by the Spanish Inquisition, involved a willow tree, a pot, a tin-coated iron lid and a lot of diligent watering and weighing, the aim being to prove or disprove a theory first proposed by Ancient Greek teachings that claimed plants grew by 'eating' soil (see page 111). His conclusion – that plants in fact grew by drinking water – was actually inaccurate, and officially rejected in the late 1600s by John Woodward, a professor and physician at Cambridge University.

Some 130 years later, in 1779, on the back of the work of English chemist Joseph Priestley, the Dutch physiologist Jan Ingenhousz (1730–1799) was still trying to crack the question about how plants grow and the effect that they had on their environment. Two and a half centuries on, although much more is known about the process of photosynthesis (see page 111), scientists are still exploring its microscopic wonders and how we can harness its life-giving powers.

What van Helmont, Ingenhousz et al. were searching for was, in fact, much bigger than the question of how plants grow. They wanted to know how the world worked and find scientific proof for how it had been created. They suspected, as we now know, that the Plant Kingdom was in part responsible for all life on Earth and, as Charles Darwin speculated in his *On the Origin of Species* (1859), that plants were also continuously evolving.

This then is botany, the study of plant life from the tiny powerhouses of plant cells to the magnitude of giant sequoia trees. It investigates what plants are, how they appear (see page 90), what they do (see page 110), what they 'know' (see page 118), how they interact with all other life on Earth and what we need to do to conserve them (see page 143). It also strives to promote consistent ways in which plant knowledge can be recorded and shared (see page 123).

As far as plants and people are concerned, Professor Kathy Willis, Director of Science at Royal Botanic Gardens, Kew, has it in a nutshell: 'A detailed knowledge of plants is fundamental to human life on Earth. Plants underpin all aspects of our everyday life – from the food that we eat, to the clothes that we wear, the materials we use, the air we breathe, the medicines we take and much more.'

Botany is, as Willis continues, 'the unique combination of beauty and science which can together provide some of the solutions for the global challenges facing humanity today' – all of which can be a lot more fun in the context of a plant-based activity you already enjoy or are passionate about from gardening, cooking and wellbeing (see chapters 3–5) to botanical art, craft or design (see chapter 6).

Opposite Observe the number and form of a tulip's petals, sepals, stamens, carpels and ovary to find distinguishing features of both the *Tulipa* genus and its plant family, Liliaceae.

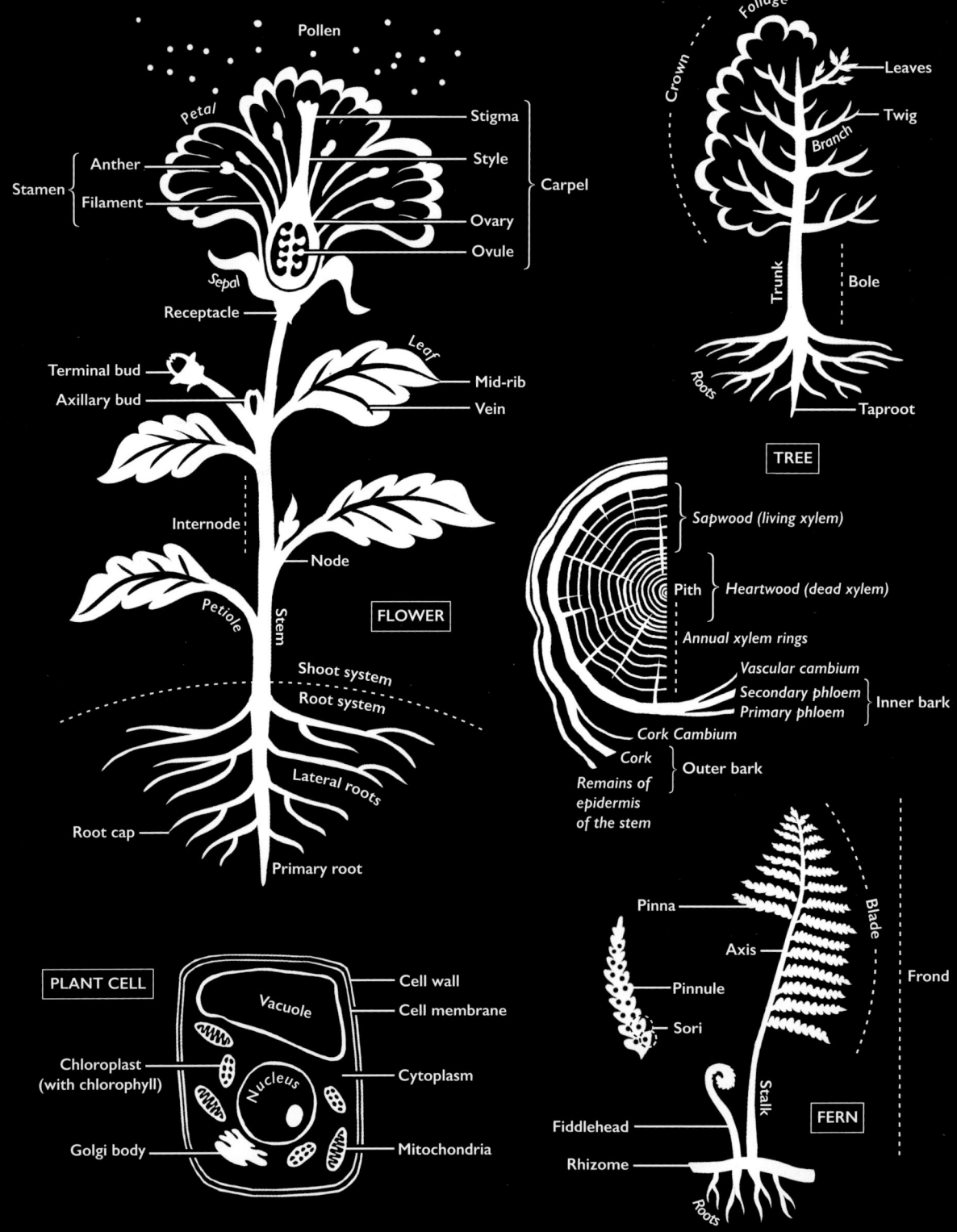

LOOKING CLOSER

From understanding why plants wilt or how sunflowers got their name to the sheer magnanimity of leaves, looking that little bit closer at the science behind plants and flowers can throw an enchanting new filter on your botanical world.

In the early third century BCE, the Ancient Greek philosopher Theophrastus set the course for future study of plants. Rather than simply list plants and their medicinal uses, his major works, *Enquiry into Plants* (*Historia Plantarum*) and *On the Causes of Plants*, also addressed how plants *worked*. It's apparent from his texts that Theophrastus was not just walking around pontificating about plants, he was closely studying and questioning their forms and functions and encouraging his followers to do the same.

Today, thanks to Theophrastus's work and subsequent centuries of botanical investigations, we now have an infinitely better understanding of why plants appear as they do, how they may have evolved and what their main processes are.

As children, most people are versed in the parts of a flowering plant, its basic life cycle and the equation for photosynthesis. Then it's over to the 'university of life'. We watch leaves change colour in the autumn, we spritz our houseplants to stop them dehydrating, we choose outdoor plants for sunlight or shade and we put bananas in paper bags with our avocados to help ripen them. We sow seeds in the spring to harvest in the summer, we're lured by fragrant sweet peas or tantalising orchids and we turn to plants for their health-giving benefits. And all the time, we're learning – often subconsciously – about how plants work.

Botanists take things one crucial step further by actively aiming to explain the science behind each of these observations.

The wilting plant, for instance, is losing water faster than it can be absorbed. Without the pulling force of transpiration (see page 112), turgidity is lost. For its part the sunflower (*Helianthus*) is so-called for its looks – the Greek *helios* meaning sun, *anthos* meaning flower – but also because its head follows the sun from east to west as it moves across the sky, possibly to enhance the process of photosynthesis and increase growth rates. And leaves, essentially a vascular plant's food factory, actually make life on Earth possible for humans by using sunlight to split water molecules and release oxygen via photosynthesis (see page 111).

In recent years botany has used advanced technology such as microscopy and genome sequencing to travel deeper still into the enquiry into plants. Botanists can now observe diverse and complex shapes of pollen (see page 108), study the DNA sequence of fossilised plants and use their expanding knowledge to help conserve as well as utilise plants. Two thousand years on from Theophrastus's first foray into classification, botany also has its very own universal 'language' (see page 124) through which scientists can discuss the many facets and applications of the Plant Kingdom.

In light of pressing issues such as climate change, biodiversity loss and human population growth, looking closer at plants and flowers has never been so necessary – but botany can also be fascinating and fun. It's amazing how even the most basic knowledge and perhaps a smattering of more-addictive-than-you-think Latin can open up a whole new window on *your* botanical world.

Opposite Five ways of looking more closely at plants, five ways of more deeply understanding not just how plants appear (see page 90) but what they do (see page 110).

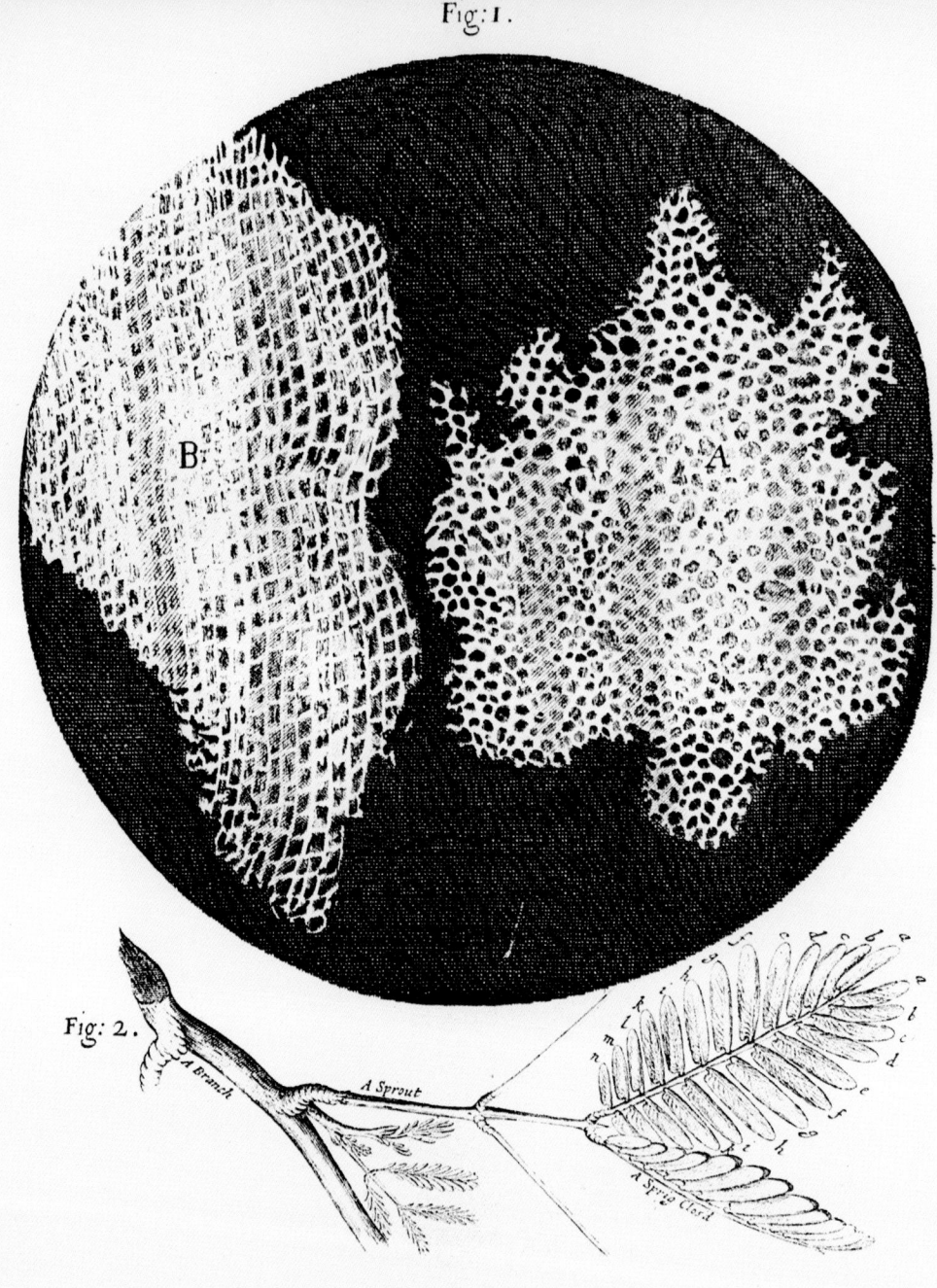

HOW PLANTS APPEAR

The appearance of plants and flowers is central to a greater understanding of them and can be divided into two complementary areas of study – plant morphology, which examines the physical form and external structure of plants and can largely be seen with the naked eye, and plant anatomy, which looks at microscopic plant structures, such as cells and tissues. Explored in tandem, the true nature of a plant and the power of its components can be revealed.

What the botanists saw

Theophrastus is often cited as the 'Father of Botany', thanks to his surviving botanical publications. But the seed for his enquiries into plants was almost certainly sown by the 'Father of Science', his teacher, Aristotle (384–322 BCE). 'If you would understand anything, observe its beginnings and its development,' said the great Ancient Greek philosopher. He was fascinated by the physical world and believed that by closely observing it, he could learn how nature worked – and by understanding how nature worked, he believed that one could understand everything.

Such natural philosophy laid the foundations for much of the botanical science that followed. In order to establish any kind of theory or conclusion about the merits of the Plant Kingdom, one had to ask nature itself for the answers, which is exactly what many of history's greatest scientific minds did.

The Renaissance era, between the fourteenth and seventeenth centuries, was a particularly heady time for the study of the natural world, eventually establishing botany as an independent science, distinct from medicine and agriculture. During this period, the advent of the printing press (*c.* 1440) made the teachings of botanical pioneers such as Aristotle and Theophrastus more accessible, thus inspiring a new raft of herbals and then floras (see page 76), while the invention of the microscope around 1590 led to more detailed enquiries into plant anatomy.

By 1665, microscopic power was so advanced that major new discoveries could be made, as illustrated by natural philosopher Robert Hooke's zoomed-in masterpiece *Micrographia* (1665). By way of a new-fangled compound microscope and illumination system, he recorded observations of thin slices of cork, perceived to be 'all perforated or porous, much like a Honey-comb'. These pores, or 'cells', as he went on to call them due to their resemblance to monks' rooms, were plant cells – or, more precisely, the cell walls in cork tissues. Hooke also examined fossils under the microscope, comparing petrified wood to a piece of rotten oak. His comparative deductions, asserting that fossils were the remains of once-living organisms impregnated by petrifying fluids, would help lay the foundations for evolutionary science.

It was another 200 years before Charles Darwin published *his* theory of evolution in his game-changing publication *On the Origin of Species* (1859). What he concluded over several decades of exploration, observation and deduction was that life forms evolved via a process of natural selection or 'survival of the fittest'. Darwin's finches usually get most of the inspirational credit for this huge scientific breakthrough, but Darwin's fieldnotes from his time on the Galápagos Islands also pay due homage to plants, including more than 200 specimens of 'all the plants in flower'.

What Darwin saw from the variety of specimens that he collected and observed was the emergence of the evolutionary tree of life, which he duly sketched when he returned. Looking at plants alongside fossils and animals made absolute sense to him, coming as he did from a rich botanical tradition: his grandfather Erasmus Darwin was an evolutionist and perceived, like Carl Linnaeus (see page 122), that plants had sexes; his father, Dr Robert Darwin, had a bountiful garden that Charles played in as a child; while his mother taught him how to identify plants from flowers.

As a young man, Charles Darwin also studied basic botany under Professor John Henslow, founder of the Cambridge University Botanic Garden (see page 75). Henslow also compiled a herbarium of British flora comprising more than 10,000 plant specimens put together by around 100 collaborators, Darwin among them. The collection was organised to emphasise variation within species and determine the limits between them, a framework Darwin continued to use when collecting his own samples. Henslow also secured Darwin's place on HMS *Beagle*, the ship that would lead him to many of his greatest discoveries.

Theophrastus, Hooke and Darwin were certainly not the only botanists to make groundbreaking discoveries about the natural world, but their *way of seeing* did have a huge impact on how other plant scientists – past and present – went on to view the Plant Kingdom.

Opposite In 1665, Robert Hooke published *Micrographia*. Within it was the first ever illustration of the cellular structure of cork – and the first use of the word 'cell' – pictured alongside a sprig of *Mimosa pudica* (sensitive plant).

IN THE FIELD

The morphological parts of plants, from roots, stems and leaves to flowers, fruits, seeds and spores, can speak volumes about what plants do to survive, and how they potentially evolved. Study these plant parts in the wild, in a garden or – as botanists often do – from herbarium sheets of pressed or dried plant specimens and you will gain further insight into the huge contribution of the Plant Kingdom to our shared world. Out 'in the field' is also the ideal place to contemplate theories, formulate ideas and draw comparisons and conclusions.

Roots

Flowers and leaves might be the crowning glories of vascular plants but they wouldn't be anything without the often hidden depths of roots. Typically relegated to the dark recesses of the soil, it's the roots that provide food, water, strength and security and thus a means for plants to survive. Which is also why, when it comes to seed germination, roots emerge first.

Embryonic roots
What these embryonic roots or radicles are looking for and can hopefully access is water and nutrients. These vital life-giving substances are then absorbed and drawn into the shoot above, with the ensuing root system providing anchorage, nutrient storage solutions and propagation potential.

Taproots and lateral roots
For most vascular plants, the root system appears in the form of a central, dominant deep-tunnelling taproot, which extends its food- and water-seeking potential via a series of lateral roots and microscopic root hairs that grow in short-lived cycles at the root tips. Root vegetables, such as carrots, are one example, delicious eating, but actually vital, nutrient storage organs for carrot plants during dormant periods. Gardeners may also recognise taproots as the thuggish, deep-boring anchors of stubborn weeds like dandelions.

Fibrous and adventitious roots
For those plants that don't display an obvious taproot, it has either died away or become part of a shallower, finer, more widespread approach. Examples include the fibrous, matted root systems of many monocots, such as grasses, lilies and palms, or the roots of ferns. These adventitious roots grow from a non-root organ, such as the stem or a leaf and are essential for regeneration and the prevention of soil erosion. They can also be found above ground as the aerial roots of epiphytes, such as orchids or ivy, the prop roots of banyan trees or mangroves, or the fine root clusters of spider plants or strawberry runners.

Opposite Species of *Narcissus*, including daffodils and narcissi, have special contractile roots that pull the dormant bulb down into the soil where the flower leaves and stem form to emerge the following spring.

Shoots

The word 'shoot' generally refers to fresh new vascular plant growth that can mostly be seen above ground. It can be used to describe the first tender shoot that emerges from a germinating seed of a gymnosperm or angiosperm, as well as parts of a plant's shoot system, including stems, buds, leaves, flowers and fruit. The shoot system plus the root system forms the whole of a vascular plant.

MAKE THE CONNECTION
Taproots and lateral roots: Taproots *Taraxacum officinale* (dandelion); *Borago officinalis* (borage) **Taproot as storage organ** *Daucus carota* subsp. *sativus* (carrot); *Pastinaca sativa* (parsnip); *Beta vulgaris* subsp. *vulgaris* (beet); *Raphanus raphanistrum* subsp. *sativus* (radish) **Lateral roots** *Quercus* (oak); *Sequoia sempervirens* (redwood); *Pinus* (pine) **Root tuber** (lateral root as storage organ) *Ipomoea batatas* (sweet potato); *Manihot esculenta* (cassava); *Dahlia* (dahlia)

Fibrous and adventitious roots: Adventitious roots (fibrous) *Cocos nucifera* (coconut palm tree); *Poa* (bluegrass/meadow-grass); *Zantedeschia* (calla lily) **Adventitious roots (aerial)** *Chlorophytum comosum* (spider plant); *Hedera* (ivy); Orchidaceae (orchid) **Adventitious roots (prop)** *Ficus benghalensis* (Indian banyan); *Rhizophora mangle* (red mangrove); *Zea mays* (maize)

Shoots: Shoots *Pisum sativum* (pea); *Vigna radiata* (mung bean); Bambusoideae (bamboo) **Shoot system** *Solanum lycopersicum* (tomato)

Stems

There's a whole lot more to stems than meets the eye. They provide support (along with roots), facilitate and elevate other vital organs, such as leaves, flowers and fruit, transport fluid between the roots and shoots via their xylems and phloems, store nutrients and produce new living tissue.

Stems also provide numerous resources for humans, including wood, cork, resin and rubber from the arborescent stems of trees; food from sugar cane and asparagus spears, cabbage buds, ginger rhizomes, cinnamon bark and bamboo shoots; and paper and textiles from the stems of conifers, flax, hemp, jute and papyrus.

Multi-tasking stems can also add ornamental interest to the garden through plants such as silver birch or corkscrew bamboo, plus they're a helpful plant identification tool if you're trying to spot the difference between a eudicot and a monocot (see pages 42–45).

Stem nodes

Look a little closer at the stem of any vascular plant and you should be able to see the nodes and, between them, what botanists call internodes. You may notice one or more leaves protruding from the nodes along with buds that may grow into leaves, cones or flowers. Adventitious roots can also grow from these nodes.

At microscopic level, stems reveal three main tissue types: dermal tissue provides waterproofing, protection and the control of gas exchange; vascular tissue handles the transport of water, nutrients and sucrose, and provides structural support; and ground tissue is the bit in between. The arrangement of these tissues varies widely among plant species.

Specialised stems

Stems also exhibit an impressive range of specialised features. Secondary xylem growth in eudicots and gymnosperms is responsible for hard and soft wood respectively. Stems can be aerial (above ground) or underground as corms, bulbs, rhizomes and tubers, and range from the main stalk to the tiny stems that hold flowers and fruit. Plus stems can cling, produce thorns, branch out in numerous ways or even imitate leaves.

MAKE THE CONNECTION

Aerial stems, buds and flower stalks:
Acaulescent (very short or concealed stem) *Liriope* (lily turf); *Viola* (viola)
Arborescent (tree-like/trunk) *Fagus sylvatica* (beech); *Cedrus libani* (cedar)
Bud (embryonic shoot with immature stem tip) *Paeonia* (peony); *Brassica oleracea* (cabbage) **Pedicel** (individual flower stalk) *Primula veris* (cowslip); *Delphinium* (delphinium) **Peduncle** (inflorescence/infrutescence stalk) *Ammi majus* (bishop's flower)

Underground stems: Bulb (short vertical underground stem with fleshy storage leaves) *Allium* (garlic/onion); *Narcissus* (daffodil); *Tulipa* (tulip) **Corm** (short enlarged underground storage stem) *Crocosmia* (montbretia); *Crocus* (crocus) **Rhizome** (horizontal underground stem) *Nelumbo nucifera* (lotus); *Zingiber officinale* (ginger); *Iris* (iris) **Tuber** (swollen underground storage stem) *Solanum tuberosum* (potato)

Branching arrangements: Opposite *Olea europaea* (olive) **Alternate** *Cotinus coggygria* (European smoke bush) **Whorled** *Pinus* (pine) **Unbranched** *Cocos nucifera* (coconut palm)

Stem adaptations: Scandent (climbing stem) *Vitis* (grapevine) **Cladode** (flattened leaf-like stem) *Ruscus aculeatus* (butcher's broom); *Opuntia* (prickly pear cactus) **Decumbent** (lying flat) *Cucurbita maxima* (pumpkin)

IN THE FIELD

Top The pads of the prickly pear cactus (*Opuntia*) are actually modified stems with reduced leaves (spines) designed to aid water storage, photosynthesis, flower production and to keep predators at bay. **Left** The branching arrangement of leaves can be incredibly useful in plant identification. Olive (*Olea euopea*), for example, has opposite branching, referring to pairs of leaves at each node.

> '*Morphology's intention is to portray rather than explain.*'
>
> Johann Wolfgang von Goethe (1749–1832), German wordsmith, poet and writer

BOTANY FOR BEGINNERS 95

Leaves

There are more words to describe leaves, it seems, than any other organ of the plant due to the huge variety in their shape, vein pattern, edging, colour, size, arrangement, texture and specialised adaptation to the environment. Leaves are also the powerhouses and metamorphising magicians of vascular plants, the evolutionary wonders that facilitate photosynthesis (see page 111) and in doing so produce food and energy for the plant, and oxygen for us.

What leaves do

In order to carry out photosynthesis, leaves have evolved to incorporate a number of key features. The greenness of leaves is produced by a green pigment called chlorophyll, which resides in the chloroplasts of some plant cells (usually the uppermost ones). By absorbing red and blue wavelengths from the sun but reflecting green ones, leaves appear green to us – and, importantly, harness the light energy they need to photosynthesise.

Many plants have large, flat leaves because this maximises the surface area for light absorption and can also maximise gas exchange through a leaf's stomata (tiny pores), which are mainly found on the underside of a leaf – carbon dioxide goes into the plant, oxygen and water out of it. Within the veins or vascular bundles of leaves lie a xylem and a phloem (see page 107), as found in a stem, which carry water and minerals to the chloroplasts, and distribute converted food energy (glucose) and oxygen around the plant for growth, storage and respiration (see page 112).

How leaves appear

Leaf architecture and behavioural patterns largely come down to survival. The thin, waxy needles of pine trees, for example, are specially adapted to brave the cold, including a sheath to protect the central leaf vein and a waxy cuticle to protect cells, save water and allow photosynthesis throughout winter, while the leaves of deciduous trees fall off to conserve energy during colder or drier periods.

Opposite The purpose of the holey leaves of *Monstera* species, such as the monkey mask plant (*Monstera obliqua*), is only theorised about – to capture more sunlight while climbing, survive high winds or help remove water?

MAKE THE CONNECTION

Stalk types: Petiolated (stalked) *Rosa canina* (dog rose); *Rheum* (rhubarb) **Sessile** (unstalked) most of the Poaceae family (grasses), this excludes a few tropical grasses, most notably bamboo

Blade divisions: Simple (single leaf blade) *Mangifera indica* (Indian mango); *Betula* (birch) **Compound** (with leaflets) *Coriandrum sativum* (coriander); *Fraxinus* (ash)

Leaf margins: Entire (smooth) *Magnolia* (magnolia); *Diospyros kaki* (persimmon) **Sinuate** (smooth-waved) *Quercus alba* (white oak); *Quercus robur* (English oak) **Dentate** (toothed) *Fragaria vesca* (wild strawberry); *Fagus grandifolia* (American beech) **Serrate** (asymmetric saw-toothed) *Hydrangea* (hydrangea); *Urtica dioica* (stinging nettle) **Lobate** (indented lobed) *Acer campestre* (field maple); *Hedera* (ivy)

Simple leaves: Elliptic (oval) *Citrus limon* (lemon); *Laurus nobilis* (bay laurel) **Lanceolate** (lance-shaped) *Salix* (willow); *Eucalyptus* (eucalyptus) **Acicular** (needle-shaped) *Pinus* (pine); *Picea* (spruce) **Ovate** (egg-shaped) *Viburnum* (viburnum); *Pyrus* (pear) **Cordate** (heart-shaped) *Viola* (viola); *Hosta* (hosta) **Hastate** (spear-shaped) *Atriplex prostrata* (spear-leaved orache) **Linear** (strip-shaped) *Calluna vulgaris* (heather); *Rosmarinus officinalis* (rosemary)

Compound leaves: Palmate (leaflets from a central point in the form of a hand) *Aesculus hippocastanum* (horse chestnut); *Schefflera arboricola* (dwarf umbrella tree); *Cannabis sativa* (hemp) **Pinnate** (leaflets arranged along both sides of mid-rib, like a feather) *Fraxinus excelsior* (Ash); *Juglans regia* (English walnut) **Bipinnate** (twice divided leaflets from central mid-rib, like a feather) *Acacia dealbata* (mimosa) **Trifoliate** (with just three leaflets) *Oxalis acetosella* (wood sorrel); *Trifolium* (clover); *Laburnum anagyroides* (common laburnum)

Flowers

Such is the multi-sensory role that flowers play in the lives of humans that it's easy to forget that all these features actually serve an even more important purpose: to ensure the reproduction and thus the survival of the plants on which they bloom.

The pollination game

Flowering plants, or angiosperms, make up more than 90 per cent of the known Plant Kingdom and comprise more than 400 known families and around 369,000 known species, but despite such diversity, they all have the same end game: to facilitate reproduction. Upon fertilisation (the moment when the male and female gametes fuse), the plant can then begin to reproduce. The way flowers do this is by encouraging the transfer of male pollen to the female ovule. This can occur within one flower (self-pollination), between flowers (cross-pollination) or both.

The parts of a flower (see diagram on page 88)

To study how flowers work, it can help to dissect a few. Complete flowers, such as hibiscus, rose, sweet pea and tulip, have sepals, petals, stamens (male reproductive parts) and pistils (female reproductive parts made of one or more carpels), arranged in whorls from the outside in. Incomplete flowers lack one or more of these structures.

Sepals are leaf-like, generally green structures usually found at a flower's base, petals are often colourful and attract pollinators, stamens comprise a pollen-producing anther on top of a filament, and pistils hold a sticky stigma at the end of a stalk-like style through which male gametes can be deposited in the ovules below via pollen tubes. The ovary is a hollow cavity that contains immature seeds called ovules. Inside those ovules are egg cells awaiting fertilisation.

Flower design is directly related to its method of pollination and environment. Some flowers rely on animals and insects to transfer pollen and thus are more attractive – brightly coloured or highly scented – while others harness the power of the wind. Flowers have also evolved to self-pollinate, cross-pollinate or both. Whatever it takes to fertilise that egg, it seems, is essentially what flowers have evolved to do.

MAKE THE CONNECTION

Pollination types: Animal-pollinated (pollen transferred by animals and insects, such as bees, birds and butterflies) *Lathyrus odoratus* (sweet pea) **Wind-pollinated** (pollen transferred by the wind) *Triticum* (wheat) **Autogamy** (pollen transferred from anther to stigma of same flower) *Galanthus* (snowdrop) **Geitonogamy** (pollen transferred from anther to stigma of different flower on same plant) *Zea mays* (maize) **Xenogamy** (pollen transferred from anther to stigma of different plant) *Cosmos* (cosmos)

Flower structures: Complete (with sepals, petals, stamens and pistils) *Tulipa* (tulip) **Incomplete** (lacking one or more basic structures) *Begonia* (begonia) **Bisexual flowers** (functional male and female organs on same flower) *Rosa* (rose) **Monoecious flowers** (functional male and female organs on different flowers but same plant) *Alnus glutinosa* (alder) **Dioecious** (functional male and female organs on different plants) *Ilex aquifolium* (holly)

Flower inflorescences: Capitulum (a dense flat cluster of small flowers or florets) *Helianthus* (sunflower) **Corymb** (flower cluster with longer lower stalks) *Cornus sanguinea* (common dogwood) **Cyme** (each branch ending in a flower with younger flowers on sideshoots) *Myosotis arvensis* (forget-me-not) **Panicle** (loose-branching cluster of flowers) *Aesculus hippocastanum* (horse chestnut) **Raceme** (flower cluster with short equal stalks at equal distances along central stem) *Digitalis purpurea* (foxglove) **Spadix** (fleshy-stemmed spike with many tiny flowers) *Arum maculatum* (wild arum/lords-and-ladies) **Spike** (a raceme where flowers develop directly from the stem) *Lavandula latifolia* (spike lavender) **Umbel** (flat-topped or spherical flowerhead, with flower stalks from central point) *Daucus carota* (wild carrot)

IN THE FIELD

Left The rare hot lips plant (*Psychotria elata*) is so named for its luscious, lip-shaped bracts, which attract hummingbirds and butterflies. **Top** Snowdrops, such as *Galanthus elwesii*, a native of the Caucasus, reproduce via self-pollination, by transferring pollen from the anther to the stigma of the same flower. **Bottom** The violet sabrewing hummingbird is an important pollinator of lobster-claw plants (*Heliconia*), banana plants (*Musa*) and passionflowers (*Passiflora*).

Top The underside of leathery polypody fern (*Polypodium scouleri*) leaves are crowded with large, rounded sori, which contain the sporangia that produce and store spores. **Right** The sensually shaped fruit of the coco de mer plant (*Lodoicea maldivica*) contains the largest seed in the Plant Kingdom. **Bottom** Seed-containing achenes from the *Taraxacum* genus, commonly known as dandelions, are carried on the wind by a feathery pappus, which functions as a parachute.

Seeds

Seeds – when you stop to think about it – are everywhere, from the mainstays of your morning coffee and toast to the seeds used to make our gardens grow. They are the givers of life and, as Kew's nuclear-bomb-proof Millennium Seed Bank confirms, a valuable resource that is absolutely vital to our survival.

The evolution of seeds
Thor Hanson was right to devote an entire book to seeds: *The Triumph of Seeds* (2015). The evolution of these remarkable structures, around 380 million years ago, was a defining moment for the Plant Kingdom, resulting in two main kinds of seed-producing plants: the flowering angiosperms (see page 40), whose seeds are enclosed in an ovary, which becomes the fruit; and the non-flowering gymnosperms (see page 37), the seeds of which are 'naked' – presented on the surface of leaves or in scaly cones.

How seeds develop
Flowering plants develop seeds upon the double fertilisation of an egg using two gametes (sperm cells). The first fertilisation process produces a zygote and then an embryo, within which is a plumule that forms the cotyledons (first leaves), a hypocotyl that forms the stem and a radicle that forms the root. The second part produces a primary endosperm and then endosperm tissue (food storage for the embryonic plant), although the microscopic seeds of orchids lack this. A seed coat (testa) is formed from the outer layers of the ovule.

Non-flowering plants are successfully fertilised by only one gamete with a second zygote often aborted or absorbed. A gymnosperm seed develops from the ovule of a female cone or cone-like structure. It also has the esssential constituents of embryo, protective covering and food storage tissues.

The art of germination
For germination to take place, seeds need water, oxygen and warmth, which explains why there are so many adaptive seed designs – they're all vying to survive.

Spores

While a seed is a well-developed multi-cellular young plant with an embryonic root, stem and leaves already formed, a spore is a single cell. It does not have an available food store or protective coat. Pollen grains are spores, so all angiosperms produce them. The spores of spore producers (see page 33), such as ferns, mosses and horsetails, can still survive in the outside world but require a moist environment, such as a damp forest floor to germinate.

MAKE THE CONNECTION
Seed producers: Angiosperms (plants with enclosed seed) *Papaver* (poppy); *Quercus* (oak) **Gymnosperms (plants with naked seed)** *Ginkgo biloba* (ginkgo); *Larix* (larch)

Seed sizes: Large seeds *Lodoicea maldivica* (coco de mer); *Cocos nucifera* (coconut palm) **Small seeds** Orchidaceae (orchid); *Begonia* (begonia)

Seed shapes: Bean-shaped *Phaseolus coccineus* (runner bean) **Square/oblong** *Trigonella foenum-graecum* (fenugreek) **Triangular** *Fagus* (beech) **Ovate/elliptic** *Citrus* x *sinensis* (orange) **Discoid** *Lunaria annua* (honesty) **Spherical** *Lathyrus odoratus* (sweet pea)

Seed dispersal by way of adapted seeds: Aril (fruit-like) *Taxus baccata* (European yew); *Litchi chinensis* (lychee) **Elaiosome (lipid-rich)** *Ricinus communis* (castor-oil plant); *Viola odorata* (sweet violet) **Hair/fibre** *Gossypium* (cotton)

Seed dispersal by way of adapted fruit (see page 102): Wing/parachute *Acer* (maple); *Carpinus* (European hornbeam); *Taraxacum officinale* (dandelion) **Barbs/hooks** *Galium aparine* (goosegrass); *Harpagophytum procumbens* (devil's claw) **Waterproof** *Rhizophora mangle* (red mangrove); *Nelumbo nucifera* (lotus)

Fruits

The word 'fruit' is used to describe a variety of edible plants such as apples, strawberries and pineapples. In botanical terms, however, fruits include many foods commonly called vegetables, seeds or nuts – tomatoes, courgettes, avocados, beans and peanuts are fruits (the edible part of the latter being the seed inside).

So what's the common denominator? The answer, in fact, is quite simple – a fruit is the mature, seed-carrying ovary of a flower, designed to help with seed dispersal. This, admittedly, leaves some room for culinary confusion – what, then, is a vegetable? Botanically speaking, this should refer to the edible leaves (lettuce), stalks (celery), roots (carrot), tubers (potato), flower buds (cauliflower) or vegetative buds (Brussels sprouts) of a plant. In common parlance, however, the word 'vegetable' is often used to describe a plant used for savoury dishes, the word 'fruit' for sweet ones – hence why rhubarb stalks are often called fruit.

In terms of edible nuts, seeds and pulses, there is similar crossover. Peas – often described as vegetables – are actually the seeds of legumes, which also makes them pulses alongside lentils and chickpeas; cashews – often described as nuts – are actually the dried seeds of drupe fruits; rice and wheat – termed as grains – are partly seed and ripened ovary; while poppy seeds are indeed seeds. It's also important to note that not all fruits are edible to us – some are the seedpods of the flowers we see in our gardens or in the wild, and some are in fact highly poisonous, such as deadly nightshade (*Atropa belladonna*).

How a fruit develops

Fruits are also divided within themselves. Simple fruits can be dry or fleshy and develop from the ripening of a single carpel or multiple carpels united in a single ovary. Most fruits fall into this category but come in a variety of seed-dispersal-aiding guises, such as berries, pomes, drupes, hips, legumes, capsules and nuts.

Aggregate fruits mature from the ripening of multiple carpels that are fused together as one but retain their separate ovaries; each ovary matures into a fruitlet, which then combine to form a whole fruit. While multiple fruit are formed from the ripening ovaries of separate flowers on an inflorescence or flowerhead. Each flower produces a fruit, which then fuse together.

Finally, come the accessory fruits, which can fall into any of the categories above – part ripened ovary, part tissue derived from another part of the flower: the strawberries, the apples and the pineapples.

MAKE THE CONNECTION
Simple fruits (fleshy): Berry *Vitis* (grape); *Musa acuminata* (banana) **Hesperidium (tough, aromatic rind)** *Citrus limon* (lemon); *Citrus x paradisi* (grapefruit) **Pepo (hard, thick rind)** *Cucurbita moschata* (butternut squash) **Drupe (stony pit)** *Olea europaea* (olive) **Pome (tough central seeded core)** *Malus pumila* (apple); **Hip (fleshy hypanthium containing achenes)** *Rosa canina* (dog rose)

Simple fruits (dry/split at maturity): Follicle (one carpel; multiple seeds; splits on one side) *Paeonia* (peony) **Legume (multiple seeds within pod)** *Pisum sativum* (pea); *Arachis hypogaea* (peanut) **Capsule (multi-carpelled ovary; multiple seeds)** *Gossypium* (cotton); **Silique/silicle (two-carpelled ovary; multiple seeds)** *Lunaria annua* (honesty)

Simple fruits (dry/don't split at maturity): Achene (small, one-seeded) *Helianthus* (sunflower) **Caryopsis/grain (one-seeded, ovary wall fused with seed coat)** *Oryza sativa* (Asian rice); *Triticum* (wheat) **Samara (winged achene)** *Acer* (maple) **Schizocarp (splits into single-seeded parts)** *Daucus carota* (wild carrot); **Loment (schizocarp legume)** *Hedysarum boreale* (sweet vetch) **True nut (fruit encased in inedible hard shell)** *Corylus avellana* (hazelnut);

Aggregate fruits: Drupelets *Rubus* (raspberry) **Achenes** *Fragaria* (strawberry) **Follicle** *Helleborus* (hellebore)

Opposite The sweet-tasting fruit known as strawberries (*Fragaria*) are in fact fleshy receptacles for the actual fruit, which are the small 'seeds', called achenes, on the outside of the strawberries.

UNDER THE MICROSCOPE

How do some petals appear velvety? How does a leaf send oxygen into the world? Does a plant have a brain or central processing unit? What can flower fossils tell us about the world we live in today? Microscopy gives us the potential to look beneath the surface appearance of things – at cells, tissues and even gene expression – to look closer at the hidden complexities of plants and flowers and the questions posed by them, and to take plant exploration into exciting new realms, perhaps even the future.

Cells

Looking at the Pathways of the Plant Kingdom (see page 24) reminds us that plants, animals and fungi are all eukaryotes – their cells are made up of membrane-bound organelles including a nucleus. So what sets plant cells apart?

Cell wall
This thick, rigid structure contains fibres of cellulose and provides protection, support and a permanent rectangular shape. Immediately inside of the cell wall is a selectively permeable membrane that surrounds a jelly-like cytoplasm. Within it are the organelles – specialised structures within the cell that perform specific and varied functions.

Nucleus
The nucleus contains the major portion of the plant's DNA, composed of thread-like chromosomes made up of nucleic acids and protein. This hereditary information, in the form of genes, directs how plants appear, develop and behave.

Chloroplasts
Oval-shaped chloroplasts are exclusive to plants. They house stacks (grana) of thylakoid discs, on which sit the photosynthesising green pigment, chlorophyll. Energy-rich molecules then move to nearby stroma, where carbon can be fixed and sugars synthesised.

Mitochondria
Rod-shaped mitochondria are relatively smaller and more plentiful. They use oxygen to release energy from sugars through aerobic respiration.

Vacuole
The largest, fluid-filled space is a vacuole. It helps keep the cell turgid, stores water and segregates waste.

Endoplasmic reticulum
This rapidly changing network of membranes connected to the nucleus is responsible for the synthesis and often storage of proteins and lipids.

Opposite Artwork of a plant cell shows a cell wall (outer band), chlorophyll-filled chloroplasts (large green ovals), a large central vacuole, a nucleus (in pink), mitochondria (small orange ovals) and peroxisomes (small blue spheres).

Tissues

Plants grow and reproduce through a process known as cell division, either via mitosis or meiosis. This allows the cells to specialise, store nutrients and be replaced if damaged. Each specialised plant cell adapts its contents to perform a particular function. Certain cells then cluster together to form dermal (exterior surfaces), ground (nutrient manufacturing and storing) and vascular (transport) tissues. Plant tissues can also be divided into two types: meristematic tissue, with actively dividing cells that lead to growth; and permanent tissue, where cells have specific roles.

Cell division
Mitosis facilitates growth and repair and produces two diploid cells (cells with a full set of chromosomes) that are identical to each other and to the parent.

Meiosis enables sexual reproduction and produces four haploid cells (cells with half the number of chromosomes) that are different to each other and to the parent. These cells become either the male pollen grains or the female egg cells.

Cell types

* **Parenchyma cells** have thin primary cell walls and large vacuoles; they carry out photosynthesis, stay alive when mature and provide the bulk of the soft part of plants.

* **Sclerenchyma cells** have thick secondary cell walls including lignin. They provide support and strength to roots, stems and vascular tissue and they die when mature.

* **Collenchyma cells** have thin primary cell walls with some secondary thickening and provide additional stretchable support, ideal for areas of new growth.

* **Water-conducting cells** are narrow, elongated and hollow. They die when mature but their cell walls remain, allowing free water flow within the xylem.

* **Sieve tube elements** are elongated, living cells that transport carbohydrates in the phloem. They rely on companion cells for missing parts.

Xylem

Vascular plants need to access water for photosynthesis and transpiration. The way they facilitate this is via a one-directional vascular tissue device known as a xylem. Xylem tissue consists of specialised water-conducting cells known as tracheary elements – either tracheid or vessel members – both of which are typically narrow, hollow and elongated. In addition, there are some fibre cells for support and parenchyma cells for the storage of various substances.

Water flow
Primary xylem is laid down at primary growth stage, when the meristem regions found at the tips of stems and roots and in buds begin to extend (meristem tissue contains unspecialised cells capable of division, growth and adaptation). The primary xylem transports water and minerals (xylem sap) to new growth, including shoots, stems and leaves, and forms one half of the vascular system in all higher plants. Xylem cells die at maturity, leaving hollow, strongly lignified cellulose tubes or lumens through which water can flow.

Wood formation
Wood is essentially secondary xylem tissue, formed at secondary growth and only in woody plants. It forms, along with secondary phloem, from lateral meristem tissue called vascular cambium. As new xylem layers are added each year, it increases the girth of roots or stems, producing the wood of tree trunks and woody shrubs, in concentric rings (as shown below).

Phloem

The second part of the vascular system, or bundle, is called a phloem, a two-way transport system made up of living cells. Its job is to carry soluble organic compounds made during photosynthesis – in particular, the sugar sucrose – to parts of the plant that need it: the roots or storage structures such as bulbs and tubers. It does this by way of sieve tube elements (strung together to form tubes), companion cells, phloem fibres and parenchyma cells.

Sap flow
Phloem sap consists of sugars, hormones and mineral elements dissolved in water. It flows from where sugars are made or stored (sugar sources) to where they are used (sugar sinks): during early development the roots; during vegetative growth the shoots and leaves; during reproductive stages the seeds and fruits. Common forms of sap in human use are maple syrup from the sugar maple (*Acer saccharum*) and the 'juice' of *Aloe vera* plants (as pictured below). You can also easily spot vascular bundles containing xylem and phloem in celery (*Apium graveolens*) stalks.

Translocation
Concentrated sugar from sugar stores causes water from the xylem to be absorbed by osmosis, creating internal pressure against the rigid cell walls. Sap then pushes through sieve element pores. As sugar is deposited into areas in need of nutrients, such as growing tissues, water exits and the pressure goes down. The pressure difference causes flow – as fast as 1m (3 ft) per hour in angiosperms.

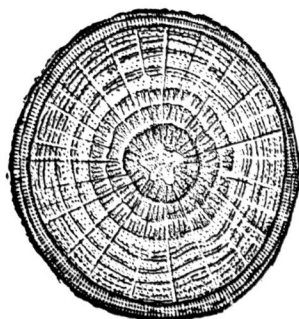

Opposite A cross-section of a three-year-old basswood or linden tree (*Tilia*) stem shows the pith (bottom) circled by three annual xylem rings and the phloem bundles at the top.

Stomata

Water goes up the xylem from the roots, sugars go down the phloem from sugar stores and the chlorophyll in chloroplasts absorbs sunlight for photosynthesis. So where do gases come in – and out? The epidermis of leaves and stems has the answer in the form of tiny stomata (pores) and their guard cells.

What stomata do
Stomata open when light, humidity, temperature, root water absorption and atmospheric conditions are right. As guard cells absorb water, the increased turgidity causes them to bow apart, creating open pores. Carbon dioxide diffuses in for photosynthesis (see page 111), usually during times of light; oxygen diffuses in for respiration (see page 112), whether it is light or dark; oxygen and water vapour diffuse out via transpiration (see page 112), mainly in the light.

Looking for stomata
Stomata are found in the sporophyte (non-sexual) generation of all land plants apart from liverworts, although their number, size and distribution varies widely – eudicots usually have more stomata on the lower surface of leaves, monocots may have the same number on both leaf surfaces, floating leaves often only have stomata on the upper side, while most trees only have them on the lower surface. A microscope should reveal them at 100X magnification (see below).

Pollen

Individual grains of pollen are too small to see with the naked eye but some of this coarse, powdery substance can form larger clusters, which *are* visible – find them on the top of the stamens of flowering plants or the male cones of gymnosperms (male cones being usually herbaceous and much less conspicuous than the often woody female cones are at maturity).

The role of pollen
Pollen can cause havoc for some humans who exhibit an allergy to one of its forms but for seed-producing plants, pollen is essential for the reproductive process. Each grain is a microgametophyte, the producer of male gametes (sperm cells). If a pollen grain lands on a compatible pistil or female cone of a gymnosperm, it has the potential to germinate by passing its sperm to an ovule containing a matching female gametophyte within its eggs.

Under the microscope
All pollen grains contain one or several vegetative (non-reproductive) cells and a generative (reproductive) cell – most *flowering* plants boast one vegetative cell only, which produces a pollen tube through which the male sperm can travel. At 100X magnification, you can see pollen's basic structure. Observed under a high-strength electron microscope, the diverse forms of this tiny substance are astounding (see below and page 334).

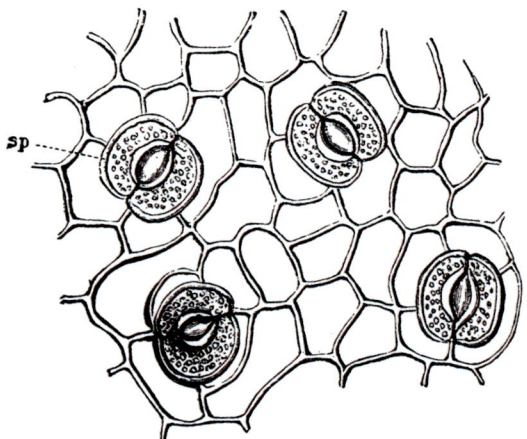

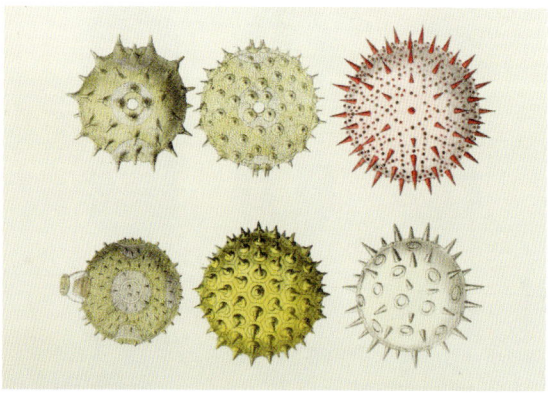

Opposite The longer male flowering catkins of the Italian alder tree (*Alnus cordata*) disperse their pollen in the wind and provide an early spring treat for bees, while pollinated shorter female catkins turn into small seed-bearing cones.

WHAT DO PLANTS DO?

Put simply, plants grow. We observe them metamorphosing through the seasons, from embryonic form to full-bodied growth to decay. We witness flowers in bud and full bloom, leaves on the trees and crumpled underfoot and pine cones opening and closing seemingly at random. We see how plant growth corresponds to climate and we experience sensory elements, such as fragrance, designed to repel and attract. What plants do is inextricably linked to their survival – and to our own, providing so much of what we need.

Photosynthesis

Photosynthesis is one of the most important processes in the world. In fact, if this vital biochemical reaction stopped happening, virtually all life on Earth would die. Most people know and accept it as a simple life-giving formula:

$$\text{carbon dioxide} + \text{water} \xrightarrow[\text{chlorophyll}]{\text{light}} \text{glucose} + \text{oxygen}$$

But it can also be beautifully explained by way of its discovery, a narrative that took more than 200 years and the experimental minds of several scientists to unfold – and that is still under close investigation to this day.

The story of photosynthesis traces back to Flemish physician Jan Baptist van Helmont and his famous willow tree experiment in the mid-1600s. By growing and carefully watering his tree in a pot of soil for five years, he found that the mass of the tree had increased by 74kg (164 lb) but that of the soil was little changed. Van Helmont concluded that water was the source of the extra mass and therefore the life-giving 'elixir'.

His deduction was wrong – largely because he ignored his own previous findings about carbon dioxide (a gas he termed *gas sylvestre*) – and he wasn't quantitative enough. His theory was subsequently disproved by the English professor and physician John Woodward some 50 years later, but van Helmont was certainly onto something – plants were increasing their mass from somewhere.

The next breakthrough came in 1771, when English chemist Joseph Priestley put a sprig of mint into a transparent closed space with a candle also burning inside. The candle burned out the air and was soon extinguished. Priestley left the plant matter in the enclosed space for 27 days and then relit the candle using a mirror and sunlight beams angled onto the candle wick. The candle then burned perfectly well in air that previously wouldn't support it.

What Priestley proved is that plants somehow change the composition of the air, a subsequent experiment involving a mouse, a transparent jar and a plant suggesting that plants restore to air whatever live animals and burning candles remove.

In 1779, Dutch physiologist Jan Ingenhousz took Priestley's work further. He also put a plant and a candle into a transparent enclosed space, which he then let stand in the sunlight for two or three days. Based on previous experiments, the air should have been pure enough to support a candle flame. What Ingenhousz did, however, was to place a black cloth over the vessel. After a few days, he tried to light the candle but nothing happened. It was a Eureka moment: plants needed light to purify the air.

By placing a small, green aquatic plant in a transparent container of water, which he then exposed to bright sunlight, Ingenhousz also noticed that gas bubbles were emitted from specific parts of the plant – around the leaves and green parts of plant stems. It was another giant leap towards cracking the code for photosynthesis and thus how plants grew.

Enter Swiss botanist, pastor and naturalist Jean Senebier in 1796, who demonstrated that plants absorb carbon dioxide and release oxygen with the help of sunlight; Nicolas-Théodore de Saussure in the early 1800s, who found that while plants need carbon dioxide, their increased mass was also due to the uptake of water; German physician Julius Robert Mayer, who in the 1840s, stated that energy can neither be created nor destroyed but rather converted – in the case of plants, light energy converted into chemical energy; and German botanist Julius von Sachs and his mid-nineteenth-century investigations into how starch is produced under the influence of light in relation to chlorophyll. More than 60 years after that, Cornelius Van Niel came up with the general equation for photosynthesis that led to the simplified formula that we use today.

All the while, and for millions of years before, green plants went about their business, drawing carbon dioxide from the air, water from the soil and using light energy from the sun to produce sugars and oxygen via the 'magical' green pigment chlorophyll – sugars and energy that would help plants grow and survive, and oxygen that would enable the existence and evolution of so many other of Earth's life forms.

Opposite Ground-creeping ornamental *Fittonia argyroneura*, or nerve plant, has striking veins of pink, white, red or silver and thrives best in filtered sunlight and high humidity as it would in its native habitat, the Peruvian rainforest.

Respiration

Respiration and transpiration don't get nearly as much airtime as photosynthesis but they're a vital team of processes that support the plant's development. Respiration creates energy, allowing growth.

Why plants respire

It's helpful, first of all, to separate breathing from respiration. Breathing refers to the muscular movement of animals that sends oxygen to respiratory organs and removes carbon dioxide from them. Respiration is a biochemical process by which organic compounds are oxidised to liberate chemical food energy. When plants respire, they use oxygen to release energy from photosynthesised sugars, creating the byproducts of carbon dioxide and water vapour.

$$\text{glucose + oxygen} \longrightarrow \text{carbon dioxide + water} \\ (+ \text{ energy})$$

How plants respire

Plants carry out respiration by taking in oxygen from the air via the stomata in leaves and stems. This goes on day and night, although at a proportionally higher level in the dark when photosynthesis stops and carbon dioxide is generally not consumed. The oxidisation of sugars then takes place within tiny intra-cellular organelles called mitochondria. The resultant cellular energy creates the building blocks of new cells, allowing the plant to grow.

Transpiration

Transpiration is both the process of water movement through a plant and the evaporation of water vapour from aerial plant parts, such as leaves, stems and flowers.

The profit of water loss

If plants need water for photosynthesis, then why lose water through transpiration? One major profit – according to the Cohesion Theory of Sap Ascent – is that transpiration causes a sucking force when water at the top of xylem channels evaporates. Like sucking on a straw, it causes a negative pressure, which lifts the xylem sap to the leaf surface. Some of the sap's water will be lost through stomata via transpiration but some water, complete with nutrients, will be used for photosynthesis. As transpiration occurs, plants cool down, making a better environment for growth.

Transpiration factors

Many plants have adapted their design to create the perfect balance of photosynthesis, respiration and transpiration. Plants with more or bigger leaves and stomata will exhibit greater transpiration. This works for larger or taller plants that need more sucking pressure to distribute water to their various parts. To reduce evaporation and procure shade, some plant species, such as cacti (pictured below), have evolved spiny, photosynthesising, waxy stems in the place of true leaves, along with the added facility to store water.

Opposite Mariposa Grove in Yosemite National Park is home to some of the world's largest giant sequoia (*Sequoiadendron giganteum*), which use transpiration to pull water and nutrients from root to towering crown.

Top The flowers of the lotus (*Nelumbo nucifera*) are bisexual and pollinated by flies, beetles or self-pollination. **Right** *Heliconia*, or false bird-of-paradise plant (due to its similarlity to *Strelitzia*), has bright, waxy bracts, inside which are nectar-rich true flowers, a favourite of pollinating hummingbirds. **Bottom** The mother of thousands succulent (*Kalanchoe daigremontianum*) uses asexual, vegetative reproduction to produce numerous plantlets that will fall and take root.

Reproduction

Plant reproduction is simply the production of new individual or offspring plants, which can be done asexually or sexually.

Asexual reproduction

Observe the stolons (runners) and plantlets of strawberry plants, or the new life that sprouts from underground storage organs, such as potato tubers, daffodil bulbs, crocosmia corms or iris rhizomes. These are all examples of the most commonly known form of asexual reproduction in plants – vegetative propagation. No pollination is required, no seeds or spores are involved and only one parent plant is needed to produce an offspring. So how does it work?

In vascular plants, vegetative reproductive parts include roots (see page 93), stems (see page 94) and leaves (see page 97), where meristem tissue is found. Undifferentiated cells in this tissue can actively divide by mitosis producing new plant growth in the form of permanent plant tissue systems. Without the fusion of gametes (female egg and male sperm) that happens with sexual reproduction, there's no mixing of genetic information. Each offspring produced is a clone of the parent plant and fellow plantlets.

This is great news when environmentally favourable traits are repeatedly reproduced. In some cases it can lead to giant clonal colonies of the same plant – duckweed and bracken, for example. Humans have also harnessed the power of vegetative reproduction as a quicker and easier way to produce many perennial plants, such as root crops, vines and fruit trees, via cuttings or grafting. It's not so great when genetically identical plants become susceptible to disease, however, which can lead to the devastation of an entire colony.

Asexual reproduction can also take place via apomixis. In flowering plants, this is asexual reproduction through seeds, created without fertilisation from the maternal tissues of the ovule. Dandelions are a good example here. While equipped with all the sexual components for pollination – pistils, stamens and an abundance of bee-attracting nectar and pollen – the ability to produce asexual seeds at the same time allows this plant to go forth and conquer.

Sexual reproduction

It's thought that sexual reproduction happened very early in the evolutionary history of eukaryotes. Today it is the most common form of reproduction for eukaryotic organisms, including humans and plants.

Although people appear to do it very differently to plants, in fact the basic biology is the same. Sexual reproduction simply means that two reproductive cells – the male gamete (sperm) and the female gamete (egg) – fuse together to create a zygote that has a mixture of genes from each parent. Each haploid gamete contains half the normal number of chromosomes of a regular cell. Therefore when the gametes fuse, the number of chromosomes is complete once again in the form of a diploid cell, and growth can begin. Mixed genes leads to greater variation and thus a better chance of survival in a changing environment.

Sexual reproduction is the way that most plants produce offspring and thus survive. Algae do it, mosses and ferns do it, conifers and ginkgo do it – but most obvious of all are the flowering plants.

We've become so used to enjoying flowers for their decorative prowess that it's easy to forget the reason for their existence. Flowers contain the reproductive organs of flowering plants (see page 98). They are designed to produce seeds from which new plants can grow – and it all happens when the male pollen is deposited onto female stigmas so that the eggs within the ovule below can be fertilised.

Those alluring, often fragrant petals are not in fact there for human appreciation (a narrative it's easy to propagate for our own satisfaction) but rather they are an extraordinarily adaptive device (see page 116) designed to attract preferred pollinators to the fold – and as our Growing & Gathering: Planting for Pollinators section shows (see page 157), along with the numerous images of flowering plants within this book, some plants will go to extraordinary lengths to do so.

Growth and decay

Whether plants reproduce sexually or asexually, or via a combination of the two, the thing that they all have in common is growth. New growth happens via vegetative propagation or when a seed or spore germinates, but either way, new tissues and structures continue to grow throughout a plant's life.

This is mainly down to the meristem cells, located at the tips of organs such as roots or shoots or between mature tissues, which divide via mitosis. Plants also retain a high number of cells that can differentiate – change from a less-specialised cell to a more-specialised one. As cells increase in number and size and differentiate along the way, shoots turn into stems, stems produce leaves, flowers bloom on the end of stems and reproductive organs can be formed.

Plants also rely on various external and internal elements and conditions to grow. Internally, a plant's genes and hormones can influence appearance, life cycle, lifespan and behaviour. Externally, growth is affected by access to water and nutrients, temperature and light to support key processes, such as photosynthesis, respiration and transpiration.

If a plant doesn't get what it needs – enough water, for example – it can die. Dead matter falls to the ground and microbes help to decompose it, providing nutrients, such as nitrogen and phosphorus, for other plants. Decomposition also releases carbon into the air, soil and water, helping to sustain the carbon cycle.

Plants also die when they reach the end of their lifespan (see page 136) – one year for annuals, two years for biennials and more than two years for perennials.

Adaptation

How do some plants grow on the slopes of cold, snowy mountains, while others survive in the stifling heat and arid terrain of the desert? Why are there so many different types of flowers, shapes of leaves or heights of trees? Why does a cactus have spines, how do ferns survive on the shady floor of a forest and why do some plants reproduce asexually? As Charles Darwin theorised in his *On the Origin of Species* (1859), it's all about the survival of the fittest – or, more technically, natural selection.

Natural selection is the process whereby plants with characteristics most suited to the environment (temperature, location, access to water or pollinators) are more likely to survive and reproduce. The genes that allowed them to be successful will then be passed to their offspring in the next generation, and so on.

Conifers, for example, have evolved to grow waxy, needle-type leaves (pictured below) that help protect from the cold and wind. While many flowers have adapted their shapes, colours or scents to attract certain pollinators or types of pollination: some orchids fool insects into thinking they are mating with one of their own kind; bees tend to like blue-purple, yellow or white flowers; some self-pollinating flowers keep their petals closed to ensure pollen falls on their own stigmas.

After thousands of years of human interaction with plants, there's also artificial selection at work, from seedless grapes and super-productive seed crops to flowers that bloom for longer or that new must-have shade of orchid for an award-winning garden. We too have learned to adapt plants.

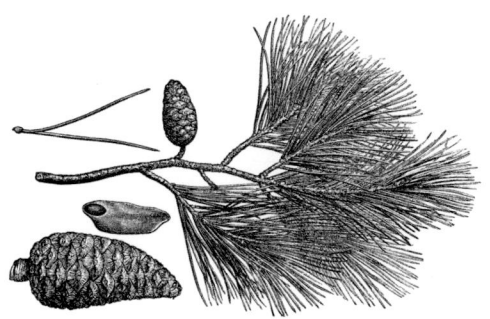

'*Let children walk with Nature, let them see the beautiful blendings and communions of death and life, their joyous inseparable unity . . . and they will learn that death is stingless indeed, and as beautiful as life.*'

John Muir (1838–1914),
Scottish-American naturalist

Top As the tiny true flowers of *Anthurium*, or flamingo flower, on its fleshy phallic spadix begin to die, so the bright red, outer waxy spathe or bract (a modified leaf) also begins to wilt and decay. **Bottom** Adaption extends from insect-mimicking flowers to flower-mimicking insects, in this case the orchid mantis (*Hymenopus coronatus*), which hides close to flowers of similar appearance (although not necessarily orchids as its name suggests) to attract food – namely flies.

What plants know

In 2013, renowned biologist Daniel Chamovitz published a fascinating book entitled *What a Plant Knows: A Field Guide to the Senses*. Highlighting the latest research in plant science, the publication presented the world of plant perception and communication – what a plant 'knows' – drawing parallels to the human senses.

Anthropomorphism does not sit well with many plant scientists, as it indicates a lack of objectivity, a veering away from the scientific need to observe behaviour rather than assume it. As such, Chamovitz is keen from the outset of his study to point out that his use of the word 'know' is unorthodox: 'Plants don't have a central nervous system; a plant doesn't have a brain that co-ordinates information for its entire body. Yet different parts of a plant are intimately connected, and information regarding light, chemicals in the air, and temperature is constantly exchanged between roots, leaves, flowers and stems, to yield a plant that is optimised for its environment . . . When I explore what a plant sees or what it smells, I am not claiming that plants have eyes or noses (or a brain that colours all sensory input with emotion). But I think this terminology will help challenge us to think in new ways about sight, smell, what a plant is, and ultimately what we are.'

Indeed, Chamovitz's idea for the book began when he realised, as a young postdoctoral fellow, that certain genes that plants used to determine if it's light or dark were also part of human DNA. As he discovered, the biology of plants and people is closer than many realise.

Seeing the light

As Charles Darwin discovered in a series of simple experiments involving seedlings with or without tips – a seedling with a lightproof tube around its middle and one with a lightproof cap on – the 'eyes' of a plant can be found at the tip of a plant's shoot rather than the part at which it is bending. It was the first step towards discovering rudimentary 'sight' in plants and describing a behaviour we now know as phototropism: how plants grow towards the light. *What a Plant Knows* goes on to explain how plants can also measure periods of darkness, use far red light to know when to 'turn off' at the end of the day and have circadian clocks just like humans. The way they do this is via photoreceptors in stems and leaves – specialised proteins that perceive different waves of light.

Smelling the way forward

Plants also appear to 'smell'. That is, they can detect volatile chemicals in the air and convert this signal into a physiological response. One olfactory receptor that has been discovered in plants is for the hormone ethylene, a small hydrocarbon gas that is found in abundance in ripening fruit and contributes to the shedding of some deciduous leaves in the autumn. Observe an avocado that ripens quicker in the presence of a banana. The banana is not telling the avocado to ripen, rather, the ethylene the banana is producing is reacting with a receptor found within the avocado. Plants also emit chemicals in response to danger or attack. These signals, in turn, can be processed by neighbouring plants.

Touch-sensitive

Some plants are well known for their rapid response to 'touch' – the insect-eating Venus flytrap (*Dionaea muscipula*) or the sensitive-leaved *Mimosa pudica*, for example. When each of these plants senses movement, the mechanical stimulation initiates an ionic cellular change that results in an electrical signal that can move between cells. This signal results in a physiological response from the plant – in the case of *Mimosa pudica*, rapidly closing its leaves.

The only way is up

Plants also grow in response to gravity: shoots up and roots down, with some leeway for wayward vines or trailing plants, of course. This upward-downward growth, known as geotropism, will continue even if a plant is turned sideways or upside down. Try it and see for yourself . . .

Opposite *A New Pitcher Plant from the Limestone Mountains of Sarawak, Borneo*, by Marianne North (see page 322), depicts *Nepenthes northiana*. Named for the artist who first illustrated it, this pitcher plant uses scent, colour, shape and secretions to attract, drown and digest insects within its vase-like traps.

BOTANICALLY SPEAKING

Life
All living organisms

Domain — *Eukaryota* (Cells with nucleus and organelles within a membrane)

Kingdom — *Plantae* (Plants)

Sub-kingdom — *Tracheophyta* (Vascular plants)

Division — *Angiospermae* (Flowering plants)

Class — *Eudicotidae* (Eudicot)

Order — *Rosales*

Family — *Rosaceae* (Rose Family)

Sub-family — *Rosoideae*

Genus — *Rosa*

Species — *Rosa canina* (Dog rose)

a) Rose bud, pale pink
b) Rose petal, ranging from white to deep pink
c) Blossom with numerous stamens and styles
d) Filament with anther
e) Style with stigma
f) Corolla with 5 separate petals and 5 sepals
g) Rose hip (false fruit) with dried sepals
h) Hip containing several fruit (achenes) inside
i) Hairy fruit with seed inside
j) Stem with hooked prickles
k) Pinnate leaf with 5–7 leaflets
l) Serrated leaflet

Don't let the Latin put you off – the scientific names of plants, the way that they are classified and their life cycles, are relatively easy to master and offer exciting new views of the botanical world.

While the common name 'rose' is perfectly adequate for many people (with a prefix of 'pink', 'tea' or 'fragrant', perhaps), botanical names such as *Rosa canina* (see opposite) or *Rosa x damascena* (damask rose) can tell us even more about a species, from how it might be related to other plants and indeed the Plant Kingdom, to the people, places or happenings involved with its discovery.

It's easy to dismiss the botanical language of plants as irrelevant outside of the scientific sphere – a jungle of Latin that can at first glance appear confusing or potentially boring. A journey behind the scenes of how this truly universal language developed, however, reveals a heady tale of plant sex, plant-collecting, fossil-hunting, technological advances and evolutionary eureka moments. It has taken thousands of years to establish and maintain, from the first plant-grouping forays of botanists such as Theophrastus (371–287 BCE) through the binomial nomenclature (two-part naming) of plants by Carl Linnaeus (1707–1778) to the gene-explorations of contemporary plant taxonomists.

Botanically speaking, a detailed or deep understanding of plant taxonomy (see page 123), nomenclature (see page 124) and plant life cycles (see page 136) is essential for botanists, plant scientists or those working within related fields such as ethnobotany, plant conservation or ecology. This usually comes via a degree in plant sciences, natural sciences or plant biology (often followed by a Masters or PhD), natural successors to the traditional 'botany degree' – now something of a dying breed sadly (the implication being that the word 'botany' denotes an old-fashioned and therefore dry subject).

For garden designers, professional gardeners and botanical illustrators, a basic grounding in botany can also be helpful. In this case it is usually assimilated into a relevant course of study or there are numerous helpful books to explore from the *RHS Botany for Gardeners* (2013) and *RHS Latin for Gardeners* (2012) to Sarah Simblet's *Botany for the Artist* (2010) – or indeed *Collins Botanical Bible* (2018).

Beyond that it's usually a question of whether an advanced knowledge of the botanical world and its specific language could enhance one's plant appreciation or view of the natural world – or even of the world at large . . .

An illuminating exercise might be to create your own herbarium collection, echoing the vital institutions of plant-based research found at Kew Botanic Gardens, in London, or the United States National Herbarium, in Washington.

By pressing or drying plant and flower specimens, it's possible to observe the minutiae of plants that you might have previously missed, while the creation of labels may require some serious sleuthing in the field, in books or online as you try to identify a plant and its name from its key features, growth patterns or habitat.

As you explore, open your mind to the backstory of a botanical name – how a plant was discovered or which plants it might be related to. Botanically speaking, you might find that you love plants even more than you did before.

Opposite Trace the taxonomy of a dog rose (*Rosa canina*) from the umbrella of life itself to its unique and ultimately descriptive species name. Now try with other species . . .

COMPELLED TO CLASSIFY

We need to talk about plants – as food, medicines, objects of beauty and as the central characters in the story of our botanical world. To do this, we need systems within which we can group and order plants, a simple way to name them and the flexibility to incorporate our findings, past, present and future. Thanks then to those people who were compelled to classify the minutiae of the Plant Kingdom over thousands of years – and those who still endeavour to do so today – a combined effort that has resulted in the universal language of plants.

The urge to order

The Plant Kingdom is so vast – nearly 400,000 known species and counting – that without some kind of universal ordering and naming system, it would be impossible for plant scientists to study and keep accurate records, or efficiently advance research valuable for agriculture, medicine or conservation.

It's doubtful that the Greek philosopher Aristotle was thinking in such a wide-ranging fashion when he compiled *Historia animalium* (*History of Animals*) in the fourth century BCE, comprising the first known classification system for animals or, indeed, for anything. Using his powers of observation, he grouped types of creatures according to their similarities – animals with blood, those that walked, those that flew and so on – a hierarchical arrangement with human beings on top, and no mention of evolution or the complex and mutable relationships thrown up by such an idea.

Aristotle also divided all living organisms into 'animals' and 'plants' with plant subgroups of 'small', 'medium' and 'large'. Although any further writings he may have made about plants did not survive, those of his pupil Theophrastus (*c.* 371–287 BCE) did: *De Historia Plantarum* (*Enquiry into Plants*) describes the anatomy of plants and classifies them into trees, shrubs, herbaceous perennials and herbs, while *De Causis Plantarum* (*Causes of Plants*) discusses propagation and growth.

Theophrastus also built on Aristotle's binomial naming system according to genus and difference, placing every organism in a family and then differentiating family members by some unique characteristic. This was game-changing and eventually highly influential but astonishingly it would be nearly 2,000 years before any further progress was made on how best to classify and name plants.

The Italian physician Andrea Cesalpino (1519–1603) finally revisited the subject of classification in *De plantis libri XVI* (1583). His interest lay in the study of potential plant medicines, for which he needed a framework for his thinking and his findings. While largely influenced by Aristotle and his successors, he introduced the *new* idea of classifying plants based on the structure of their fruits and seeds, in turn inspiring Swiss botanist Gaspard Bauhin (1560–1624) to thus describe 6,000 different species of plant in *Pinax theatri botanici* (*Illustrated Exposition of Plants*) in 1623.

Bauhin's classification system was not particularly innovative but it did group plants according to their 'natural affinities'. More importantly, it grouped them according to genus and species – a word first given its biological definition by English naturalist and author of *Historia Plantarum* (*The History of Plants*, 1686–1704), John Ray. Adding a species name was a brilliant idea that absolutely laid the foundations for the plant nomenclature of today, but it would take the left-field thinking and ultimate logic of Swedish botanist Carl Linnaeus to help put everything into a viable order.

By the time Linnaeus took his first steps into botany, there was a confusing profusion of classification systems in use. At this time global exploration was rife and new plants were being discovered in rapid succession. His answer to this conundrum was the publication of his most groundbreaking work, *Systema Naturae* (*The System of Nature*) in 1735 – or, as the subtitle of its tenth edition read: *System of nature through the three kingdoms of nature, according to classes, orders, genera and species, with characteristics, differences, synonyms, places.*

The book organised plants and animals from kingdoms all the way down to species. Linnaeus was also the first to consistently use binomial nomenclature (see page 124) combined with a hierarchical system of taxonomy, as presented in his 1753 publication *Species Plantarum* (*The Species of Plants*). 'God created, Linnaeus organised', was one of his favourite claims.

When Charles Darwin published *On the Origin of Species* in 1859, ideas about creationist hierarchy and strict Linnaean taxonomy were also challenged. Evolutionary 'Tree of Life' representations became popular in the nineteenth and twentieth centuries, using fossils to try to understand the links between organisms through the ages. Today, this work continues using advanced DNA analysis. With all the methods combined and an agreed taxonomic system largely now in place, more than one million of an estimated 8.7 million species of living plants, animals and other organisms have so far been formally described.

Opposite Carl Linnaeus, as illustrated by Robert J. Thornton in 1807. Thorton's homage to Linnaeus's sexual system includes *The Temple of Flora*, one of the most poetic, sumptuous botanical books ever made.

COMPELLED TO CLASSIFY

The need to name

The often overlooked beauty of binomial nomenclature is that many features associated with common names (such as allusion to appearance, use, mythology, habitat, history or discovery) are intrinsic parts of botanical names, too. The fact that botanical names are written in Latin (though daunting at first) bestows the additional superpower of making these names universal. Thus French-speaking plant lovers can discuss specimens with those from India; plant scientists can discuss their findings with paleobotanists; gardeners can more easily choose preferred species for their plots; lists can be compiled of rare or protected plants; and publications can be produced for a global audience.

Of course, there's certainly a charm in referring to plants by their common names – names that conjure up childhood walks through the countryside when such names were first learned, or that beautifully describe how a plant looks or behaves in language that is easily understood – but it can also be immensely confusing and sometimes even dangerous. *Anthriscus sylvestris*, for example, has numerous nicknames (cow parsley, mother die, wild chervil, wild-beaked parsley and keck among them), some of which are also used to describe other plants as well. Its lacy, umbelliferous flowers are also easy to confuse with Queen Anne's lace or wild carrot (both common names actually refer to *Daucus carota*), edible parsley (*Petroselinum crispum*) and carrot (*Daucus carota* subsp. *sativus*), plus species of skin-irritating hogweed (*Heracleum*) and the highly toxic, absolutely not edible hemlock (*Conium maculatum*).

Before the universal adoption of Carl Linnaeus's system of binomial nomenclature (using a genus name followed by a species name), plant names could also be extremely lengthy: *Plantago foliis ovato-lanceolatis pubescentibus, spica cylindrica, scapo tereti* meaning 'plantain with pubescent ovate-lanceolate leaves, a cylindric spike and a terete scape', for example. In the binomial system this plant is now referred to as *Plantago media*, in line with the current governing body for botanical plant names, the International Code of Nomenclature for algae, fungi, and plants (ICN).

Following in the footsteps of the *Lois de la nomenclature botanique* – the 'best guide to follow for botanical nomenclature' – established in 1867 at the International Botanical Congress, in Paris, its guidelines are vital for the standardisation of botanical names. These can then be shared through resources, such as the International Plant Names Index (IPNI), which shows who named the plant and when, or The Plant List (www.theplantlist.org), a working database of all the known plant species in the world, created as a response to the Global Strategy for Plant Conservation (GSPC) to help conserve biodiversity.

As per the descriptive language used by early botanists, ICN-governed names are in Latin, although often heavily drawn from Greek or other languages. Each assigned name then includes an initial-capped genus name (see page 130) followed by a lowercase species name (see page 133), the whole written in italics. Subspecies, variety, form, cultivar and hybrid (see page 134) may also be included, each written according to the ICN code. It's also helpful to know that the genus name is a noun and the species name usually an adjective. This is where the romance lies – scientific names, like common names, tell stories.

The word *Rosa*, for example, probably stemmed from an Ancient Greek word for the rose plant, while *canina* (see diagram on page 120) translates as dog, from the belief that the root was effective against the bite of a mad dog. *Rosa* x *damascena* is a hybrid of *Rosa gallica* (French rose), *Rosa moschata* (musk rose) and *Rosa fedtschenkoana* (a mountain rose named after Russian botanist Olga Fedtschenko) that was introduced to Europe from Damascus, Syria.

Meanwhile the story behind the naming and indeed creation of the hugely popular *Rosa* 'Madame A. Meilland', or Peace rose, stems from cuttings sent abroad by French horticulturist Francis Meilland during World War II. Successfully propagated and then introduced as a cultivar in 1945, the Peace rose – a pale-yellow, pink-tinged hybrid tea rose – was given to all delegations of the inaugural meeting of the United Nations in San Francisco. The accompanying note read: 'We hope the "Peace" rose will influence men's thoughts for everlasting world peace.'

Opposite *The Flowering Plants and Ferns of Great Britain* (1846) illustrates similarities between cow parsnip (*Heracleum maximum*), species of hartwort (*Tordylium*), coriander (*Coriandrum sativum*) and hemlock (*Conium maculatum*).

Kingdom

Classification is all about establishing similarities and difference. Thus the highest category that plants might fall into is 'life'. It is composed of living matter that shows certain attributes such as growth, responsiveness, metabolism, energy, transformation and reproduction. As proposed by American microbiologist and biophysicist Carl Woese (1928–2012), who carried out extensive work on evolutionary relationships by studying genetic codes and sequences, there are then three domains: Bacteria, Archaea and Eukarya. What sets eukaryotes apart are cells with a membrane-bound nucleus and other organelles, which contain genetic material, and within this domain are several kingdoms, of which the Animal Kingdom, Fungi Kingdom and Plant Kingdom are the most well known. How eukaryotes are divided today differs according to various scientists. In the United States, textbooks usually refer to six kingdoms; in the United Kingdom it's five.

Division

In botany, the rank referred to as 'division' is pretty much equivalent to phylum. The International Code of Nomenclature for algae, fungi and plants accepts both. The word 'phylum' was originally used by German biologist and artist Ernst Haeckel (1834–1919); the word 'division' by German botanist August W. Eichler (1839–1887), who first classified plants to reflect the concept of evolution. Eichler's system was the first to divide the Plant Kingdom into 'non-floral' plants (Cryptogamae) and 'floral' plants (Phanerogamae), to separate Phanerogamae into angiosperms and gymnosperms, and then angiosperms into monocots and dicots. Eichler's work laid the foundation for today's evolving systems and now includes various subdivisions. See the diagram on page 24 for examples of common divisions and subdivisions: land plants (embryophytes), vascular plants (trachaeophytes) and non-vascular plants (bryophytes), for example.

Class

The concept of the taxonomic rank class was first introduced by French botanist Joseph Pitton de Tournefort (1656–1708) in his *Eléments de botanique* (1694). It was then utilised by Swedish botanist Carl Linnaeus in *Systema Naturae* (1735) to divide his three proposed kingdoms: Animals, Vegetables (plants) and Minerals. The classes into which he split the Animal Kingdom were based on morphological features and are still largely in use today: Mammalia, for example. The classes he used for plants, however, were based on Linnaeus's Systema Sexuale, an artificial key that did not stand the test of time. Indeed, the rank of class is sometimes omitted from modern classification systems in favour of unranked clades – groups of organisms with a common ancestor and all its lineal descendants. Clades are not just grouped, they are nested within each other, branching out as a family tree to reflect evolutionary history. This, of course, is made all the more possible and illuminating by advanced microscopy and genetics.

Order

The rank order appears between class and family, with the spectrum widened to include superorder and suborder. Orders were also used by Linnaeus to subdivide classes into smaller, more comprehensible groups, which is exactly what is done today, just in different ways according to the classification system in use. Some groups have also been set up to try to establish a consensus on the taxonomy of various plants rather than using separate systems. The Angiosperm Phylogeny Group (APG), for example, reflects new knowledge about flowering plant relationships discovered through studies into their evolutionary history. Their latest system, APG IV, included 64 recognised orders, which were then subdivided into various clades and 416 families. The clade monocots includes the order *Liliales*; the clade eudicots, the order *Ranunculales*; the clade magnoliids, the order *Laurales*. Respectively, they play host to the lily, buttercup and laurel families.

COMPELLED TO CLASSIFY

Top left *Wild Flowers* by illustrator Katie Scott (see page 314) from the wonderful *Botanicum* (Big Picture Press, 2016) shows various species of the division angiosperm. **Top right** A pistachio plant by artist Claude Aubriet for Joseph Pitton de Tournefort, a pioneer in systematic botany – particularly genera – and author of the influential *Eléments de botanique* (1694). **Left** German biologist, philosopher and artist Ernst Haeckel (left) coined many words and phrases, including phylum, ecology, phylogeny, stem cell and the kingdom Protista.

> 'Natural bodies are divided into three kingdomes of nature: the mineral, vegetable, and animal kingdoms. Minerals grow, Plants grow and live, Animals grow, live, and have feeling.'
>
> Carl Linnaeus, 'Observations on the Three Kingdoms of Nature', Nos 14–15, *Systema Naturae* (1735)

BOTANY FOR BEGINNERS 127

Family

The taxonomic term *familia* was first used by the French botanist Pierre Magnol in *Prodromus historiae generalis plantarum, in quo familiae plantarum per tabulas disponuntur* (1689), referring to 76 tabled plant groups. Carl Linnaeus subsequently used the term in *Philosophia Botanica* (1751) to denote major groups of plants such as trees, herbs, ferns and palms. The rank of 'order' and 'family' were then used interchangeably for a time, with various botanists, organisations or even countries preferring one term over the other.

Still, plant families do enjoy relative stability today. They are one of the eight major taxonomic ranks, located after order and before genus. Their naming is now governed by various international codes – thus all of them end with the suffix '-aceae'. A family is named after one of its members (its genus or species), which gives rise to many well-known family names used in everyday language. The 'rose family' (Rosaceae), for example, stems from the genus *Rosa*; the parsley family (Apiaceae) stems from the genus *Apium* and the related species commonly known as parsley (*Petroselinum crispum*). The latter illustrates how plant names can change as new botanical discoveries require a plant to change groups. Parsley is also known by the synonyms *Apium crispum* Mill. and *Apium petroselinum* L, as it was once assigned to the *Apium* genus along with celery (*Apium graveolens*).

Plant families are hugely important to botanists and plant scientists but a working knowledge of them can also be really useful for related activities such as gardening. Because plant families have shared characteristics, many of which can be seen with the naked eye, it's a good place to start with plant identification. Flowers of the Campanulaceae, or bellflower, family are usually bell-shaped with five flower parts, often in shades of blue. Plant families often share behavioural features, too, such as preferred habitats, life cycles, pollinating systems or chemical components, which can be handy for finding beneficial attributes, such as hardiness, medicinal properties or propagation methods. For a full list of the world's known plant families – 642 at last count – see The Plant List (www.theplantlist.org). Or pick up a plant families book, head out into 'the field' and see how many revealing connections you can make.

MAKE THE CONNECTION

Commonly known plant families: Apiaceae (carrot/celery/parsley family); Asteraceae (daisy family); Boraginaceae (forget-me-not family); Brassicaceae (cabbage family); Fabaceae (bean family); Campanulaceae (bellflower family); Caryophyllaceae (carnation/pink family); Ericaceae (heather family); Cucurbitaceae (gourd family); Geraniaceae (geranium family); Iridaceae (iris family); Lamiaceae (mint/deadnettle family); Lauraceae (laurel family); Liliaceae (lily family); Magnoliaceae (magnolia family); Malvaceae (mallow family); Musaceae (banana family); Oxalidaceae (wood sorrel family); Papaveraceae (poppy family); Polemoniaceae (phlox family); Primulaceae (primrose family); Ranunculaceae (buttercup family); Rosaceae (rose family); Saxifragaceae (saxifrage family); Solanaceae (potato family); Violaceae (violet family)

Opposite and Right *Chrysanthemum* species and globe artichokes (*Cyanara cardunculus* var. *scoymus*) belong to the Asteraceae (daisy) family, sharing a compound inflorescence (flower head) made up of lots of smaller flowers.

Genus

The genus name (plural genera) of a plant precedes the species name in the binomial (two-part name): the *Acer* of *Acer nigrum* (black maple); the *Allium* of *Allium canadense* (wild onion); the *Helianthus* of *Helianthus annuus* (sunflower); the *Camellia* of *Camellia sinensis* (tea). It describes a group of plants within a family that have even closer similarities or shared characteristics and lineage. Thus if the two-part botanical name of a plant shares its genus with another plant or plants, you can assume that they are closely related, even if it's not obvious from their looks.

Within the potato or nightshade family (Solanaceae), for example, resides the genus *Solanum*. This genus includes some possibly surprising relatives: potatoes (*Solanum tuberosum*), tomatoes (*Solanum lycopersicum*) and aubergines or eggplants (*Solanum melongena*). It also includes *Solanum nigrum* (black nightshade), which is the most likely candidate for a plant known as strychnos during Pliny the Elder's times (23–79 CE). He used the word 'Solanum' to describe a comforting or sun-loving plant (the Latin *sol* meaning sun, *solamen* meaning a comfort), giving an idea of how long some genus names have been around. The genus *Solanum*, along with other familar genera, was firmly established by Carl Linnaeus in 1753 and now boasts up to 2,000 species – a particularly prolific group.

Today's taxonomists use molecular data to confirm genus status, which means that some established names have significantly changed. The tomato, for instance, was at one time given its own genus, *Lycopersicon*, by English botanist Philip Miller (1691–1771). With the help of genetics, however, its original name provided by Carl Linnaeus was restored, as the genus *Solanum* was proved to be correct.

How a genus gets its name is now subject to strict codes of nomenclature. Some derive from historic references, while others are made up from scratch where a new genus is established. Either way, there's often a romantic notion in there somewhere. The genus *Theobroma*, nominated by the ever-present Linnaeus, translates as 'food of the gods' after the Greek *theos* (god) and *broma* (food). *Theobroma cacao* (cocoa tree), the best-known species, gives us chocolate. More recently, a newly identified genus of fern – the *Gaga* genus – was named after the musician Lady Gaga, in recognition of her 'fervent defence of equality and individual expression' and inspired by her similarly shaped costume at the 2010 Grammy Awards.

MAKE THE CONNECTION
Commonly known genera: *Allium* (includes cultivated onion, garlic, spring onion, shallot, leek and chives); *Solanum* (includes potato, tomato and aubergine); *Eucalyptus* (includes some of the tallest trees in the world); *Euphorbia* (includes the Peking spurge, candelabra tree and poinsettia); *Ranunculus* (includes buttercups, spearworts and water crowfoots); *Brassica* (includes broccoli, cauliflower, cabbage, turnip and rapeseed); *Rosa* (includes wild and ornamental roses in four subgenera); *Quercus* (includes white oak and European oak); *Salvia* (includes common sage, sacred white sage and chia); *Mentha* (includes garden mint and peppermint); *Tulipa* (including the lady tulip, eyed tulip and garden tulip); *Phalaenopsis* (moth orchids, some of the most popular cultivated orchid species); *Helleborus* (includes 'Christmas rose' hellebore)

Opposite and right Species in the bulbous genus *Allium* share a strong onion or garlic scent, linear leaves and umbels of star-like flowers, including the shallot (*Allium cepa*), opposite, and the ornamental *Allium hollandicum*, right.

Species

The English naturalist John Ray was the first to give the concept of species a biological definition in his *Historia Plantarum* (1686): ' . . . no surer criterion for determining species has occurred to me than the distinguishing features that perpetuate themselves in propagation from seed. Thus, no matter what variations occur in the individuals or the species, if they spring from the seed of one and the same plant, they are accidental variations and not such as to distinguish a species . . . Animals likewise that differ specifically preserve their distinct species permanently; one species never springs from the seed of another nor vice versa.' In short, species bred true and did not change.

For basic botanical purposes, species is the basic unit (smallest taxon) of plant taxonomy and in most cases denotes organisms that breed with each other (the potential for hybrids blurs the boundaries somewhat). Species ranks below genus and, in common parlance, refers both to the specific epithet that follows the genus in most two-part botanical names and to the full two-part name.

Thus the striking magnolia species also known as bull bay or southern magnolia (*Magnolia grandiflora*) is written out as the genus *Magnolia* followed by the specific epithet *grandiflora*. The specific epithet is an adjective, as most species names are, derived from the Latin words *grandis*, meaning 'big' and *flor*, meaning 'flower'. Other species in the Magnolia genus therefore need different combinations of names to set them apart: *Magnolia stellata* (star magnolia) or *Magnolia macrophylla* (bigleaf magnolia), for example.

Once species names from across the Plant Kingdom start to become more familiar, it's not unusual to come across the same specific epithet being used to describe different genera. *Hydrangea macrophylla* shares its second name with *Magnolia macrophylla* but the two are distinct species validated by the precedence of their genus name. Often such epithets refer to structure (*arboreus* = tree-like), form (*caespitosus* = tufted), habit (*altus* = tall), colour (*alba* = white) or habitat (*alpinus* = alpine) but some reference people (*hookeri* = after William J. Hooker) or places (*japonica* = from Japan). Not only are species names vital for the often ultimate identification of a plant, they can be a fun way to learn more about the plants, too. Just remember to keep the genus name initial-capped, the species name lowercased and both italicised when put together to impress fellow species-loving colleagues or friends.

MAKE THE CONNECTION
Commonly known specific epithets: (the second part of the species name): *annua* (annual); *arvensis* (of the field); *borealis* (from the north); *digitata* (leaves like a hand); *elegans* (elegant); *floribundus* (free-flowering); *grandiflora* (large-flowered); *grandis* (large); *leiocarpus* (with smooth fruits); *mexicana* (from Mexico); *multicaulis* (with many stems); *nanus* (dwarf); *nocturna* (nocturnal); *odoratus* (fragrant); *officinalis* (with herbal uses); *orientalis* (eastern); *peregrinus* (exotic); *perfectus* (complete); *purpuratus* (purple); *racemosus* (with flowers in racemes); *rudis* (wild); *sanguineus* (blood-red); *sempervirens* (evergreen); *simplex* (unbranched); *stellaris* (star-like); *sylvestris* (found wild); *tinctorius* (used for dyeing); *undulatus* (wavy); *variegatus* (variegated); *virens* (green); *viridescens* (becoming green); *vulgaris* (common)

Opposite *Magnolia* is an ancient genus of more than 200 species, named after the French botanist Pierre Magnol. **Right** The Orchidaceae family, meanwhile, boasts 20,000–30,000 different species, including 60 or so kinds of moth orchid.

Variations on a theme

Variation within species is not uncommon, although some species, such as the monkey puzzle tree (*Araucaria araucana*), do exist on their own. Botanists deal with variation by further dividing species. Some variation is also manmade, with new plants assigned names accordingly.

Subspecies
A subspecies is usually denoted where geographically separate populations of a species exhibit recognisable genetic differences, but not enough to be a distinct species. Subspecies have the potential to successfully interbreed with each other.

MAKE THE CONNECTION
Coincya monensis subsp. *monensis* (Isle of Man cabbage), endemic to the British Isles; *Daucus carota* subsp. *sativus* (carrot), domesticated from wild carrot

Variety (Varietas)
Varieties also have a slightly different natural botanical structure, but the geographical element has less of a bearing. Varieties of distinct appearance may exist in proximity to each other and interbreed naturally to create hybrids, a habit exploited by plant breeders.

MAKE THE CONNECTION
Camellia sinensis var. *sinensis* (Chinese tea); *G. elwesii* var. *monostictus* (greater snowdrop)

Form (Forma)
Form ranks below variety and usually designates plants within a species that display very minor alternative characteristics, such as colour or leaf shapes. They may occur anywhere within a species range.

MAKE THE CONNECTION
Rosa rugosa f. *alba* (Japanese rose with white flowers); *Athyrium angustum* f. *rubellum* (lady fern with red stems, 'Lady in Red')

Cultivar
The word 'cultivar' stands for cultivated variety, where plants are selected for desirable characteristics that are maintained during propagation. For added clarity, the International Code of Nomenclature for Cultivated Plants (ICNCP) governs that cultivars produced since 1959 are denoted using modern language and quotation marks rather than additional Latin.

MAKE THE CONNECTION
Malus domestica 'Braeburn' (Braeburn apple), cultivated in 1952; *Monstera deliciosa* 'Variegata' (variegated Swiss cheese plant)

Trade designation
Plant breeders often designate selling or marketing names for their plants. These names reject quotation marks, instead using a different font (the system used by the Royal Horticultural Society) or small capitals (ICNCP).

MAKE THE CONNECTION
Rosa 'KORbin' (iceberg), white floribunda cultivar bred by Reimer Kordes in 1958

Hybrid
A hybrid is the offspring of two plants of different species or varieties. Hybridisation can happen naturally in the wild or by means of human intervention. Most occur between species of the same genus, but it is also possible between genera of the same family.

MAKE THE CONNECTION
Helleborus x *sternii* (Stern's hybrid hellebore), hybrid of *H. argutifolius* and *H. lividus*; *Citrus* x *limon* (Meyer lemon), hybrid of lemon and mandarin, native to China

Heirloom
This refers to an established, multi-generational cultivar. Many species have kept traits through open pollination (non-hybrid) to breed true, or by grafts and cuttings.

MAKE THE CONNECTION
Solanum lycopersicum (yellow pear tomato), first grown in Europe in 1805

COMPELLED TO CLASSIFY

1. Apple, Golden Sweet.
2. "   Talman Sweet.
3. "   Bailey Sweet.
4. "   Sweet Bough.

Top There are more than 7,500 known cultivars of apple (genus *Malus*), some of which were illustrated by painter Alios Lunzer for the *Brown Brothers Continental Nursery Catlog* of 1909. **Bottom** Columbine (*Aquilegia*) varieties, hybrids and cultivars are numerous, including 'Crimson Star' – stunning planted *en masse*.

> *'To such an extent does nature delight and abound in variety that among her trees there is not one plant to be found which is exactly like another; and not only among the plants, but among the boughs, the leaves and the fruits, you will not find one which is exactly similar to another.'*

Leonardo da Vinci, Italian Renaissance polymath (1452–1519)

BOTANY FOR BEGINNERS 135

Areca Catechu L.

THE CYCLE OF LIFE

The life cycle or lifespan of any living organism provides an illuminating insight into its evolutionary development. Members of the Plant Kingdom are no exception, metamorphosing between sexual and asexual phases of reproduction in a process called the alternation of generations (see page 33). Some types of plants also go through seasonal life cycles, influenced by external factors, such as temperature and access to water or daylight. This includes annuals, biennials and perennials but also the further divided hardy, tender and herbaceous plants.

Annual

Annuals complete their life cycles – that is they grow, germinate, flower and set seed – all within one year, and then die. Summer annuals germinate during spring or early summer and mature by autumn of the same year, while winter ones germinate during the autumn and mature during the following spring or summer. These plants are often low-growing and are important ecologically as their vegetative cover helps alleviate soil erosion (although they can also be considered weeds).

For some annuals, the seed-to-seed life cycle is even shorter than a year. *Arabidopsis thaliana* (thale-cress) completes its life cycle in as little as six weeks. This ephemeral annual also enjoys a 'Jekyll and Hyde' lifestyle, both as a widely considered weed that easily colonises wasteground and a widely studied botanical specimen – the first plant to have its genome sequenced. It's even being studied by NASA in space.

In cultivation, many food plants are grown as annuals, including virtually all domesticated grains. They're also great choices for injecting short-lived but high-impact colour into gardens or landscapes, with many annuals blooming all summer long. Indeed, many of the world's most riotous displays of wildflowers are self-seeding annuals that burst forth year after year.

MAKE THE CONNECTION
Lathyrus odoratus (sweet pea), *Citrullus lanatus* var. *lanatus* (watermelon); *Stellaria media* (chickweed); *Zinnia elegans* (common zinnia); *Oryza sativa* (Asian rice); *Triticum aestivum* (wheat); *Nigella damascena* (love-in-a-mist); *Zea mays* (maize); *Papaver rhoeas* (field poppy); *Centaurea cyanus* (cornflower); *Rhinanthus minor* (yellow rattle)

Biennial

The enigmatic biennials take two years to complete their biological life cycle. In the first year, they germinate and produce vegetative growth such as leaves, stems and roots. Then there's a dormant period, during the colder months of winter, for example, and the following year they produce further vegetative growth, then they flower, set seed and die. Under extreme conditions – such as untimely exposure to cold temperatures or drought – a biennial may complete its life cycle more rapidly. Thus many biennials act and are treated as annuals.

MAKE THE CONNECTION
Digitalis purpurea (foxglove); *Allium cepa* (onion); *Daucus carota* subsp. *sativus* (carrot); *Matthiola incana* (Brompton stock); *Petroselinum crispum* (parsley); *Campanula medium* (Canterbury bells); *Myosotis sylvatica* (wood forget-me-not); *Lunaria annua* (honesty), pictured below; *Brassica oleracea* (cabbage)

Opposite This 19th-century plate of betel palm (*Areca catechu*) shows various stages of this perennial's life cycle, including the areca nut, which is often chewed wrapped in betel leaf (*Piper betel*).

Perennial

Perennials live for more than two years, perennial meaning 'through the years'. Some become dormant from year to year, sometimes dying back below the ground. During favourable seasons, they come back and if seasonal conditions remain ideal, they will often flower and fruit for many years to come.

Many perennials live for much longer than the two-year minimum. Peonies (*Paeonia*), for example, can last for decades, while the bristlecone pine (*Pinus longaeva*) can exist for millennia. The oldest known member of bristlecone (also the oldest non-clonal organism on Earth) is more than 5,000 years old.

When used in common parlance, usually within the realm of gardening or landscape design, perennials often refer to herbaceous perennials. They exclude woody plants such as trees, shrubs and sub-shrubs. Botanically speaking, perennials cover all these groups and include many well-known species found in the wild, from evergreen and deciduous trees to bulbs and meadow flowers to woody and herbaceous shrubs.

MAKE THE CONNECTION
Dahlia (dahlia); *Foeniculum vulgare* (fennel); *Echinacea purpurea* (purple coneflower); *Aster alpinus* (Alpine aster); *Cichorium intybus* (common chicory); *Phlox* (phlox); *Achillea millefolium* (yarrow); *Iris* (iris); *Narcissus* (daffodil); *Sedum* (sedum); *Verbena bonariensis* (purpletop verbena); *Miscanthus sinensis* (maiden grass); *Medicago sativa* (Lucerne); *Asparagus officinalis* (garden asparagus); *Rheum rhabarbarum* (rhubarb); *Armoracia rusticana* (horseradish)

Herbaceous

The terms 'herbaceous' and 'woody' refer to the nature of various plants. Herbaceous plants don't have woody stems above ground, but beyond that, they can be annuals, biennials or perennials and display life cycle behaviour to match. There are a wide variety of herbaceous flowering plants in the world.

For gardeners, herbaceous borders often translate as swathes of herbaceous perennials that bloom from spring to autumn and die back each year, while for botanists, herbaceous plants (sometimes 'herbs' for short) are just non-woody ones: grasses, spring bulbs, meadow flowers, to name a few.

Meanwhile, for most people herbs are simply plants with leaves, seeds or flowers that are exploited for their medicinal, aromatic or flavoursome properties. Either way, the verdant and sensory appeal of 'herbs' provides layers of lush coverage for much of our world.

MAKE THE CONNECTION
Musa acuminata (banana); *Aquilegia* (columbine); *Paeonia* (peony); *Salvia sclarea* (clary sage); *Valeriana officinalis* (valerian); *Origanum vulgare* (oregano); *Hosta* (hosta); *Dryopteris filix-mas* (male fern); *Lathyrus latifolius* (perennial peavine)

Woody

Woody plants are usually perennial, as a longer life cycle is required to grow the wood from secondary xylem (see page 107). They include obvious specimens, such as trees, but also woody shrubs and lianas – the first angiosperms were thought to be massive woody plants. Today's woody species provide year-round height and sculptural interest.

MAKE THE CONNECTION
Hedera helix (ivy); *Wisteria sinensis* (Chinese wisteria); *Lavandula angustifolia* (English lavender); *Rosmarinus officinalis* (rosemary); *Hydrangea macrophylla* (hydrangea); *Ilex aquifolium* (holly); *Fagus* (beech); *Jasminum officinale* (jasmine)

Left Banana and plantain plants (genus *Musa*) are technically giant herbs, their apparent stem made up of the bases of huge herbaceous leaf stalks. **Top** Rosemary (*Rosmarinus officinalis*) is a woody shrub, the leaves and flowers of which make a wonderful culinary herb. **Bottom** Coneflower (*Echinacea purpurea*) 'Green Edge' and 'Rubinglow' are teamed with switchgrass (*Panicum virgatum*) 'Rehbraun' in a perennial border by landscape designer Piet Oudolf at RHS Wisley.

THE CYCLE OF LIFE

THE CYCLE OF LIFE

Hardy

Plant hardiness is all about a plant's ability to survive adverse growing conditions – cold, heat, drought, flooding or wind, for example. It's also dependent on the plant's location. In the United Kingdom, the Royal Horticultural Society (RHS) has published a set of applicable hardiness conditions with advice on how to care for plants. It ranges from 'Very hardy' (can survive in colder than -20°C/-4°F) through 'Hardy – average winter' (-20 to -15°C/-4 to 5°F), 'Half-hardy' (-5 to 1°C/23 to 34°F), 'Tender' (1 to 5°C/34 to 41°F, requiring frost-free glasshouse in winter) to 'Heated glasshouse – tropical' (warmer than 15°C/59°F), with several ratings in between.

In the United States, there are a range of conditions from baking desert to frozen tundra, for which the United States Department of Agriculture (USDA) has developed a scale and map of hardiness zones. This, or versions of it, has now also been adopted by other countries. Checking where you live can therefore help when you are planning planting schemes. However, often the best measure of hardiness is tried-and-tested experience, either through received wisdom passed down via generations or by trying things out yourself.

Tender

When a plant is described as tender, it usually means that it doesn't thrive in cold temperatures. The advice is to bring these plants in during winter or wrap plants that are not easily moved in sheets of fleece. If in doubt, look on seed packets, plant labels or hardiness scales for advice before choosing a plant.

Many tender, sub-tropical or tropical plants make fabulous house or conservatory plants, while those that are deemed drought-tolerant or drought-resistant have the bonus of enduring hot summers – something to think about in relation to climate change.

MAKE THE CONNECTION
Hardiness (temperate zones): Tender
Astrophytum ornatum (monk's hood cactus) **Warm temperature** *Ocimum basilicum* (sweet basil); *Aloe vera* (Barbados aloe) **Sub-tropical** *Spathiphyllum wallisii* (peace lily) **Tropical** *Monstera deliciosa* (Swiss cheese plant) **Drought-tolerant** *Chamaedorea elegans* (parlour palm, pictured)

MAKE THE CONNECTION
Hardiness (temperate zones): Very hardy
Calluna vulgaris (heather); *Cornus alba 'Sibirica'* (Siberian dogwood) **Hardy (very cold winter)** *Acer palmatum* (Japanese maple); *Viburnum furcatum* (forked viburnum)
Hardy (cold winter) *Hedera hibernica* (Atlantic ivy); *Cyclamen alpinum* (Alpine cyclamen)
Hardy (average winter) *Crocus laevigatus* (smooth crocus); *Rosmarinus officinalis* (rosemary)
Half-hardy *Dahlia* (dahlia); *Lathyrus odoratus* 'Matucana' (sweet pea 'Matucana')

Opposite Plants such as common heather (*Calluna vulgaris*), which grows on the wild open moorland around Stanage Edge in England's Peak District, need to be hardy enough to withstand potential cold and wind.

1. Emily Henderson. 2. Blanche Burpee. 3. Purple Invincible, F₁. 4–11. The various F₂ types obtained by self-fertilising F₁. 4. Purple Invincible. 5. Duke of Westminster. 6. Painted Lady. 7–9. Corresponding dark winged types. 7. Purple, with purple wings. 8. Duke of Sutherland. 9. Miss Hunt. 10 and 11. F₂ whites. Notice that there is no *hooded* red.

THE ART OF SCIENCE

Botany for Beginners offers a tiny glimpse of the botanical world so far explored. Hopefully it inspires curious minds to journey even further into the Plant Kingdom and to view the science behind plants and flowers as every bit as beautiful as the specimens it aims to explain. As the English gardener and architect Sir Joseph Paxton (1803–1865) once said: 'Botany – the science of the vegetable kingdom, is one of the most attractive, most useful, and most extensive departments of human knowledge. It is, above every other, the science of beauty.'

Mapping our botanical future

In 1866, the Augustinian friar and scientist Gregor Mendel (1822–1884) published a paper that would pave the way for modern plant science. Previously, over a period of eight years, he patiently cultivated and cross-bred more than 30,000 common edible pea plants (*Pisum sativum*) and then systematically compared seven easy-to-see characteristics: pea shape, pea colour, flower colour, flower position, plant height, pod shape and pod colour.

His findings led to what is now known as Mendel's Laws of Inheritance, whereby each inherited trait is defined by a gene pair, one half passed on from each parent; genes from different traits are sorted separately from one another so the inheritance of one is not dependent on the other; and an organism with alternate forms of a gene will express the form that is dominant – most notably, in Mendel's case, white flowers over purple ones.

Mendel's findings were monumental but failed to gain the immediate acclaim they deserved. This was in part because they appeared at odds with the accepted theory of the time that inherited traits were blended and because they didn't, as yet, correlate with Charles Darwin's heralded theory of evolution, published only a few years earlier in 1859. Mendel was also backwards in coming forwards on the publicity front, devoid of Darwin's dogged persistence or Carl Linnaeus's sensationalist showmanship.

It would be another three decades before Mendel's work received the attention it merited, simultaneously 'rediscovered' by three independent botanists – Hugo Marie de Vries, Karl Franz Joseph Correns and Erich Tschermak von Seysenegg – and then widely championed by English biologist William Bateson (1861–1926). The words 'genetics' (from the Greek word for descent) and 'gene' (to describe a Mendelian unit of heredity) were subsequently coined by Bateson and the Danish botanist Wilhelm Johannsen respectively – and with that a new botanical era was born. Genetics would now dominate the scene.

Fast-forward to the present day and terms such as chromosome, DNA, genome and gene expression are now commonplace. Technology is so advanced that we've successfully mapped 99 per cent of the entire human genome, not to mention the sequence of thousands of plant specimens with targets for a representative of every life form on Earth. We've cloned sheep, genetically modified crops, built huge DNA databases – and the supersized computers needed to power them – developed true-blue plants, recreated a model of the 'world's first flower' using data from over 800 living plant species, produced the most accurate evolutionary family trees yet and even sequenced a genome in space. So far so sci-fi, with the jury still out on whether genetic engineering provides vital solutions to an uncertain, increasingly populated, under-resourced future, or it could unleash devastating consequences as a result of people playing God with Nature.

Those who stand behind such advances argue that genetic modification could deliver the disease-free, hardier crops we need to secure enough food, medicines and fuel for a population rising by around 83 million people a year (as per the *World Population Prospects: The 2017 Revision* published by the UN Department of Economic and Social Affairs in 2017) – crops that can grow in an environment that is rapidly evolving as a result of climate change, including increasing temperatures and CO_2 emissions, and more widespread drought and fires.

Those who stand against these advances counter that GM crops provide foods that are unsafe for our health, can produce sterile seeds within a short space of time and monopolise the market so that smaller producers get pushed out. Plus biofuels created with plants may not be a 'green' solution after all. The counter movement stands with alternative initiatives, such as organic farming and permaculture.

Either way, botany – the science behind plants and flowers – is much bigger news than people often give it credit for. Let's hope we can use our knowledge wisely to map a positive future for both people and our botanical world – plant-loving superheroes come forth and be counted.

Opposite Plate V from the 1909 edition of *Mendel's Principles of Heredity: a Defence* (1909) by William Bateson illustrates Mendel's Laws of Inheritance, a concept that would lay the foundations for the modern study of genetics.

'It seems to me that the natural world is the greatest source of excitement; the greatest source of visual beauty, the greatest source of intellectual interest. It is the greatest source of so much in life that makes life worth living.'

David Attenborough,
British broadcaster and naturalist (1926–)

CHAPTER THREE

GROWING
&
GATHERING

YOUR BOTANICAL WORLD

We all have our own version of the botanical world, the way in which we personally relate to the Plant Kingdom. Growing or gathering your own plants – for food, as ingredients for recipes or remedies, for the home, inspiration or simply because of a love of nature – can deeply enhance that experience.

Whatever your relationship to plants, growing or gathering will at some stage have something to do with it. Perhaps cooking is your thing or concocting herbal remedies? Any plant-based ingredients you use will have been grown and gathered at some point. The same goes for crafting with plant-based materials or experimenting with floral design – and, of course, the most obvious 'growing and gathering' activities: gardening and foraging.

This chapter explores the latter two elements in more detail from planting suggestions that can directly benefit humans (see page 150) or nature (see page 156), to introductory guides on foraging for food (see page 162), provisions (see page 170) and naturecraft (see page 174). Then, if you're interested in knowing more about a particular topic, there's an extensive Further Reading list at the end of the book (see page 392) with easy-to-navigate sections to help you find the perfect companion for *your* botanical world.

Or, for instant gratification, turn to chapters 4, 5 or 6 for carefully curated collections of delicious botanical recipes, remedies or nature-inspired art, design and craft, compiled with the help of an international selection of experts. Hopefully there's something here to stimulate a deeper connection or new interaction with plants, or perhaps revive a latent one.

If there's one message to take away from the remainder of *Collins Botanical Bible*, it's that there's no set way to commune with nature. Nor do you need to know everything there is to know about plants to enjoy growing, gathering or any related pursuits – although deepening your understanding of the natural world can enhance your experience.

The plant-inspired experts featured in the coming chapters, from grow-your-own aficionados to wild food educators to botanical artists, will confirm that the best way to learn about the many facets of the Plant Kingdom is out 'in the field' – whatever that chosen field may be. Read more about our experts' 'botanical worlds' next to each feature-length recipe, remedy or profile to get a sense of what has inspired them to feed, heal or create with plants. Quotes accompany these contributions, and offer a fascinating read, perhaps even more so when read collectively.

There are no set rules that say you *have* to find a plant-loving niche, however. For many people, exploring and harnessing the botanical world is a basic instinct, an intuitive or spiritual engagement – which is no surprise when you consider that growing and gathering have been part of the human experience for tens of thousands of years, from the first hunter-gatherers through the Neolithic Revolution to variables of farming and foraging in present-day times – while for others, the Plant Kingdom serves as a wonderful prism through which to simply observe what it means to be alive – to grow, gather, survive, thrive and ultimately connect with the natural world. Enjoy the journey.

Opposite 'Fennel (*Foeniculum vulgare*)' by Sonya Patel Ellis (see page 386) for *Herb Garden at The Marksman* (2016) pays homage to this wonderful wild and cultivated culinary herb.

Just a little bit of faith in a plant's ability to grow and *your* ability to nurture it.

The year is 1913. The place, a desolate valley somewhere in an Alpine region of Provence, France. A lone hiker in search of water is saved by a middle-aged shepherd who takes him to a spring he knows. Curious about the shepherd, the hiker stays for a while. He learns that the shepherd, named Elzéard Bouffier, is a widower who has decided to restore the landscape by single-handedly growing a forest, tree by tree. He does this by making holes in the ground with a pole and dropping into each one acorns gathered from miles away.

The hiker returns to the valley in 1920, shell-shocked and depressed from fighting in World War I. He is greeted by young saplings taking root in the valley, with streams running through it, a result of dams made by Bouffier higher up in the mountains. The young soldier makes a full recovery amid the peace and beauty of the newly-verdant valley, and thence returns to the same idyllic spot each year.

Bouffier continues to plant his trees over four decades and swaps shepherding for beekeeping to better fit and serve the environment. By the end of the tale, the valley pulsates with life, receives official protection (its existence widely believed to be a bizarre natural phenomenon) and is peacefully settled. The valley becomes a 'Garden of Eden' for more than 10,000 human inhabitants, attracted by the sense of wellbeing and happiness that the natural landscape affords, who themselves plant vegetables and flowers, cabbages and rose trees, leeks and snapdragons, celery and anemones, as well as fields of rye and barley.

This story, published by French author Jean Giono in 1953 as *The Man Who Planted Trees*, is a fiction. But even so, it says much about the act of growing plants and the benefits this can bring to both people and environments. It also serves as a reflection of Giono's long-standing love of the natural world, further illustrated by his wish that the book should remain copyright-free so that it could be distributed at will.

Bouffier's ecological success story also highlights how simple growing can be. If you haven't grown your own plants before, follow in Bouffier's footsteps and maybe start with one thing and see what happens – a pot of rosemary, a fruit tree or a houseplant, perhaps. Live with your plant a little and let it teach you what it needs to stay happy and healthy. A dehydrated, over-shaded or nutrient-depleted plant will soon let you know how it's feeling via the state of its leaves, flowers or overall growth. Trust in the process and allow yourself to make mistakes – and have fun.

It may not be possible to surreptitiously plant a whole forest in today's tangle of red tape (although there are some amazing stories of people who have) but neither do you need a huge private garden to tend your own patch. Some of the most beautiful green spaces are cultivated in pots, on balconies, up walls or in communal plots. Let our planting suggestions – in collaboration with landscape designer Miria Harris (see page 379) – be your guide and find your own reasons to nurture. As the English architect and landscape designer William Kent once said: 'Garden as though you will live forever.'

Opposite There are so many ways to engage with the botanical world. Find those that most appeal to you, then use this book's index to find further inspiration and ideas.

GROWING FOR YOU

Consider the green spaces that you feel the strongest connection to, or the features of your favourite gardens, parks or the countryside that you take most pleasure in. Now think about the aspects of your life that you could potentially grow yourself – plant-based foods, cut flowers, culinary or therapeutic herbs, or materials for creative projects such as print-making, pressed botanicals or styling your home. The elements that you find most rewarding, inspiring or uplifting should be the ones that you choose to grow for yourself.

Planting for wellbeing

Gardening is widely accepted to bring a range of health benefits from feel-good factor hormones stimulated by the physical act of digging and planting, to the emotional wellbeing inspired by being outdoors and around nature. The planting suggestions that follow take things one step further by actively choosing plants for their known therapeutic, nutritional, sensory or environmentally enhancing qualities. Such plants could be brought together in the form of a traditional or dedicated herb garden. Or, as London-based landscape designer Miria Harris suggests, 'woven into herbaceous borders alongside swathes of successively blooming flowers and evergreen shrubs to maintain colour, vibrancy and uplift throughout the seasons'.

A fantastic recent example of planting for wellness was expert organic gardener Jekka McVicar's Ayurveda-inspired show garden: 'A Modern Apothecary', which she created for the RHS Chelsea Flower Show in 2016. Within the peaceful curves and symmetry of her award-winning garden lay borders of scented flowers and healing herbs: mint for tea; flax for the joints; skin-soothing calendula; memory-boosting rosemary and relaxing lavender – a garden for all ages in which to reflect and escape the stresses of modern life.

MAKE THE CONNECTION

Love Common myrtle
Myrtus communis
Aphrodite's sacred herb of love is still used in wedding bouquets, its verdant leaves producing a multi-tasking essential oil said to aid respiratory and skin conditions.

Digestion Fennel
Foeniculum vulgare
The aniseed flavour and digestive medicinal properties of fennel run through its delicate yellow flowers, seeds, feathery leaves and swollen stem base.

Relaxation Lavender
Lavandula angustifolia
Best known for its fresh, clean, relaxing aroma, lavender also has anti-inflammatory and antiseptic properties, as well as a wide range of culinary uses.

Stress relief Thyme
Thymus vulgaris
This drought-tolerant star of the herb garden is a delicious component of *bouquet garni* and can be brewed into a mood-boosting, stress-busting tea.

Happiness Portland Rose
Rosa 'Comte de Chambord'
An exceptionally perfumed rose with bundles of old-world character. Enjoy the warm, pink, full-petalled, fragrant flowers in your garden.

Perspective Silver birch
Betula pendula
Wellbeing means looking after the environment as well as yourself. A silver birch tree provides ornamental beauty, natural shade and is amazing at filtering pollution, too.

Nutrition Blueberry
Vaccinium corymbosum
Growing your own berries is a fun and easy way to add some positive nutritional value to your garden. Blueberries are a great source of the vitamins K and C.

Peace and purification Peace lily
Spathiphyllum
Bring the peace and greenery of the outdoors in with a houseplant. The peace lily is easy to care for and can help remove harmful toxins from the air.

Opposite Fill your garden with the wonderfully relaxing and uplifting scent of lavender (*Lavendula*). To harvest the aroma simply snip off stems just before flowers open, hang to dry and then fill sachets with the dried buds.

Planting for inspiration

Nature inspires on so many levels, providing the theme, focus, motif, backdrop or materials for countless creative works as well as inspiring many of science's most important Eureka moments. From Claude Monet's Giverny to Carl Linnaeus's botanical garden at Uppsala, plants provide a rich tapestry of ideas and inspiration.

'Green spaces provide areas in which to stop, think, and dream up new ideas. A garden full of tantalising fragrance, colour and texture is the perfect foil to the hustle and bustle of daily life,' advises landscape designer, Miria Harris. Indeed, many of her designs have been commissioned by fellow creative types dreaming of a space where they can switch off and be re-inspired.

Harris achieves this by creating year-round interest, via plants that continue to enthral through autumn and winter – the late-flowering blooms, evergreen shrubs, foliage that turns from green to red or gold, textural tree trunks, vibrant stems or sculptural seedheads. Sensory elements are also key – think deeply-scented flowers to induce feelings of nostalgia, the rustle of grasses or trees to aid relaxation, or herbs for mood-enhancing dishes, tinctures or teas. Each design is personal, attests Miria, so plant for your personality and allow your inspiration to run wild.

MAKE THE CONNECTION

Wild abandon Love-in-a-mist
Nigella damascena 'Miss Jekyll Alba'
The white variant of this lacy border lovely is ironically named for colour-loving gardening guru Gertrude Jekyll. Scatter seeds and let nature take its course.

Instant exoticism Chinese wisteria
Wisteria sinensis
Claude Monet's garden at Giverny starred in many of his most famous paintings, its iconic bridge arching under a canopy of fragrant wisteria.

Escapism and romance Common jasmine
Jasminum officinale
So-called for the heavenly fragrance of its blooms, plant this 'Gift from God' near a bedroom window for the sweetest of dreams.

A plant for all seasons North American redbud
Cercis canadensis 'Forest Pansy'
Year-round colour comes via purple-pink buds in spring, then heart-shaped leaves on elegant grey stems that go from green to purple to golden-bronze.

Peace and harmony Olive
Olea europaea
This all-powerful, peace-bearing symbol of the Ancient World – the 'tree of eternity' – is also drought-tolerant and can yield delicious fruits.

Sculptural interest Bearded iris
Iris germanica
From religious icon and royal banner to a favourite subject of the artist Georgia O'Keeffe, the regal iris naturally stands out from the crowd.

Creative flow Common grapevine
Vitis vinifera
The rambling grapevine, or rather the fermentation of its luscious fruit, has probably inspired more creative flow than any other plant. Pay homage up a wall or in a conservatory.

Think pink Chinese meadow rue
Thalictrum delavayi
Placing striking pink flowers such as Chinese meadow rue against an inky background instantly contemporises the colour, first named after the original pink *Dianthus plumarius*.

Opposite In this detail of Chinese meadow rue (*Thalictrum delavayi*) from a garden designed by Miria Harris, the elegant pink petal parasols and creamy stamens are a dramatic foil to the black backdrop of the house.

'It is a golden maxim to cultivate the garden for the nose, and the eyes will take care of themselves.'

Robert Louis Stevenson, Scottish author, poet and travel writer (1850–1894)

Top Plant *Geum* 'Totally Tangerine' en masse, as shown in this garden designed by Miria Harris, for a prolific profusion of orange blooms through spring and summer. **Right** Afford plants bright, bold backdrops à la Jardin Marjorelle, Morocco or Frida Kahlo's garden, Mexico (see page 348). **Bottom** Take conservatory chic one step further with House of Hackney's iconic Palmeral print – paying homage to former Hackney nursery garden Loddiges, once home to the largest hothouse in the world – paired with a potted palm.

Planting for style

As the great twentieth-century artist Henri Matisse once said, 'There are always flowers for those who want to see them.' Indeed it's hard to find an era, of the last three millennia at least, that didn't look to the Plant Kingdom for its stylistic flourishes and motifs: from the floral adornments of Ancient Egypt, Greece and Rome through the richly-embroidered fabrics introduced by the Silk Road, to the influence of Tulip mania, the botanically-rich Arts and Crafts Movement, late 1960s 'flower power' and most recently, the cornucopia of plant-inspired designs on offer today. Green is the new black, suggested colour experts Pantone in 2017.

What's particularly attractive about current and hopefully lasting trends for all things botanical is that plants and flowers are so universally appealing. Plants are beautiful by nature: they bring people together, inspire wellbeing and happiness, and are hugely diverse so there's lots of room for personal preference, self-expression and various modes of application.

'In terms of garden design, planting for style runs the gamut of choosing stylish plants for statement beds or borders, to growing specimens that could be used to style up your life, your home, or even your dinner', says landscape designer Miria Harris (whose husband, Michelin-starred chef Tom Harris, uses freshly-picked heritage herbs to garnish his dishes; see chapter 4). Growing style, it seems, is entirely possible.

MAKE THE CONNECTION

Complementary colours Black-eyed susan
Rudbeckia fulgida var. *sullivantii* 'Goldsturm'
Place the yellow sun-ray petals of rudbeckia against a swathe of purple asters to create colour harmony and flow.

Floral design Japanese anemone
Anemone x *hybrida* 'September Charm'
Japanese anemones make great cut flowers, holding their shape in the vase and supplying a constant supply of ready stems if left to spread and colonise via their rhizomes.

Drama and intrigue European Smoke bush
Cotinus coggygria
The mid-height smoky blooms of oval-leaved cotinus filter sunlight for a soft-focus aspect. Combine with grasses and perennial flowers for year-round atmosphere.

Showstopping swathes Avens
Geum 'Totally Tangerine'
This Chelsea Flower Show favourite flowers from late spring to autumn; its citrus-orange petals are also attractive to bees and butterflies. Plant in swathes for a high-impact design.

Botanical artistry Male fern
Dryopteris filix-mas
The shuttlecock fronds of the male fern go well with hard landscaping. They're also the perfect specimens with which to make your own botanical wall art.

Exotic accents Tulip
Tulipa
Tulip mania makes absolute sense when these majestic, joyful beauties burst through in spring. Keep your patch personal with the pick of numerous varieties.

Modernist moves Agave Americana
Agave Americana
Place boldly-shaped desert succulents or cacti against brightly-coloured backdrops for a mid-century Californian or Mexican vibe. Agave is also super-low maintenance in sun-drenched climes.

Conservatory chic Bamboo palm
Dypsis lutescens
Jump on the houseplant bandwagon with an ornamental bamboo palm or two, adding instant style and quite literally freshening up tired spaces – the leaves actively filter toxins from the air.

GROWING FOR NATURE

Growing for nature is all about choosing plants or planting combinations that could positively impact the environment or its wildlife. The most obvious way to do this is by selecting specimens that attract pollinators. Better still, you could weave such plants into a wider planting scheme that promotes biodiversity. Think about specimens that provide shelter to a variety of wildlife, are heirloom or native species, can tolerate a range of conditions for year-round greenery, or return nutrients to the soil. The Earth will thank you for it.

Planting for pollinators

Pollinators are the caretakers of much of the Plant Kingdom, ensuring the fertilisation and thus reproduction of pollen-bearing plants by transferring male sperm to the female egg of the same species. The majority of flowering plant species rely on such pollinators to make seeds that will become the next generation of plants. Thus it's vital that the insects – the bees, butterflies, moths and hoverflies – birds and animals that provide such a service have ample incentive to fulfil their role. That incentive is nectar, and there is a lot that gardeners can do to help provide it.

As Josefina Oddsberg, co-founder of Bee Urban Sweden (www.beeurban.se) explains: 'Planting flowers for bees is all about letting your garden burst with flowers rich in pollen and nectar throughout the season. In early spring, that's crocus, scilla and allium, during high summer, herbs such as lemon balm, mint and thyme are perfect, in late summer bees love heather, wild oregano and orpin. Make sure you use organic seed and plants, and avoid hybrids of any kind, as these are often sterile. If you want to keep it simple, choose aromatic plants such as lavender and herbs, and bushes and trees that give fruit and berries. They are always a good choice for our pollinating friends.'

MAKE THE CONNECTION

Easy pickings Argentinian vervain
Verbena bonariensis
Verbena's nectar-rich, luminous purple flowers on tall elegant stems bloom from summer into autumn offering easy, inter-season, perennial pickings for bees.

Wild style Coneflower
Echinacea purpurea
Sterile outer petals draw a host of insects into the central cone of tiny, nectar-filled florets, while the seedheads are loved by birds.

Go wild Common poppy
Papaver rhoeas
Try throwing some wildflower seeds around, then wait for the big reveal. The black-on-red centre of the field poppy is a magnet for bumble and honey bees.

Honey-making shrubs Strawberry tree
Arbutus unedo
So-named for its strawberry-like autumnal berries beloved by birds, this glossy evergreen also contributes to honey production via its late creamy blossoms.

Fragrant attraction Guelder rose
Viburnum opulus
A good, mid-height screening shrub that fills the air with scent from an abundance of white or pale-pink flowers; a draw for humans and pollinators alike.

Nature and nurture Switch grass
Panicum virgatum
Butterflies love to lay their eggs in swishing clumps of frothy-flowered switch grass, providing caterpillars with protective cover and an instant leafy feast. (Though it will attract pollinators, switch grass itself is wind-pollinated.)

Early offerings Early crocus
Crocus tommasinianus
Bumble bee queens are often starving after a long winter hibernation and need all the nectar they can get. Lay on an early spring feast for them with a clump of these crocuses.

Companion planting Borage
Borago officinalis
Plant blue starry borage around monoecious vegetable crops, and bees, butterflies and hoverflies will help ensure that pollination, and thus fruiting, takes place.

Opposite Help support local bee communities and thus the wider issue of worldwide bee decline – and the production of many vital food crops – by introducing bee-friendly, nectar-rich flowers that bloom across the seasons.

Planting for biodiversity

Planting for pollinators is one step towards protecting and promoting the biodiversity of the Plant Kingdom. Without our bees, birds and butterflies, a monumental proportion of plant species would simply die out. At the same time, with nectar sources in decline, whole species of pollinators are under threat.

Planting for biodiversity forces you to look at the bigger picture – how plants provide wildlife habitats or enrich soil, for example – which can only be a good thing. The visual effects can also be stunning, as the influential Dutch landscape designer, nurseryman and author Piet Oudolf illustrates through his uber-naturalistic gardens for projects such as the High Line in New York and Hauser & Wirth Somerset.

Piet's design philosophy is to encourage beautiful, dreamy planting that offers personal health benefits from the very act of gardening and connecting with nature, but also plays a crucial role in conserving vital ecosystems and thus the environment. Pollinators and other beneficial insects should have food, shelter and protection throughout the year; thus plants are chosen for their seasonal life cycles, for their leaves, seedheads and silhouettes as well as their flowers.

Landscape designer Miria Harris includes similarly naturalistic elements in her gardens: 'It's all about taking a layered approach and giving nature a helping hand. Look after the soil, work with the seasons and plant with sustainability in mind. The wildlife should follow.'

MAKE THE CONNECTION

Food for free Woodland wild strawberry
Fragaria vesca
Every garden should have wild strawberries – little sweeties for us, and food for wildlife including small animals and birds. An excellent choice for edging.

Autumn nuts Filbert
Corylus maxima 'Purpurea'
If you don't mind hazel-loving wildlife such as squirrels and jays in your garden, plant a purple filbert. You can eat the nuts, too, while admiring the broad foliage.

Uncertain climes Silver grass
Miscanthus sinensis 'Silberfeder'
Include plants that can perform well in a range of conditions. This super grass is incredibly drought tolerant but will also perform well in soggier climes.

Pond life Bulrush/cattail
Typha minima
A pond feature can help attract a variety of wildlife, including frogs, newts and insects. This least-invasive cattail species filters water and provides safe-haven spots.

Decorative and generous Great burnet
Sanguisorba officinalis
The raspberry flowers of this old-world herb morph into dark seedheads come winter – a treat for birds. An extensive root system also helps prevent erosion.

Replenish soil Lupin
Lupinus 'Masterpiece'
Lupins, like other members of the legume (Fabaceae) family, have the power to fix nitrogen in the soil. This one has violet-purple spires with orange-red flecks.

Positive pest-control Prostrate rosemary
Rosmarinus officinalis 'Prostratus Group'
This evergreen shrub deters pests naturally, so plant prostrate rosemary around your vegetable patch to deter bean beetle, carrot fly and mosquitoes, and to attract beneficial insects, too.

Shade and shelter Tufted hair grass
Deschampsia cespitosa
Provide natural shade and shelter for wildlife, and food for larval butterflies with this silvery-flowering grass. It's also ideal for stabilising poorly draining soil.

Opposite Lindsey Carr's *Garden* beautifully captures the biodiversity of a wetlands habitat. Find more of the Glaswegian artist's botanical world online: www.littlerobot.org.uk ; Instagram @lindseycarrart.

REASONS TO GATHER

REASONS TO GATHER

There are many reasons to give foraging a try, but one stands out above the rest: creating the space, time and impetus to explore 'gatheredness' and thus reconnect with the wild has immeasurable mental and spiritual benefits.

'Most of us are still related to our native fields as the navigator to undiscovered islands in the sea. We can any afternoon discover a new fruit there which will surprise us by its beauty or sweetness. So long as I saw in my walks one or two kinds of berries whose names I did not know, the proportion of the unknown seemed indefinitely, if not infinitely, great.' So begins Henry David Thoreau's epic literary work *Wild Fruits,* an unfinished text that was finally published in 2000.

Thoreau, one of history's finest nature writers, began to formulate the idea for the book in the early 1850s. By 1859 his concept of wildness – in short, 'in Wildness is the preservation of the world' – was sufficiently honed to begin writing *Wild Fruits* in earnest. Into the book went lyrical accounts of his wanderings through the nature of New England, along with stores of botanical knowledge and advice on how to identify, gather and eat the 'wild fruits' that he came across on his travels: wild apples, cranberries, huckleberries and chestnuts among them.

Richard Mabey, author of the more contemporary wild food classic, *Food For Free* (1972), later said in *Weeds* (2010) that Henry Thoreau's *Wild Fruits* '... celebrated that mysterious quality of "gatheredness" that clung like a savour to foraged wildings'. In his book *The Perfumier and the Stinkhorn* (2011), Mabey also refers to *Morel Tales* (1998), by the American sociologist Gary Alan Fine, about the culture of wild-mushroom hunting in the United States, in which interviewees talk about the same quality of 'gatheredness' that Thoreau sought to promote. As Mabey explains, it is that which 'makes wild foods taste different from shop-bought ones, and about the ecstasy of discovery'.

Gatheredness, then, is something that we should all potentially aim for, or at least try, not just in the gathering or foraging of wild foods such as fruits, nuts, seeds and roots but also materials for herbal remedies, natural dyes, botanical pressings, nature-crafting or, indeed, whatever sparks your interest. Do ensure that you adhere to any relevant countryside codes, as some plants, plant parts, or even whole areas of landscape are off limits in order to protect wild species or avoid biodiversity loss.

Even if you have no interest in foraging food or materials, the wild is the perfect place to simply gather inspiration, letting ideas form as you walk through nature. For just being in nature is a tonic, especially valuable in the constant buzz and virtual reality of today's 'Digital Age'.

It's also amazing how much plant knowledge can be subconsciously gathered simply by regularly immersing yourself within a landscape, more still by looking closer with the help of a site- or interest-specific plant identification guide – for inspiration see our Further Reading list (see page 392), or search through the various recipes and remedies in chapters 4 and 5.

At the end of the day, the more frequently you gather, and the better company you keep, the greater the gatheredness you should find.

Opposite There are many reasons to gather – for food (see page 162), provisions (see page 170) and naturecraft (see page 174) for example – but also to connect with nature.

FORAGING FOR FOOD

With so many of our food sources brought to us quite literally on a plate, it's easy to forget that food is all around us. It's there in the hedgerows in the form of berries and edible flowers. It covers the forest floor in the guise of wild garlic or the nuts and seeds that drop to the ground in the autumn. There's a veritable salad bowl of fresh leaves growing wild, not to mention some of the tasty roots hiding underground. There's much to learn here, from long-forgotten dishes to grassroots-survival food, and the folkloric history is fascinating too.

Fruits

If you've ever been blackberry picking in the countryside, then you've been foraging for fruit. This delicious ingredient of jams, jellies, pies and crumbles is easy to identify, a breeze to pick – if you mind the thorns and keep to the higher reaches – and instantly edible if you can't wait until you get them home. Wild strawberries are similarly straightforward fare.

More advanced foragers will know that nature holds further fruity treats in the form of apples, plums, peaches and cherries from trees that once formed part of an orchard and are now in public countryside, or their wild counterparts, the crab apples or wild cherries.

The temperate climates that support the growth or cultivation of such fruit may also play host to an abundance of 'berries', where the environment is right, such as the marsh-loving cranberry or hedgerow-dwelling sloe, elderberry or rosehip. More tropical climes may render fruits to match – from lemons, oranges and olives to mangoes, bananas and coconuts (the latter being hard-cased drupes, not nuts!).

Wherever you are in the world, it's worth checking which fruits are foragable and where you have permission to pick. That way you can get fruit for free without upsetting the ecology of your picking spot. You could also try growing your own fruits so that you can 'forage' at will, or even sign up to share your harvest with likeminded, land-sharing souls.

MAKE THE CONNECTION

Blackberry
Rubus fruticosus
Unlike its raspberry relative, the blackberry keeps its torus (core) when picked. The juicy fruitlets can then be used in jams (see page 218), jellies, desserts and wine.

Wild strawberry
Fragaria vesca
The strongly-flavoured woodland or wild strawberry is thought to have been eaten by humans since the Stone Age. Collect by threading onto straw-like grass.

Wild cherry
Prunus avium
The main ancestor of the cultivated cherry can be identified by two red glands on the leaf stems and white blossom. Birds love the fruits, so get there fast.

Crab apple
Malus sylvestris
High in the setting agent pectin, sunset-yellow crab apples make great jams and jellies when combined with blackberries, rowan berries, rosehips or summer herbs.

Bilberry/whortleberry
Vaccinium myrtillus
Like its cultivated cousin the American blueberry (*Vaccinium corymbosum*), this wild red-to-blue, deeply staining version is great in cakes, muffins and crêpes.

Sloe
Prunus spinosa
In Britain, purple-black sloe drupes are traditionally harvested after the first frosts, then used to create sloely-does-it infused gin (see pages 243 and 246) or vodka.

Rosehip
Rosa rugosa/Rosa canina
There are numerous uses for vitamin-C–rich rosehips, from the Swedish rosehip soup, *nyponsoppa*, to herbal tea, syrup (see page 218) and wine.

Cranberry
Vaccinium oxycoccus
Native to cool temperate regions of the Northern Hemisphere, the marsh-loving wild cranberry ripens in the autumn. Its festive cousin *Vaccinium macrocarpon* is native to North America only.

Opposite Late summer and autumn, as this 1870 engraving of blackberries and acorns illustrates, is a wonderful time for gathering wild fruits and nuts. Take children along and get them involved in the wonders of nature's bounty too.

Nuts and seeds

As described earlier in the book (see page 102), the word 'nut' is commonly used not only to refer to 'true nuts' (fruits encased in an inedible hard shell, such as the hazelnut, chestnut and acorn) but also the seed of a cone (pine nut), the seed of a legume (peanut) or the seed of drupes (walnut, pecan and almond).

It's quite an interesting concept to get your head around when you're out foraging, not least as a reminder that while the parent plant of the 'nut' actively throws its wares down in a bid to spread its seeds far and wide, humans may not be the carrier they were looking for. Remember, therefore, to leave some for the squirrels or birds, not only as direct food for them but also for potential transportation and germination.

In the United Kingdom, edible nuts include the sweet chestnut, hazelnut, pine nut and acorn. In the United States, the hickory (including the pecan) and black walnut may be added to the list. Plus there are less well-known survival foods such as beech masts (mast-year dependent and if the squirrels don't get there first). Nuts can also be fun for kids to naturecraft with (see page 179).

On the 'seed' front, worthy plants such as fennel do grow wild, so it's worth going on a hunt to see if you can find a runaway colony. Or grow your own crop; this stalwart perennial will reward you with liquorice-like treats (in the form of seeds and pollen) year after year.

MAKE THE CONNECTION

Common hazel
Corylus avellana
The nuts of all hazels are edible, including the filbert (*Corylus maxima*) and the American hazel (*C. americana*). Eat them while green, before the squirrels do, or dry them to eat later.

White walnut
Juglans cinerea
Native to America, feral across Europe, the 'English walnut' can be picked 'wet' in summer or later in autumn, when the green husks split to reveal ripe brown nut cases.

Acorn
Quercus
Follow in the footsteps of early humans and Native American peoples and brew up a cup of roasted acorn. Be sure to leave some acorns behind for the animals, though.

Sweet chestnut
Castanea sativa
A high-carbohydrate food source for thousands of years; remove spiny cases and then roast the delicious inner nuts over an open fire or in the oven (see page 217).

Pine nut
Pinus
Around 20 pine species produce large enough seeds to eat. Harvesting them from the cones and shell is time consuming, but just think of all that home-made pine nut pesto (see pages 190 and 206).

Fennel seeds
Foeniculum vulgare
If you're lucky enough to find a rogue fennel colony and the anise-like seeds are fully ripe, get harvesting. Freeze your loot in a jar for a week to kill off any bugs. Find a recipe idea on page 221.

Hickory nuts
Carya
The pignut (*Carya glabra*), shagbark (*C. ovata*), shellbark (*C. laciniosa*) and pecan (*C. illinoiensis*) hickories all produce soft, sweet, edible nuts. Crack when dry.

Linden fruits
Tillia
After the sweet-tasting flowers (great as a relaxing tea) – come linden fruits (often called 'seeds'). Grind blossoms and immature fruits together with grapeseed oil to make a tasty chocolate substitute.

FORAGING FOR FOOD

' *The bitter-sweet of a white-oak acorn which you nibble in a bleak November walk over the tawny earth is more to me than a slice of imported pineapple.* '

Henry David Thoreau, American naturalist and philosopher (1817–1862), in *Wild Fruits*, finally published in 2000

Top Linden (*Tilia*) flowers make a calming, stress-busting tea or leave to ripen to harvest small nutlets, which contain one or two tiny, chocolate-tasting fruits when young. **Bottom** Sweet chestnuts (from *Casteanea sativa*) and acorns (from *Quercus* species) have both been cooked and eaten since ancient times. The chestnut's spiky protective husk and acorn's cupule (cup-shaped base) are also great for naturecrafting with kids (see page 179).

GROWING & GATHERING 165

Althaea rosea Cav.

Edible flowers

Even if you come home empty-handed, looking for edible flowers is a great way to up the ante on your plant identification skills. There are so many similar-looking flower species out there, it's super important that you really, truly know how to sort the edibles from the potentially fatals, and the can-picks from the protected or endangered.

On that note, the best way to tap into nature's edible flower store is to go slow. Foraging experts far and wide advise that you first familiarise yourself with the weeds, herbs, bushes and trees in your neighbourhood and try to understand plants as part of a larger ecosystem by studying them as communities *in situ*. After that, get to know one plant at a time – possibly one plant per week or even one per month. That way you have the time to be sure that it's the right plant, do a tolerance test for any adverse reactions and learn about the best way to eat your bounty.

Most flowers have their most intense oil concentration and flavour when harvested after flower buds appear, but before they open. It's also advisable to harvest early in the morning, after the dew dries but before the heat of the day. For home-grown edible flower patches, check that pre-bought plants have not been treated with chemicals.

MAKE THE CONNECTION

Marshmallow
Althaea officinalis
Native to Europe, Western Asia and North Africa, the root was originally used to make marshmallow candies. For easy sweet pickings, use the flowers in salads.

Elderflower
Sambucus nigra
Scour the hedgerows and pollution-free roadsides for midsummer elderflowers. The fragrant white blossoms make a beautiful pressé or novel fritters.

Wild garlic
Allium ursinum
Ramsons, bear leek, buckrams – there are many names for this bulbous plant. There are also many ways to eat them, including pickling the stalked buds, pounding the bulbs or leaves into wild pesto (see pages 190 and 206) or making kimchi (see page 193).

Hollyhock
Alcea rosea
The 'English country garden' relative of the mallow and hibiscus produces flowers in a range of vibrant colours. Use fresh or crystallised as a showy garnish.

English daisy
Bellis perennis
You can harvest these innocent beauties from a clean patch of lawn or meadow. Pickled daisy buds work well as caper substitutes, while flowers can prettify salads. Test first as some people react.

Sweet violet
Viola odorata
Look for violets in early spring in shady woodland areas. The deep purple, vitamin-C-rich flowers impart a distinct sweet flavour to syrups, jellies and desserts.

Hibiscus
Hibiscus
If you live somewhere warm enough for hibiscus to thrive, harvest some petals and steep them in hot water to make a tangy, colourful drink. Garnish with full blooms.

Prickly pear
Opuntia ficus-indica
This spiny desert-dweller is often shunned as food. That's a shame, as the despiked pads, super-sweet 'tuna' fruits and bright, crunchy flowers make delicious snacks.

Opposite A chromolithograph of *Alcea rosea* (common hollyhock) from volume 1 of Franz Eugen Köhler's *Medizinal-Pflanzen* or *Medicinal Plants* (1887) illustrates deep maroon, almost black petals – quite the garnish.

Stalks, leaves and roots

Eating your greens is more than possible in the wild, although many of nature's edible leafy offerings are an acquired taste after the perfectly-honed, cultivated varieties that we're used to. It's worth tempering those tastebuds, though, especially when you see just how many forageable plants there are – many of them right on your doorstep or masquerading as garden weeds.

In the temperate climates of Europe or North America, such edible weeds potentially include nettle, dandelion and the less well-known Alexanders, cleavers and great plantain. Edible leaves, stalks or roots, however, can be found growing worldwide and in a variety of biomes, with many wild foods offering a fascinating glimpse into the origins of many of today's cultivated foods – yam tubers, beets, lotus roots or spinach leaves, for example – and what our hunter-gatherer and early agrarian ancestors may have foraged or later cultivated to survive.

If you're tempted to gather any of these plant parts do note that many countries, including Britain, have legally-binding countryside codes that prohibit digging up any kind of root unless you have the landowner's permission. It's also dangerous to eat any wild food you can't absolutely identify. For peace of mind, consult an expert first, and always forage with care.

MAKE THE CONNECTION

Dandelion
Taraxacum officinale
Not-so-weedy dandelions provide leaves for salads, soups or stir-fries, flowers raw or fried and roots as a root veg, or dried and roasted as a coffee substitute.

Sugar maple
Acer saccharum
Native to large swathes of North America, the sugar maple has long been pillaged for its sugary sap. Tap trees in spring and then boil the watery sap to make rich syrups, delicious on pancakes or porridge.

Marsh samphire
Salicornia europaea
Also known as beach asparagus, crow's foot greens, sea beans and pousse-pierre, cook up and strip off the salty edible flesh of marsh samphire for a real taste of the sea. It's also great with fish.

Wild burdock
Arctium
Native to the Old World, but widely introduced worldwide, burdock root is cooked and eaten in parts of Asia. Spring leaf stalks and leaves can also be eaten raw.

Stinging nettle
Urtica dioica
Side-step the sting by blanching leaves in water and tap into the spinach-meets-cucumber taste, rich in vitamins A and C. Great in soups, in wild pesto (see pages 190 and 206) or as tea.

Chickweed
Stellaria media
Native to Europe but naturalised in parts of North America, this stubborn weed is also an edible spring green with a similar taste to an iceberg lettuce.

Alexanders
Smyrnium olusatrum
Known to Theophrastus and Pliny the Elder, Alexanders has been replaced in many dishes by similar-tasting celery. Hunt for it in coastal areas of its native Europe or its adopted Britain.

Sea beet
Beta vulgaris subsp. *maritima*
Native to coastal Europe, northern Africa and southern Asia, the wild ancestor of beetroot, sugar beet and chard has a lovely spinachy saltiness.

Opposite *Taraxacum officinale* (dandelion) from volume 1 of Franz Eugen Köhler's *Medizinal-Pflanzen* or *Medicinal Plants* (1887) illustrates the spectrum of its edible parts including leaves, stems, flowers and root.

FORAGING FOR PROVISIONS

Our hunter-gatherer ancestors would not just have been scouring the wild for food. They would also have been on the look out for medicinal herbs, materials to make shelters and later, as life became more civilised, provisions that might further enhance their lives such as cut flowers for the home. Thankfully, some of the traditional wisdom required to identify, gather and make use of such provisions has not only survived, it continues to inspire.

Medicinal herbs

From Ayurveda to the monasteries of the Middle Ages and to hedgerow herbalism, plants as medicines provided historic civilisations and cultures with the means with which to survive, endure, strengthen, grow and potentially modernise. Enquiries into the properties of plants for medicinal purposes also revealed their non-medicinal properties, such as culinary ingredients and natural dyes.

Today, herbalism continues to be practised in many parts of the world, by some indigenous peoples, but also as an increasingly accepted form of holistic healing in places where Western medicine usually prevails. In commercial form at least, preparations containing herbs such as echinacea, evening primrose and milk thistle are commonplace.

The number of herbalists trained to create bespoke preparations from freshly sourced plants and flowers also appears to be rising with demand, as Chapter 5: 'Botanical Remedies' and the Further Reading section (see page 392) reveal. In this case spot-on plant identification and handling is key. Administering the wrong plant or part could have serious consequences, as could treating someone who should avoid certain herbs or ingredients (children, pregnant women, or those with serious medical conditions for example). If you do wish to try out a remedy for therapeutic reasons or for an ailment, always consult an expert first.

MAKE THE CONNECTION

Pot marigold
Calendula officinalis
Found growing wild in southern Europe, the bright orange flowers of calendula have anti-inflammatory properties that are well known to help soothe skin.

Lady's mantle
Alchemilla vulgaris
Native to Europe and beyond, the dew-collecting leaves of lady's mantle combined with the tiny chartreuse flowers make a soothing 'monthly' tea that reduces menstrual bleeding.

Evening primrose
Oenothera biennis
The seeds of this sun-loving, fatty-acid-rich plant were sent home by early English settlers of America as early as 1614. Collect petals and make your own essential oil.

Feverfew
Tanacetum parthenium
Native to Eurasia but now found worldwide, feverfew tea is said to help relieve or prevent migraines. Test it before you drink it all, though, as fresh leaves can be bitter.

Juniper
Juniperus communis
Juniper is famed as the primary flavour of gin (see page 246), although only a few species yield edible berries so use a guide when out foraging.

Yarrow
Achillea millefolium
The genus *Achillea* nods to Achilles, who reportedly used yarrow to treat battle wounds. Herbalists still use it today to treat numerous, hopefully less-bloody, ailments.

Ribwort plantain
Plantago lanceolata
A common weed of cultivated land, a tea from plantain leaves is said to be a highly effective cough medicine, while the brown flowers taste like mushrooms.

Goosegrass
Galium aparine
Kids love the sticky stems of goosegrass, but the bruised plant can also help relieve nettle stings. Or try weaving the stems together to make a rough sieve or basket.

Opposite Women gather medicinal plants in a 15th-century copy of a manuscript by the Andalusian Arab physician Albucasis (936–1013) entitled *Observations on the nature of different alimentary and hygenic products*.

Cut flowers

Cut flowers is a tricky subject on the foraging front. Although picking armfuls of wildflowers feels like a romantic pastime – certainly one that was undertaken with enviable whimsy in the past – encouraging the masses to participate in such an activity could seriously undermine the ecology of some environments.

Flowers, after all, produce seeds, from which the next generation of plants will grow. Many of these seeds are encased in fruits, which various wildlife depend on for their survival. Plus, removing flowers in their prime also takes away valuable nectar for pollinators, along with the pollen needed for such plants to reproduce.

A good proportion of tempting, bouquet-worthy flowers also grow in protected wildflower areas, are covered by countryside codes, or are on the endangered list. Make sure to do your research so you know what is permissible in your area.

Even so, it's worth listing some of the gorgeous wildflowers that you could pick if you were allowed to. It may be that a friend doesn't mind you foraging for them on their land, or you could always grow your own seasonal, wildflower garden. That way you would have gorgeous cut flowers or foliage available all year round and could avoid buying expensive, mass-produced or imported flowers at the same time. Plus you would be planting for pollinators (see page 157) and biodiversity (see page 158), adding value to the habitat of your local birds and bees.

MAKE THE CONNECTION

Cow parsley
Anthriscus sylvestris
Very common in northern Europe, and invasive in North America, a foraged bunch of cow parsley adds a touch of wild to bouquets without usually causing ecological harm.

Honesty
Lunaria annua
Native to the Balkans and Southwest Asia, and naturalised through the temperate world, look for pretty pink flowers in the spring and silvery moon-like seedheads in the summer.

Corn poppy
Papaver rhoeas
The bright red petals of this iconic symbol of remembrance drop off easily in flower arrangements, so it's worth waiting a little and including dried seedheads instead.

Cornflower
Centaurea cyanus
Widely naturalised outside its native European range, the cornflower delivers a lovely pop of bright blue to bouquets, works well as a culinary garnish, and looks pretty in the garden too.

Ox-eye daisy
Leucanthemum vulgare
Migration laws means nothing to this widespread (often invasive) flower – a mainstay of the wildflower meadow and an ancient symbol of patience. A must for a homegrown wild patch.

Daffodil
Narcissus
Spring is definitely here when the daffodils rise again, trumpets blaring in brilliant shades of sunshine yellow. Be careful not to 'steal' these from private or protected land.

Purple coneflowers
Echinacea
Native to eastern and central North America, and the epitome of 'prairie planting' style, the rayed petals and chunky cones of echinacea make great cut flowers. Seedheads can also add interest.

Holly and ivy
Ilex/Hedera
Holly and ivy hold true as enduring symbols of winter. Collect shiny evergreen leaves and berries to create lovely foundation wreaths and garlands.

'*I paint flowers so they will not die.*'

Frida Kahlo, Mexican artist (1907–1954)

Top Mexican artist Frida Kahlo (see page 348), as this 1950s portrait shows, is renowned for wearing flowers in her hair, often gathered from her garden to complement the traditional Mexican clothing she also favoured. **Bottom** A vase of wildflowers can be a wonderful memento of a walk in nature – just make sure you have permission to pick them first, and steer clear of endangered species. For daily reward, create a cutting garden of the wild blooms you love the most.

FORAGING FOR NATURECRAFT

Before the Industrial Revolution, 'craft' meant a skilled, handmade trade that often involved working with naturally-sourced materials. The end result of such activities was either a useful or beautiful object – or, as Arts and Crafts Movement proponent William Morris philosophised, ideally both. More recently, there has been a noticeable revival of plant-based artisan pursuits, from working with natural dyes to herbaria-style pressed botanicals. Foraging for materials to craft with is a great way to get inspired, connect with nature and can also keep costs down.

Pressed botanicals

A recent trend for all things botanical has led to a renewed interest in the art of pressing flowers. But rather than the stylised, rather fusty Victorianesque arrangements produced in flower pressing's last heyday in the 1970s and 1980s, more recent trends are for naturalised mounts that reference botanical illustrations or herbaria, or super simple designs that let flowers, leaves, seedheads and plant silhouettes do the talking – see Chapter 6: Inspired by Nature for examples, including naturalistic artworks produced for my own pressed botanical brand, The Herbarium Project (see page 386), which led to the making of this book.

If you're attempting more earnest forays into pressed botanicals – beyond placing flowers on blotting paper between the pages of books, which is a lovely way to begin – invest in or make your own large-format press in order to arrange larger specimens in the naturalistic style. Other simple tips for successful pressing include: collecting on dry days (avoid dewy mornings); inspecting specimens first for insects or damage; pressing for three weeks to ensure samples are fully dry; and committing to a bit of trial and error – some specimens do press better than others or need specific handling. It's also fun to grow your own plants or flowers, or persuade others to let you pick on their land in exchange for a bespoke commission. Remember, though, to always make sure you don't pick anything that is protected.

MAKE THE CONNECTION

Lady fern
Athyrium filix-femina
Native throughout most of the Northern Hemisphere, lady fern leaves press flat easily for simple, graphic mounts on card or under glass.

Wild carrot
Daucus carota
Pressed individually, the (also edible) flowers of the tightly packed umbel make beautiful inserts for resin jewellery, especially against a dark, velvety background.

Cosmos
Cosmos bipinnatus
Native to scrub and meadowland in Mexico and much of North and South America, pink or candy-striped cosmos petals keep their colour well against their dark feathery leaves.

California poppy
Eschscholzia californica
Native to the United States and Mexico but found in wildflower meadows elsewhere, this poppy's long seedheads contrast beautifully with their papery orangey-yellow petals (see opposite).

Ginkgo
Ginkgo biloba
This ancient plant is native to China but is now grown worldwide. The unique fan-like leaves work really well pressed and encased in papier-mâché lanterns.

Crocosmia
Crocosmia x *crocosmiiflora*
The warm-hued orange crocosmia is too pretty to be called invasive, as are its other names: montbretia, falling stars and coppertips. Another great colour keeper when pressed.

Columbine
Aquilegia
Around 60–70 species of this dove-shaped beauty reside in the meadows, woods and mountains of the Northern Hemisphere. Pressing reveals lovely detail, plus there are lots of colours to play with.

Japanese maple
Acer palmatum
All acers have beautiful leaves, but those of the starry Japanese maple are particularly fine, especially in autumn. Mount leaves in a graphic pattern for added drama.

Opposite *California Poppy* (2013) by Sonya Patel Ellis (see page 386) from the exhibition *Hortus Uptonensis* (2015) at West Ham Park, London uses the art of pressed botanicals to look closer at the minutiae and ephemerality of nature.

Natural dyes

The majority of natural dyes are made from plant sources such as roots, berries, bark, leaves, flowers and wood, with natural dyeing a commonplace practice until the 1850s, when William Henry Perkin kickstarted a synthetic dye revolution with the chance invention of purple-hued mauveine – Perkin was actually trying to synthesise the anti-malarial drug quinine.

In the face of mass production, artisans such as William Morris (see page 369) and other members of the Arts and Crafts Movement tried to keep the use of natural dyes alive, preferring the muted colours and subtle variability of the shades they produced. Although it wasn't enough to hold back the tide of industrialisation, such efforts did thankfully help preserve a record of natural dyeing ingredients and techniques. Natural dyeing traditions have also been upheld by various indigenous peoples around the globe.

Such resources are now invaluable to those seeking to recapture the old ways, colours and shades of plant dyes. Foraging for dye stuffs and making your own dye can be great fun, too, from colouring eggs with red onion skins (earthy shades), red cabbage (purple-blue) or turmeric (yellow) to replicating ancient colours with madder (red), weld (yellow-green), woad (blue) and indigo (indigo-blue). Keep a colour chart so that you can more easily replicate favourite shades.

MAKE THE CONNECTION

Golden tickseed/calliopsis
Coreopsis tinctoria
Common to Canada, northeast Mexico and the United States, and naturalised in China, the yellow-red blooms yield a yellow, orange or brown dye.

Dahlia
Dahlia
Native to Mexico, all dahlia (except the white ones) can produce dye. Use an alum mordant (dye fixative) for warm yellow and orange, or an iron mordant for green.

Sunflower
Helianthus annuus
Sunflowers were first domesticated in the Americas, with Native American peoples traditionally producing a purple-black dye from the seeds and a yellow one from the flowers.

True indigo
Indigofera tinctoria
Naturalised in tropical and temperate Asia, as well as in parts of Africa, the famous deep blue dye is obtained through the processing of the plant's leaves. The colour is even lovelier faded.

Weld
Reseda luteola
Native to Eurasia and one of the earliest known dyes, weld delivers all shades of yellow. Harvest the yellow-green flowers and leaves before the fruit sets.

Common madder
Rubia tinctorum
Madder root has been used since ancient times as a vegetable red dye, an alternative to cochineal, which is made from insects feeding on red prickly pear (*Opuntia*) fruit.

Woad
Isatis tinctoria
This now-noxious weed dates back to ancient times, when it was cultivated throughout Europe to produce an indigo-blue dye and a tattoo ink.

Lady's bedstraw
Galium verum
Lady's bedstraw roots were traditionally used to make orangey-red dyes – for Scottish tartans, for example – especially where its fellow red-dye-making madder would not grow.

Left True indigo (*Indigofera tinctoria*) as illustrated in G.T. Wilhelm's *Encyclopedia of Natural History* (1817) was an original source of indigo dye. **Top** A 19-century illustration of weedy Dyer's woad (*Isatis tinctoria*), one of the three staples of the European dyeing industry along with weld and madder. **Bottom** North American native Golden tickseed or calliopsis (*Coreopsis tinctoria*) was traditionally used to ward off lightning as well as being exploited for its potential yellow or red dye.

Top Pussy willows, referring to the super-soft furry catkins of several *Salix* (willow or sallow) species, appear long before the leaves and are a wonderfully sensory sign of spring. **Right** Pretty posters of botanicals such as this illustration of false fruits (see page 102) are a great way to draw children into the wonders of the Plant Kingdom. **Bottom** Woody cones from pine (*Pinus*) conifers inspire numerous crafting projects for kids and adults alike, including these lovely painted 'cone-flowers'.

Naturecrafting with kids

Naturecrafting is a lovely way to introduce children to the wonders of the Plant Kingdom and rediscover seasonal childhood joys for yourself.

Spring, for example, is a great time to sow new seeds, study first shoots and leaves, collect 'confetti' from blossom trees or enjoy the excitement of bulb plants emerging from their winter hibernation.

Summer brings a riot of flowering plants each with its own special name and fascinating backstory – it's never too early or too late to sow the seed for plant identification. Late summer also brings the harvest of various edible crops, which can be grown at home or harvested from a pick-your-own farm. Plus this is the season to try flower pressing (see page 175) or making cyanotypes (see page 80).

Autumn holds the potential for drifts of sunset-coloured leaves (from deciduous trees), interesting seeds and nuts and cones (from coniferous trees), with which to create collages, sculptures, characters or games. In winter, or in colder climes, you could tempt kids into the fresh air by organising a bracing foraging expedition in search of evergreen foliage, berries or hardy flowers to weave into festive wreaths or garlands. The only rule – apart from teaching kids about the usual countryside codes of course – is to have lots of fun.

MAKE THE CONNECTION

Horse chestnut
Aesculus hippocastanum
In Britain, shiny brown horse chestnuts are threaded onto pieces of string ready for battle in a game of 'conkers'. Threaded together in larger groups, they also make excellent 'caterpillars'.

Snapdragons
Antirrhinum
Make a snapdragon puppet show by animating the dragon's mouth between finger and thumb. In autumn the blooms turn into rattles of skull-like seedheads.

Willow
Salix
Fuzzy 'pussy willows' provide a sensory springtime treat, while the bendy branches of white willow (*Salix alba*) are ideal for weaving and sculpture.

Common reed
Phragmites australis
Country rambles can turn up all sorts of creative treasures. If you're in or near a wetlands, look for naturally hollow reeds, which can then be used as primitive ink pens.

Sycamore
Acer pseudoplatanus
Paint the wings of nature's helicopters in different colours, then throw them into the air and see which takes the longest to come down. Hours of free fun.

Dock leaf
Rumex obtusifolius
Eager young naturecrafters may well fall victim to a clump of stinging nettles, so show them how to find a dock leaf (hopefully nearby) with which to naturally soothe the pain.

Pine cone
Pinus
Collect fallen pine cones in winter and hang them on ribbons for pretty seasonal decorations, make animal sculpture, or watch as they open and close depending on the weather.

Clover
Trifolium repens
Organise a many-leaved clover hunt – the Guinness World Record set in 2009 was 56 leaflets – or make beautiful indentations in air-drying clay using the flowers, stalks and leaves.

LIVING WITH NATURE

What is nature? Is it, as is most often the Western view, those parts of the physical world that are not human or manmade: the plants, animals and landscapes? Or do you believe, as many indigenous peoples of the world do, that humans are very much a part of nature, intrinsically connected to the land and its spirits? Either way, the more we live with nature – urge ourselves to get outdoors and spend time with plants, flowers and the landscape – the greater our chances of connecting with and ultimately respecting it. Nature is, after all, our home.

Growing home together

'Nature is not somewhere to visit, it is home,' wrote the American poet and conservationist Gary Snyder in a collection of nine captivating essays entitled *The Practice of the Wild* (1990). The quote is often shared minus a reference to Snyder's insightful investigation into the human relationship with nature, which is a shame. Weaving together aspects of mythology, history, anthropology, etymology and spirituality, plus Snyder's own personal experiences and interpretations of 'wilderness' and 'wildness', this is environmentalism at its most inspiring – a lecture-free, beautifully lyrical exploration of what it means to live in harmony with nature. It is also a vital reminder of our collective responsibility to safeguard our 'home'.

The Practice of the Wild came 35 years after the famous Six Gallery reading of 1955, at which fellow poet Allen Ginsberg read *Howl*, and Snyder *A Berry Feast*, a poem he had written while staying on the Warm Springs Reservation in the summer of 1950. Native American culture shapes much of Snyder's work; he grew up as neighbour to the Salish peoples of Puget Sound, and witnessed their close relationship to the land. In 1969, having spent more time absorbed in the Native American culture and that of Zen Buddhism – especially the principle of *Ahimsa*, or non-harming – he published *Four Changes*, a widely-circulated environmental treatise that warned of overpopulation, pollution and consumption, particularly of fossil fuels.

Some of Snyder's green-living solutions were a little far out for those not involved with the counterculture movement, but others laid the foundations for the kind of 'environmentally-friendly' advice that is finally infiltrating the collective psyche: carpooling, using your own shopping bags, opting for natural fertilisers and using less water among them. While his 'total transformation' of 'the five-millennia-long urbanizing civilization tradition into a new ecologically-sensitive, harmony-oriented, wild-minded scientific/spiritual culture' still has a way to go, there is at least a new global awareness of what needs to be done, including how to conserve 'wilderness'.

A Western concept largely denoting a place of pristine natural beauty untouched by Man, ideas about wilderness were particularly prevalent in the writings of the nineteenth-century American author Henry David Thoreau, the American poet and essayist Ralph Waldo Emerson and Scottish-American naturalist and 'Father of the National Parks' John Muir.

Muir was particularly vocal about the preservation of the wilderness, and his observations are wonderfully poetic, describing not just the spiritual beauty of the wild landscape, but also the devastating effects of deforestation caused by the over-logging and the ravage of meadow ecosystems by 'hoofed locusts' (grazing sheep). His adventurous, literary activism certainly struck a chord with the masses, and his legacy includes the creation and extension of Yosemite National Park, the co-founding of the Sierra Club (inspiration to visionary landscape photographer Ansel Adams, see page 339) and, not least, the cornerstone of the modern-day conservation movement.

Today, the International Union for Conservation of Nature (IUCN; www.iucn.org) and its associated database Protected Planet (www.protectedplanet.net) count more than 160,000 protected areas dedicated to the long-term conservation of nature. In total these areas cover over 12 per cent of the world's surface and include national parks, nature reserves, terrestrial and marine zones, and sustainably managed public and private landscapes. It's a huge and vital achievement, yet not without debate: the first national parks sought to protect 'the wilderness' but in doing so displaced huge numbers of indigenous communities who had lived in and shaped such ecosystems for millennia – peoples who called the wilderness home; while the idea of humans *protecting* the wild somehow places humans outside of that very wildness. Can we really conserve nature if we retain a stance of being separate from it?

This is a fine line that needs treading carefully but surely one we're collectively capable of. Hopefully, as Snyder puts it in his poem *For the Children*, we can 'Stay together learn flowers go light' – a motto for growing and gathering if ever there was one.

Opposite The Ansel Adams Wilderness in the Sierra Nevada, California, near Yosemite National Park and Mammoth Lakes has been inhabited by Miwok, Monache, Mono, Washo and Shoshone peoples for thousands of years.

'Live in each season as it passes; breathe the air, drink the drink, taste the fruit, and resign yourself to the influence of the earth.'

Henry David Thoreau,
Walden: Or, Life in the Woods (1854)

CHAPTER FOUR

BOTANICAL RECIPES

NATURE'S LARDER

Connect with nature as you cook – by growing, gathering or observing the provenance or season of ingredients – and let the Plant Kingdom reward you, not just with a wealth of delicious ingredients but also the inspiration with which to prepare an array of plant-forward feasts.

Looking for an easy or instant route into the botanical world? Look no further than nature's larder, or indeed the plant-based food or drinks that you consume most regularly: fruit, vegetables, coffee, tea, rice, bread, wine or juice perhaps. Now think about how much you really know about where these foods or ingredients come from, their provenance – be it plant or region of origin – or seasonality.

Take courgettes, for example. Shop-bought, they would most commonly be described as 'long, dark green vegetables with firm pale flesh'. Grow your own, and you'll soon be aware that the fruit of the courgette plant (*Cucurbita pepo*) ripen around midsummer from bright orange flowers and can be harvested as baby fruit, left to grow to 'standard size' or until they are mature enough to be called a marrow. You can also eat the flowers, traditionally found in Italian marketplaces, ready to be stuffed, lightly battered and fried to a crisp (see page 204).

If you really get into growing your own courgettes you may end up planting yellow or stripy varieties that seem to taste infinitely better than the ubiquitous green ones. You might also find such choices in local farmers' markets or speciality shops and end up devoting disproportionate amounts of time to courgette-based conversations or debates. Should you fry them, sauté them, add them to cakes or bakes, eat raw in a salad or stewed down in a soup? Do they taste better thinly sliced on a pizza, cut lengthways in a stir-fry or stuffed whole? What's the best courgette-flower stuffing – ricotta and basil, or do they taste better with feta and mint?

The humble courgette serves as an inspiring example of just how versatile nature's larder, and thus a plant-forward diet can be – a concept that an increasing number of people are latching onto. Indeed, plant-based foods are no longer relegated to the side dish or vegetarian or vegan option. Today, they are just as likely to be the gourmet stars of the show.

This botanical shift is in part due to the ingredient-elevating work of plant-forward chefs such as Yotam Ottolenghi, René Redzepi, Jeremy Fox, Amanda Cohen, Alice Waters and Jeong Kwan (see www.plantforward50.com). But credit must also be given to those less lofty legions of plant-loving pioneers who tirelessly inspire others to grow their own produce, forage for wild food, eat sustainably, get to know the botanical world more intimately or, as author and activist Michael Pollan so eloquently put it in his *In Defence of Food* (2008), simply 'Eat Food. Not too Much. Mostly Plants.'

It's a sage mantra to bear in mind as you hopefully try some of the delicious recipes to follow. Plus eating 'mostly plants' is really not so hard considering the huge generosity of the Plant Kingdom itself. If Claire Ptak's Fig Leaf Ice Cream (see page 201) doesn't get you, then surely the earthy pull of Tom Harris's Orange, Fennel Seed and Almond Cake (see page 221) or the intrigue of John Rensten's Magnolia Flowers Lightly Pickled in Elderflower Vinegar (see page 197) will.

Opposite Garden nasturtium (*Tropaeolum majus*), as depicted by James Sowerby for *Medical Botany* (1832), are entirely edible, including the flowers and seedpods (see page 191).

Tropaeolum majus

Published by W. Phillips, August 1.st 1809.

EAT THE SEASONS

There's nothing new about eating the seasons – before the advent of refrigerators, preservatives and reliable long-distance transport, it was simply the intuitive thing to do. Today it's an active choice that can bring flavoursome rewards.

Once upon a time, seasonal eating was the only option. That which could be eaten fresh would be picked straight from the tree, plucked from a shrub, gathered from the ground or a hedgerow, or cultivated to be harvested with relative immediacy.

Over time, humans also developed ways to store or naturally preserve edible produce to avoid wasting seasonal gluts and to ensure that food was available during less-bountiful periods. Jams, jellies, marmalades and conserves all owe a debt to the art of survival, preceded by smoking, dehydrating, fermenting, salting and pickling when sugar was still a luxury commodity. In fact, sugar was first introduced to Europe as a medicine rather than a food, 'a kind of honey found in cane, white as gum ... used only for medicinal purposes', as described by Pliny the Elder in his *Naturalis Historia* (77–79 CE).

Today, there's a new importance attached to eating seasonally, imbued with factors such as the desire to protect the environment, deal with climate change, support local economies, go organic, and conserve ancient plant species. There's also a case to be made for the superior flavour, texture and seasonally beneficial nutrients of freshly picked, peak-harvest fruit, vegetables or legumes. This, coupled with the idea that eating seasonally, and thus exploring and exploiting the natural qualities of plants, has the potential to bring you closer to the botanical world, the cycle of life and, indeed, your own biorhythms.

Eating the seasons is also one of the most satisfying and sociable ways to explore the botanical world, as illustrated by some of the delicious recipes within this chapter and the plant-loving gourmets who have so generously provided them.

Interested in growing your own seasonal produce? Turn to recipes by Mark Diacono (see page 198) or Matt and Lentil Purbrick (see page 237). Intrigued by the idea of foraging for seasonal produce? Find recipes, tips and advice from a host of wild food educators including Robin Harford (see page 193), John Rensten (see page 197), Sarah Witt (see page 210), Pascal Baudar (see page 238), Dina Falconi (see page 206) and Adele Nozedar (see page 226).

Recipes have also been kindly donated by a seasonally-inspired selection of chefs, cooks and mixologists, among them Tom Harris (see pages 221 and 233), Bernadett Vanek (see page 194), Matt Hoyle (see page 212), Claire Ptak (see page 201), Nina Olsson (see page 209), Kitty Travers (see page 214), Skye Gyngell (see page 222), Gill Meller (see page 225), Paola Gavin (see page 234), Alan McQuillan (see page 244), Amy Zavatto (see page 246), Charisse Baker (see page 253) and Rawia Bajtalmal Edwards (see page 254).

As chef Tom Harris attests, 'Seasonality, locality and historical connection can be harnessed and conveyed in so many interesting and fun ways, from adapting traditional recipes that once absolutely relied on seasonal or local produce, to designing dishes that use less-well known edibles, such as the tops of turnips or radishes as well as the roots.'

Eating the seasons is in our nature – roots, shoots, fruit, flowers, leaves and all.

Opposite Temperate climate-dwellers can use the wheel to find fruits, vegetables or legumes through the seasons. Or make your own seasonal calendar to match your region.

SPRING

Spring is a time of renewal and rejuvenation, when many plants begin to grow and flower or are sown for seasons to come. The transformation from bare branch or soil to a crown or carpet of green can be remarkably swift once the weather allows. Suddenly, culinary delights are everywhere, from fields full of pea shoots to parks brimming with cherry or magnolia blossom, to forest floors peppered with wild garlic, nettle or wood sorrel – a time for eating tender shoots and leaves, baby roots and greens and early edible flowers such as violets.

What to eat now
The spring edit

Globe artichoke
(*Cynara cardunculus* var. *scolymus*)
Edible in bud form, the artichoke's tastiest morsels include the fleshy leaf bases and the inner heart.
Goes well with aioli, lemon, harissa

Pea (*Pisum sativum*)
The young, sweet peas of spring sum up the freshness of the season, as does a garnish of pea shoots.
Goes well with butter, mint, potato

Rhubarb (*Rheum rhabarbarum*)
Hot on the heels of the first red-green rhubarb stalks of spring come a host of delicious desserts.
Goes well with orange, ginger, strawberry

Spring greens (cultivar of *Brassica oleracea*)
What the first cabbages of the year lack in heart they more than make up for in vitamin-rich dark leaves.
Goes well with cannellini beans, hazelnut, soy

New potatoes (*Solanum tuberosum*)
No need to peel these thin-skinned, sweet-tasting baby potatoes. Eat whole, cooked or in salads.
Goes well with chives, dill, soured cream

Asparagus (*Asparagus officinalis*)
Asparagus shoots should be harvested young before they turn woody, when they can be eaten raw or cooked.
Goes well with Hollandaise sauce, Parmesan, tarragon

Nettle (*Urtica dioica*)
Try tender leaves of nettle as a spinach-meets-cucumber leaf vegetable, or as a nutritious soup.
Goes well with eggs, cheese, garlic

Radish (*Raphanus raphanistrum* subsp. *sativus*)
For a crisp-meets-mellow taste of spring, serve raw roots with salted butter along with warm wilted tops.
Goes well with balsamic vinegar, risotto, anchovy

Spring onions (*Allium cepa* var. *cepa*)
Also known as scallions or salad onions, use raw in salads, wilted in stir-fries or as a base flavouring.
Goes well with lime, romesco sauce, chilli

Wild garlic (*Allium ursinum*)
The soft, pointed, aromatic leaves are at their best in spring, as are the pungent edible flower buds.
Goes well with pine nuts, asparagus, mozzarella

> **Sensory notes of spring**
> Fresh • Green • Sweet • Delicate

Opposite Spring onions (*Allium*), chives (*Allium schoenoprasum*), asparagus (*Asparagus officinalis*), peas (*Pisum sativum*) and the last Brussels sprouts (*Brassica oleracea*) are lovely candidates for a verdant spring feast.

Try four things
The spring edit

Bouquet garni
Makes 1

3 sprigs of parsley
2 sprigs of thyme
1 bay leaf
1 leek leaf (the outer layer of the leek removed in one piece)

The term *bouquet garni* first appeared in the classic *Le Cuisinier François* (1651), by François Pierre Sieur de La Varenne, a book that was to introduce a distinctly modern way of cooking. It also included this delicate flavouring for broth, stews, soups and sauces – previously, potent mixes of spices and herbs were usually used to mask unpleasant flavours.

1. The traditional way to make a 'garnished bouquet' (the rough translation of *bouquet garni*) is to envelop parsley, thyme and a bay leaf within a length of leek leaf. The leaf can then be pressed shut and tied with twine at both ends to secure the herbs inside.

2. Alternatively, tie the stems of the herbs together with string or enclose them in a piece of cheesecloth or muslin.

3. You can also experiment with bespoke additions to your *bouquet garni*, including orange peel, basil leaves, salad burnet, rosemary, peppercorns, savory, tarragon or any vegetable stalks.

Wild pesto
Makes 1 small pot

1 small bunch of wild garlic or nettle leaves, washed
50g (½ cup) grated Parmesan cheese
50g (½ cup) toasted pine nuts or walnuts
1–2 tbsp extra virgin olive oil
1 clove garlic
Dash of lemon juice
Salt and freshly ground black pepper

Pesto, or *pesto alla genovese*, as it's formally known, originates in Genoa, the capital of Liguria, in Italy. Pesto literally translates as 'to pound' – traditionally this was done with a pestle and mortar, to crush the key ingredients of garlic, European pine nuts, coarse salt, basil leaves, Parmigiano-Reggiano (Parmesan cheese), *pecorino sardo* (a hard cheese made with sheep's milk) and olive oil. This wild version is a delicious alternative that can be made with freshly foraged nettle or wild garlic leaves.

1. Simply blitz everything in a blender and serve with pasta. Any leftovers can be stored in an airtight jar topped with a layer of olive oil.

Poor man's capers
Makes 1 small pot

50g (½ cup) nasturtium seedpods
120ml (½ cup) water
2 tbsp salt
175ml (¾ cup) white wine vinegar
2 tsp sugar
2 bay leaves
2 small sprigs of thyme

Typically, capers are the pickled flower buds of the sun-loving caper plant (*Capparis spinosa*). A fun alternative is to pickle the small round seeds of the nasturtium (*Tropaeolum majus*).

1. Simply divide each seedpod into three seeds, wash and place in a jar. Bring the water and salt to the boil in a pan to make a brine; once boiling, turn off the heat. Pour the brine over seeds, cover and leave to soak for 3 days, then strain the brine into a jug using a fine sieve.

2. In a small pan, bring the vinegar, sugar, bay and thyme to the boil. Return the seeds to the jar, pour over the pickling liquid and let everything cool. Cover and refrigerate for 3 days before use. Your 'poor man's capers' should then keep for 6 months.

Rhubarb compote
Serves 3

700g rhubarb, trimmed and cut into chunks
75g (⅓ cup) soft brown sugar
Grated zest and juice of 1 orange
1 tsp freshly grated root ginger

Derived from the Latin *compositus*, meaning mixture, the word 'compote' historically referred to a medieval concoction of fruit in syrup that was believed to balance the effect of humidity on the body, and was often served towards the end of a meal. Today, it usually refers to whole pieces of fruit in sugar syrup, often seasoned with spices or aromatic additives such as vanilla, orange peel, cinnamon, cloves, ground almonds, grated coconut, candied fruit or raisins. Rhubarb compote was particularly popular between the two World Wars, its tart flavour beautifully complemented by orange juice and/or ginger.

1. To make it, preheat the oven to 180°C/350°F/gas mark 4. Arrange all the ingredients in a shallow ovenproof baking dish and cook for 30–40 minutes. Serve hot or cold with yogurt or a dessert.

Robin Harford, Eatweeds, Exeter

Wild garlic kimchi

Makes 1.5 litres (6½ cups)

Love kimchi? Try this exquisite and delicious wild garlic version. But be warned – it has a kick like a mule.

1kg (2¼ lb) wild garlic leaves and stems
75g (¾ cup) gochugaru (coarse Korean red pepper powder)
3 tbsp freshly grated root ginger
150g (1⅓ cups) grated daikon (mooli)
2 tbsp dried sea lettuce sprinkles (optional)
2 tbsp sea salt
1 tbsp ume shiso/ume plum seasoning (vegetarian options), or fish sauce

Note
If you're not familiar with kimchi, see page 230 for an explanation and a traditional recipe.

1. Wash the wild garlic leaves and stems, then shake dry and roughly chop.

2. In a large bowl, combine and thoroughly mix the gochugaru, grated ginger, grated daikon, sea lettuce (if using), sea salt and ume shiso or fish sauce.

3. Add this paste to the chopped wild garlic leaves and, using your hands, thoroughly squeeze and press the wild garlic until everything is covered with the paste and there is a decent amount of liquid coming from the mixture.

4. Place the kimchi into a sterilised clip-top 2-litre (64-oz) Kilner jar, press down the leaves and close the lid. Allow to sit for a week before taste-testing.

5. The kimchi will keep for at least 6 months without refrigeration, longer if placed in the fridge. If you live in a cold climate, you can also store outdoors, in the shade.

Opposite Prettily-flowered, broad-leaved wild garlic (*Allium ursinum*) is easily identified by its location – usually in damp, deciduous woodlands – and its distinctive garlicky smell. Find carpets of it from early spring to mid-summer.

MAKE THE CONNECTION
Find out more about Robin's botanical world at www.eatweeds.co.uk; @robinjharford

BOTANICAL RECIPES
The spring edit

MY BOTANICAL WORLD...
'*Foraging is a deep journey into your soul, it is not to be taken lightly, as the latest fad or craze. For once the plants have "got you" there is no turning back. As I journeyed deep into this plant world, the green wall slightly thinned, becoming ever so translucent, almost gossamer-like, until I perceived the hedgerow through a green veil. Then, one day, the Wild Redeemer blessed me, that veil parted, and I entered a wonderland of plants, imagination and mystery. Journey well.*'

Bernadett Vanek, Born Under the Sun, California

Pea and nettle gnudi with a braised fennel almond crema

Serves 1

Sweet, crunchy pea tendrils are insanely delicious and stunning in salads or as a garnish. The delicate flavour and texture of early pea shoots also combines beautifully with almond ricotta, and that other spring messenger, the stinging nettle, to create lovely bite-sized dumplings called gnudi. Stinging nettle has to be blanched to lose its bite, but after that the nutrient-rich leaves also work well in teas, soups, sauces and oils. Add a nutty almond cream and a touch of zingy lemon juice to the pea tendrils and blossoms and you have a dish that's fresh, vibrant and full of springy pea flavour.

Large bunch of stinging nettles (enough to make 55g/¼ cup of blanched nettles)
150g (1 cup) peas
135g (1 cup) whole almonds
Nutritional yeast, to taste
Zest and juice of 1 lemon
1 clove garlic (optional), smashed
1 large fennel bulb (or 2 small ones), sliced, reserving the fronds for garnish
1 tbsp olive oil, plus extra to drizzle
A little vegetable stock (optional)
240ml (1 cup) almond milk
Pinch of cayenne pepper and ground nutmeg (optional)
Pea shoots and blossoms, to garnish
Salt and freshly ground black pepper, to taste

1 For the almond ricotta, soak the almonds overnight in a bowl of cold water. The next day, drain and remove the skins, then blend in a food processor, adding water in small amounts until you have a ricotta-like texture. Add the nutritional yeast slowly to achieve the desired 'cheesy' flavour, plus salt and a few drops of the lemon juice to taste. Flavour with the smashed garlic, if desired.

2 To blanch the stinging nettles and peas, place each in small separate pans of boiling salty water for 1 minute. Scoop out with a slotted spoon and transfer immediately to bowls of ice-cold water to retain their green colour.

3 To make the gnudi, place the blanched peas (keeping a few back for a garnish) and nettles plus the almond ricotta in a food processor. Mix until soft but it holds its form. Shape into quenelles using two small teaspoons, or roll into small balls.

4 To make the crema, sauté the sliced fennel in the tablespoon of olive oil. Add a little water or stock (if using) until covered and cook on a very low heat until soft. Once cooked, place in a high-speed blender and slowly add the almond milk until you have created a creamy sauce. Season to taste with salt, black pepper, the lemon zest and most of the remaining juice. A touch of cayenne pepper and ground nutmeg also work well with this dish.

5 Prepare the garnish by drizzling the pea shoots, pea blossoms and reserved blanched peas with olive oil and the remaining lemon juice.

6 To plate, place a swirl of crema on a shallow plate, then top with the gnudi and the garnish. Keep the ingredients close to each other for a natural yet stylish effect.

MAKE THE CONNECTION
Find out more about Bernadett's botanical world at www.bornunderthesun.com; @born.under.the.sun

Above Use slightly sweet and crunchy edible pea (*Pisum sativum*) blossoms and delicate shoots and tendrils to garnish your dish for a spring-fresh taste and a wonderfully naturalistic, seasonal finish.

MY BOTANICAL WORLD . . .

'*I love collecting ingredients from the markets and farms around my home in sunny Pasadena, California, growing my own or foraging in the nearby countryside or on my travels. Fresh, local, seasonal, organic produce always tastes superior and inspires the most wonderful flavour combinations and natural plating arrangements.*'

MY BOTANICAL WORLD...

'*I study wild food, pick wild food, obsess about wild food and love sharing what I have learnt. I set up Forage London to give city dwellers like myself a chance to enjoy and discover some of the amazing wild foods that grow all around us.*'

John Rensten, Forage London

Magnolia petals lightly pickled in elderflower vinegar

Makes 1 jar

Spring in a jar. These petals remind me of ginger and also of celery with a bitter hint of chicory – a delicate and complicated flavour indeed. A wild salad has as many surprises as it has ingredients, so some thin slices of magnolia are a great addition. There are more than 200 varieties; mostly in the UK they are deciduous trees and a few evergreens, but I find the taste and texture very similar in all the species I have so far tried. I recently used dried and powdered magnolia petals as an excellent substitute spice when making my son some gingerbread biscuits. They were a terrific success, all consumed the same day.

Magnolia can also be used as the base of a floral wine or a fragrant sorbet, but I think there are many more, untapped possibilities for this wonderfully complex ingredient. I find it flowering abundantly in the city from as early as mid-February, so as well as including it in my wild chai recipe, I have it in mind to experiment with it as a seasoning for fish cakes, maybe even home-made burgers or sausages.

3 tbsp elderflower vinegar (or 2 tbsp white wine or cider vinegar)
2 tbsp sugar
Magnolia blossoms

1. For this super recipe I used last year's sweet-scented elderflower vinegar, made simply by leaving half a dozen elderflower sprays in vinegar for a week or so or until it takes up their wonderful fragrance. You could just use a white wine or cider vinegar, but it will need diluting quite a bit so as not to overpower the blossoms.

2. Play with the quantities, but in my version gently heat the elderflower vinegar or white wine vinegar and sugar together with about 6 tablespoons of water and then leave to cool before adding to roughly enough petals to fill a standard-sized glass jar (450g/16oz). They are ready to eat just 12 hours later and, surprisingly, these delicate flowers keep for ages – the colour gets lost but the taste does not. Alternatively, pickle them in rice vinegar and soy (a great recipe for this version is on Robin Harford's brilliant website www.eatweeds.co.uk). I add them to salads, serve them with cheese or cold meats, or just munch a few each time I walk past the jar.

Recipe from *The Edible City: A Year of Wild Food* by John Rensten (Boxtree, Pan Macmillan, 2016)

Opposite Deliciously-petalled *Magnolia discolor* (now *Magnolia liliiflora*), as illustrated by this beautiful stipple engraving by Pierre Joseph Redouté, has been cultivated for centuries in China and Japan, and is now found worldwide.

MAKE THE CONNECTION
Find out more about John's botanical world at www.foragelondon.co.uk; @foragelondon

Mark Diacono, Otter Farm, near Honiton, Devon

Garlic scape mimosa

Serves 4 as a lunch or starter

I first made this with asparagus, and its simplicity and flavour make it one of those recipes I keep coming back to throughout the year – it works equally well with bamboo shoots, chard stalks or green beans. Here, garlic scapes replace the asparagus. As garlic grows, it throws up a stalk, at the end of which a flower develops – scapes are the top 15–20cm (6–8 in) of those stalks. Harvesting them before the flower has matured directs the plant's energies to the garlic bulb rather than to creating a flower and gives you one of my favourite spring treats into the bargain.

3 medium eggs
400g (14oz) garlic scapes
4 tsp English mustard
2 tsp honey
1 tbsp white wine vinegar
3 tbsp extra virgin olive oil
2 tbsp capers, chopped
Small handful of chives, chopped
2 or 3 chive flowers, broken into florets
Small handful of dill, chopped
Zest of ½ lemon, finely grated
A couple of good pinches of paprika
Sea salt and freshly ground black pepper

1. Place the eggs in a saucepan of cold water, bring to the boil and immediately turn off the heat. Leave the eggs in the water for 12 minutes. Run cold water into the pan for a minute or so to stop the cooking process.

2. Steam or simmer the scapes until just tender – this can take 3–6 minutes depending on their thickness.

3. Make the dressing in a small bowl – whisk the mustard and honey into the vinegar, add a couple of decent pinches of salt, then whisk in the olive oil.

4. Peel and grate the eggs – it's a slightly messy palaver, but the texture is worth it.

5. Arrange the scapes on a large plate, lay the grated egg on top. Scatter over with the capers, chives, the chive flowers, dill and lemon zest, dust with paprika, season with salt and pepper, and splash with dressing.

© Mark Diacono, 2014, *A Year at Otter Farm*. Used by kind permission of Bloomsbury Publishing Plc.

MAKE THE CONNECTION
Find out more about Mark's botanical world at www.otterfarm.co.uk; @mark_diacono

Opposite Garlic scapes are the tender stalks and bulbil-pods of hardneck garlic *Allium sativum* var. *ophioscorodon* (as opposed to softneck garlic *Allium sativum* var. *sativum*). The late spring season is short so grab while you can.

MY BOTANICAL WORLD...
'Cooking and growing draws people together: other gardeners, cooks, growers and enthusiasts. Clichéd as it sounds, when you are out there growing and eating, you become more aware of the little things – the changing light, the first blossom, the buzzard passing.'

MY BOTANICAL WORLD...

'*My parents and grandparents taught me how to cook. Food was at the centre of our lives and signified joy. We were always in the kitchen helping or scrumping for the best wild and backyard produce. Today, my palate calls for growing my own lemon verbena, rosemary, mint and rose geranium in pots in our back seating area at Violet. I add herbs to our quick strawberry jam; the pears in the neighbouring council estate are picked and poached and turned into tarts. Fig leaves foraged from the Hackney area also infuse our custard and enhance the flavour of our figs when we roast them to put in our Almond Polenta Muffins.*'

Claire Ptak, Violet Bakery Cafe, London

Fig leaf ice cream

Makes 1.5 litres (6½ cups) or 15 scoops

Fig trees can be found in many back gardens in London and often overflow onto the street. Seldom do the figs turn into much; the fruit tends to stay under-ripe throughout the season, unless the fig tree has the good fortune to grow inside a walled garden that provides it with much-needed warmth. This suits me just fine because I am mostly interested in those large green fragrant leaves. For a wedding rehearsal dinner for two of our most regular customers, Rick and Caroline, my friend and chef Joe Trivelli prepared a wonderful Italian Irpinian menu and I made this ice cream with roasted fig leaves.

10 new spring fig leaves
350g (1½ cups) whole milk
175g (⅔ cup) caster sugar
4 egg yolks
650g (2½ cups) double cream

Note
You will need an ice-cream maker for this recipe.

1. Preheat your grill to high. Lay the fig leaves out flat on a baking tray. Place the tray on the highest rack under your grill and leave the door ajar. After a few minutes you will start to smell the wonderful heady aroma of the fig leaves warming up and then start to singe under the flame. Let them take on a little bit of colour before you take them out.

2. In a heavy-bottomed pan, warm the milk, sugar and fig leaves until just beginning to bubble. This won't take long, so while it's heating up, put your egg yolks into a bowl and whisk to break them up. Measure the cream into a large container or bowl and set aside.

3. When the milk is ready, temper the yolks by pouring a little of the milk onto them, whisking as you go. Now pour the tempered yolks back into the remaining warm milk in the pan. Stirring continuously, heat until the mixture starts to thicken at the bottom of the pan, checking it now and again by bringing your stirring spoon out of the pan. Pour the custard mixture into the cold cream and whisk well to prevent the custard cooking any further. Cover and put in the fridge for at least 1 hour to cool.

4. Once the ice-cream base has cooled, pour it through a fine sieve to remove the leaves and any eggy bits. Pour into your ice-cream maker and churn for about 20 minutes, following the manufacturer's instructions. Freeze for 1 hour before serving. This will keep for 3 to 4 days in the freezer before it starts to get icy.

Recipe from *The Violet Bakery Cookbook* by Claire Ptak (Square Peg, 2015).

MAKE THE CONNECTION
Find out more about Claire's botanical world at www.violetcakes.com; @violetcakeslondon

Opposite The luscious 'fruit' of the common fig (*Ficus carica*) is actually an infructescence called a syconium, in which many tiny flowers and – if pollinated by a specialised wasp – multiple one-seeded drupelets are born.

SUMMER

Summer is packed full of ready-made treats from stone fruits such as cherries and peaches to pick-your-own strawberries, as well as crisp fresh peas, lettuce and baby courgettes, or juicy tomatoes. Indeed, it's quite possible to prepare a delicious, summer feast completely raw and fuss-free and, thanks to the warming weather, take the dining experience outside. For the season on a plate, flavour or embellish with fragrant edible flowers, aromatic herbs, peak-ripe fruit or pickled vegetables served up with a sparkling cordial or whimsical cocktail.

What to eat now
The summer edit

Broad bean (*Vicia faba*)
Although best plucked from young and tender pods, this ancient crop is often dried to extend the season.
Goes well with dill, winter savory, chilli

Cherry (*Prunus* cultivars)
Early summer is peak season for cherries. Try sweet varieties raw or fuller-flavoured sour ones in cooking.
Goes well with vanilla, chocolate, goat's cheese

Fennel (*Foeniculum vulgare*)
Fennel just keeps on giving, via aniseedy pollen, aromatic seeds, feathery leaves and crunchy bulbs.
Goes well with chicory, sumac, pear

Peach/Nectarine (*Prunus persica*)
Fleshy-fruited peaches and nectarines hail from the same species; gorgeous fresh or in summer puddings.
Goes well with almond, honey, raspberry

Strawberry (*Fragaria* x *ananassa*)
This wonderful accessory fruit is a mainstay of summer sauces and puddings, or is just perfect on its own.
Goes well with black pepper, balsamic vinegar, banana

Blackcurrant (*Ribes nigrum*)
These small, piquant berries are rich in vitamin C and are great raw or cooked into sweets, savouries, preserves or drinks.
Goes well with fennel, mint, mango

Courgette (*Cucurbita pepo*)
Experiment with green, gold or stripy fruit or try stuffing and frying the gorgeous golden flowers.
Goes well with olive oil, garlic, ricotta

Fig (*Ficus carica*)
Another ancient crop, eat these naturally honeyed morsels fresh or dried, in savoury or sweet dishes.
Goes well with cardamom, cheese, rosemary

Sorrel (*Rumex acetosa*)
Add a tangy kiwi flavour to salads, soups or sauces with this versatile, naturally sour, leafy herb.
Goes well with potato, peanuts, spinach

Tomato (*Solanum lycopersicum*)
If you cultivate one fruit, make it a tomato – a key sauce and salad ingredient and the taste of summer.
Goes well with basil, avocado, coriander

> **Sensory notes of summer**
> Fruity • Crisp • Herbaceous • Floral

Opposite Try growing your own tomatoes (*Solanum lycopersicum*) in a garden, greenhouse or even indoors – they have perfect flowers capable of self pollination to produce the lush berry-type fruits.

Try four things
The summer edit

Stuffed courgette flowers
Makes 16

16 courgette flowers
300g (1½ cups) ricotta
100g (1 cup) grated Parmigiano Reggiano (Parmesan cheese)
Handful of chopped basil
1 lemon
120ml (½ cup) soda water or sparkling mineral water
100g (⅔ cup) self-raising flour, sifted
Sunflower oil for deep-frying
Sea salt and freshly ground black pepper

Stuffed courgette flowers are often considered a delicacy. In parts of Italy, however, these deep-fried edible flower morsels are still served up as street food – the Italians having developed the long green 'zucchini' variety of squash in the nineteenth century after it was brought there from the Americas. Either way, they are a super tasty, fun treat.

1. Wash the flowers (non-fruiting male ones, ideally) gently under cold running water, blot dry and remove the inner pistil, stamen and any outer leaves with a knife.

2. For the stuffing, blend the ricotta, grated Parmesan and the basil, and season with the zest and juice of a quarter of the lemon, plus salt and pepper to taste.

3. For the batter, add the soda or sparkling water to the sifted flour in a bowl and whisk to the consistency of single cream.

4. Pipe or spoon the filling into each flower cavity, then twist the ends of the flower together, coat in the batter and quickly dip into a pan of boiling sunflower oil, 10-cm (4-in) deep (hot enough to crisp a cube of bread) – or bake in the oven for 15 minutes at 180°C/350°F/gas mark 4.

Summer fruit coulis
Serves 2

200g (1½ cups) raspberries
100g (1 cup) redcurrants
1 tbsp icing sugar
1 tbsp lemon juice

A coulis, being a smooth, strained purée of fruit or vegetables, is usually served as an accompaniment to a dish, as a prettifying plating-up device or a burst of complementary flavour. A summer fruit coulis is the perfect way to use up peak harvest gluts of berries such as raspberries, strawberries and redcurrants and can be served cold or hot on desserts, ice creams or breakfast bowls.

1. For ease, whizz all the ingredients together with a blender. When the mix resembles a thick purée, strain it through a fine sieve until you have a shiny, seed-free liquid.

Once you've got the hang of this one you can also try inventing your own coulis. Fruit ones usually taste better with a little sugar and can be refined with citrus or spices. Savoury coulis can be made by sautéing vegetables, seasoning with aromatic herbs or spices, adding liquid such as stock and simmering for 20 minutes before blending and straining.

Pickled cucumber

Serves 2

2 tbsp pickling salt
2 tbsp sugar
500ml (2 cups) distilled white vinegar
500ml (2 cups) cold water
8–10 small pickling cucumbers
1 tsp mustard seeds
1 clove garlic, peeled
5 dill crowns or a handful of dill fronds

Dill-pickled cucumbers are often known as gherkins, although these technically refer to a variety of dwarfed cucumber known as the West Indian or burr gherkin (*Cucumis anguria*). If you're thinking of growing your own cucumbers for pickling, go for a similarly small and thick-skinned variety, such as the kirby cucumber.

1. To pickle them, first make the brine by dissolving the salt and sugar in vinegar in a small non-reactive saucepan over a low heat, before whisking in the cold water.

2. Stuff sliced cucumbers into a large sterilised jar, add the mustard seeds, garlic and dill (also possible to grow yourself) and fully cover the contents with the cooled brine.

3. Cover and let sit in the fridge for around 24 hours. The pickles should keep in the fridge for up to one month. Don't be alarmed if your garlic clove turns blue – it's a perfectly harmless chemical reaction caused by the acids in the vinegar.

Crystallised flowers

20–30 an hour

Edible flowers (whole or petals) such as viola, violet, rose, nasturtium, lavender, cherry, apple, primrose, borage, jasmine, cornflower, sage, thyme
1 egg white
50g (¼ cup) caster or 50g (½ cup) icing sugar

Crystallised flowers are a lovely way to prettify a cake, bake, dessert or beverage. They're also the perfect way to dress up less-distinctive-tasting blossoms or to preserve seasonal favourites for use later in the year. For peace of mind, grow your own edible flowers, to ensure that your chosen blooms are definitely edible and chemical-free. It's also wise to collect them on a dry, sunny day when the petals are fully open and as dehydrated as possible, and remove any unwanted bugs or stalks.

1. To crystallise, use tweezers to dip your flowers or petals first into a saucer of lightly beaten egg white, followed by a saucer of sugar – caster sugar for a more crystallised effect, icing sugar for matt. Use a fine paintbrush to tease the mixture into any folds. Any excess sugar can then be gently shaken off before laying flowers on baking parchment in a warm, dry place for 24–48 hours.

2. Your candied flowers are now ready for use or can be stored for several weeks between parchment sheets in an airtight container. For candied flowers that last longer, replace the egg white with gum arabic at a ratio of 1 teaspoon to 1½ tablespoons of water, vodka or distilled organic rosewater.

Dina Falconi, Foraging & Feasting, Hudson Valley, United States *Makes about 200g (1 cup)*

Wild green pesto

Versatile and nutritious, wild green pesto is a gourmet delight and an herb class favourite. I enjoy serving it as a spread on bread. (In this manner, it was delightfully received as an appetiser at a fundraising dinner which took place in a castle on the Hudson River. I appreciate the humble weeds being honoured and bringing in the big bucks.)

Pesto, typically added to pasta, tastes great topping whole grains and burgers, and in soups, sauces and dressings. Make it creamy by mixing it with equal parts of yogurt, crème fraîche or cream. In this recipe we use any wild greens that are eaten raw. While some friends make pesto from only dandelion leaf (very bitter) or nettle (which generally needs cooking to chill out the sting), I suggest using half pungent, aromatic plants and half milder-tasting ones. If you are adding bitter-flavoured plants, include small amounts – about one half cup per batch. I usually triple all the ingredients in this recipe for a bigger batch.

50g (½ cup) hard grating cheese, such as Pecorino Romano or Parmigiano, cut into chunks
75g (½ cup) nuts or seeds (preferably soaked and dried); some good choices are sunflower seeds, green pumpkin seeds, hazelnuts, almonds, pecans and pine nuts
1–2 cloves garlic (less garlic is used here than in typical pesto recipes, so you can taste the flavour of the herbs)
150g (3 cups) packed wild greens (edible raw) of choice, coarsely chopped if leaves are large
75–195ml (⅓–¾ cup) cold-pressed olive oil
Sea salt

MAKE THE CONNECTION
Find out more about Dina and Wendy's botanical world at www.botanicalartspress.com; @foragingandfeasting

1 Place the cheese in a food processor and process until well pulverized.

2 Add the nuts or seeds and garlic and process until medium-fine ground.

3 Add the wild greens and process until the mixture is well minced.

4 With the food processor running, add the olive oil until the desired consistency is reached: 75ml (⅓ cup) oil produces a thicker consistency – nice for spreading on bread, crackers, etc. For a thinner, looser sauce add another 60–120ml (¼–½ cup) of olive oil.

5 Taste first, then add salt if desired.

Wild green variations
Here are four good plant combinations for making wild green pesto:

* 75g (1½ cups) violet leaf
  50g (1 cup) bergamot leaf
  25g (½ cup) garlic mustard leaf or flowering top

* 75g (1½ cups) wood sorrel leaves and tender stems
  50g (1 cup) star chickweed leaves and tender stems
  25g (½ cup) bergamot leaf

* 85g (1¾ cups) sheep's sorrel leaves and tender stems
  50g (1 cup) dayflower leaf and tender stems
  12g (¼ cup) ground ivy leaf and tender stems (can include flower, too)

* 75g (1½ cups) day lily shoots
  62g (1¼ cups) field onion greens
  12g (¼ cup) garlic mustard leaves
  (this combination is available only in early spring)

Garden variations

Here are two good plant combinations for making 'tame' green pesto:

* 75g (1½ cups) rocket/arugula
  30g (1 cup) parsley
  25g (½ cup) watercress

* 45g (1½ cups) French sorrel leaf
  30g (1 cup) chervil
  15g (½ cup) hardy marjoram

Cultivated additions

Of course, pesto is most commonly made with basil, but other cultivated plants work surprisingly well and taste great mixed with wild plants. Try adding 50g (½ cup) leaves. Use one or a combination of the following fresh culinary herbs per batch of wild green pesto: winter or summer savory, oregano, thyme, sweet or perennial hardy marjoram, tarragon, coriander.

From the book *Foraging & Feasting: A Field Guide and Wild Food Cookbook* by Dina Falconi; illustrated by Wendy Hollender.
www.botanicalartspress.com

Above An 18th-century illustration of rocket or arugula (*Eruca sativa*), a wild and widely cultivated herb that is rich in vitamin C and potassium. Eat the peppery leaves but also the flowers, young seed pods and mature seeds.

BOTANICAL RECIPES
The summer edit

MY BOTANICAL WORLD . . .

'*To forage means to dance with the land. It means responding to and resonating with ecosystems: acknowledging the gifts that the earth offers us; learning how to use wild plants for food, medicine, clothing, and more. This resource – a treasure found often right in our own backyards and literally at our fingertips – is so vast it's astounding. The trick is to be able to see it, to train our vision to see beyond what we normally see so we can comprehend what actually has been there all along. In doing so, an exhilarating and reassuring world opens up for us.*'

MY BOTANICAL WORLD...
'*I use edible plants like dill and chives to evoke cherished memories from my childhood summer days in Sweden. Dill is our garlic in northern Europe, and it sometimes overwhelms me emotionally for a split second, just as I recognise the taste in my mouth, feeling it deep down in my being, my origins, the flora and fauna of my youth. For my cookbooks and my food styling and writing, I often connect flavours with places and events, it adds an emotional dimension to the tasting of food. Cooking mostly vegetarian or vegan, I mainly use botanical ingredients like herbs and alliums to create interest in dishes. It is my palette.*'

Nina Olsson, Nourish Atelier, Netherlands

Herbed potatoes

Serves 4

Roasted baby potatoes with a chive and dill pesto sauce tastes like Nordic summer. In Scandinavia we would make a salad like this for Midsummer's feasts.

Baby potatoes don't need peeling and only need a little wash or scrub, if necessary. You can also make this salad with older potatoes. Peel or scrub the potatoes thoroughly before roasting, and extend the cooking time by 5–10 minutes for older potatoes.

1kg fresh potatoes, halved and scrubbed
Olive oil
Salt and freshly ground black pepper
A couple handfuls of baby spinach
½ red onion, diced
A handful of pine nuts

For the chive and dill pesto sauce
1 clove garlic
2 handfuls of chives
1 handful of dill
1 handful of parsley
50ml (¼ cup) crème fraîche (optional)
3 tbsp extra virgin olive oil
1 tbsp grated Parmesan cheese or nutritional yeast
2 tbsp lemon juice

1. Preheat oven to 220°C/425°F/gas mark 7.

2. Spread the potato halves on a baking sheet, drizzle with olive oil and sprinkle with a little salt and pepper. Roast for around 20 minutes, turning occasionally, until tender and golden brown. Watch the potatoes to avoid burning them in the last few minutes.

3. Meanwhile, blend all the pesto ingredients together in a food processor to a fine, loose paste.

4. Remove the potatoes from the oven and place in a big serving bowl. Add the baby spinach and sprinkle with the red onion and pine nuts. Toss everything together and drizzle with half the pesto sauce – keep the rest for serving at the table.

Recipe from *Bowls of Goodness* by Nina Olsson, with permission of Kyle Books (2017)

Vegan note
Serve with vegan crème fraîche and rawmesan instead of Parmesan.

Opposite The potato plant (*Solanum tuberosam*) was domesticated thousands of years ago in the Andes region of South America to which it is native, before travelling to Europe with the Spanish in the 16th century.

MAKE THE CONNECTION
Find out more about Nina's botanical world at www.nourishatelier.com; @nourish_atelier

Sarah Witt, High Desert Test Kitchen, California

Skillet mesquite cornbread

Makes 1 (26-cm/10-in) pan

Most people think of smoked meats and barbecued potato chips when they hear the word 'mesquite'. But the mesquite tree (*Prosopis glandulosa*), a member of the Fabaceae family, is actually one of the most prolific food producers of the American desert.

This thorn-adorned, feathery-leaved tree puts out long, nutritious pods that ripen in midsummer. When they're straw-coloured and brittle – easily snapped in half and slightly sweet on the tongue – they're ready to harvest. The entire pod can be ground whole, and makes a fine powder that can be used as a gluten-free addition to baked goods, or as a substitute for sugar in sauces and salad dressings. The flavour is subtly sweet and earthy, with deep caramel undertones. I liken it to carob (*Ceratonia siliqua*), and find that it pairs well with orange, chocolate and cream.

Mesquite definitely merits a spot on the dessert spectrum, but it can easily work in a savoury capacity as well. Play around with the flavours a bit – substitute rye for the whole wheat, or add seasonings that suit your palate. I throw in a few teaspoons of local sages I've dried, and when I desire heat, I amp up the bread with cayenne or chilli. Make it your own.

170g (1 cup) cornmeal
120g (½ cup) whole wheat flour
60g (½ cup) mesquite powder
1 tbsp baking powder
1 tsp salt
A few grinds of black pepper
55g (¼ cup) cold butter, cubed
2 eggs
90g (¼ cup) honey
250ml (1 cup) goat's or cow's milk

1 Preheat oven to 220°C/425°F/gas mark 7. Lightly grease a medium-sized, 26-cm (10-in) cast-iron frying pan/skillet.

2 Sift all the dry ingredients into a large bowl. Using your hands, massage the butter cubes into the flour mixture until you have evenly distributed the ingredients, resulting in pea-sized crumbles (although it's okay if you have some large butter chunks).

3 Whisk the eggs with the honey and milk, and stir into the flour mixture until just combined.

4 Pour into the prepared pan and bake in the oven for 18–20 minutes, or until an inserted toothpick comes out clean. You may need to adjust baking time and temperatures if you use a different cooking vessel.

5 Serve with butter and a sprinkling of salt or drizzle of honey.

Notes
If you don't live in an area where mesquite trees grow, you can purchase mesquite powder through numerous online outlets. If you do have access to fresh pods on trees, always harvest before summer rains and never collect from the ground, as fallen pods can acquire harmful molds or fungus. You can grind the whole dried pods in a blender, but if you can locate a communal hammermill, that's preferable. Check out www.desertharvesters.org for more information on processing your own pods.

MAKE THE CONNECTION
Find out more about Sarah's botanical world at www.sarahwitt.net/hdtk; @thesarahwitt

Opposite Honey mesquite (*Prosopis glandulosa*) pods can be ground into meal or flour, while the wood lends a unique flavour to foods cooked over it. Mesquite honey made by nectar-supping bees is another desert treat.

MY BOTANICAL WORLD . . .

'*Skillet mesquite cornbread was the first recipe I developed after taking a native cooking class run by the Cahuilla Indians at the Malki Museum in the Morongo Reservation, California. Mesquite was prominently featured that afternoon as we communally cooked, finding its way into acorn breads and dumplings that accompanied venison stew, cholla bud succotash, tepary bean salad and chia candy. The respectful relationship my teachers had with their wild resources inspired me to find that well of gratitude within myself — that desire not only to take, but to nurture the plants that I have come to know.*'

Matt Hoyle, Nobu 57, New York

Hand roll

Makes 1 roll

This recipe – a mixture of your favourite ingredients, dressed in a sesame and soy vinegar and rolled in nori (seaweed) – includes ingredients from several groups of the plant kingdom from seeds and grains to roots, vegetables, herbs and flowers. Miso (used in the crunchy topping) and soy sauce (in the sauce) are both made from fermented soy beans. Korean cooks I have met tell me that this (Jang) is the basis of their cuisine. Japanese cuisine also utilises these plant-based products to give the depth and umami that underpins it. It is interesting that where these fermented vegetable sauces and pastes are used, dairy is almost non-existent in the cuisine.

1 sheet nori (1 sheet per roll)
Selection of pickles, vegetables, herbs, leaves, edible flowers, seeds and grains
Furikake, to serve (see notes)

For the sauce
66g (½ cup) sesame seeds
100ml (scant ½ cup) rice vinegar
50ml (¼ cup) soy sauce
16g (1 tbsp) lemon juice
18g (1 tbsp) sesame oil

MAKE THE CONNECTION
Find out more about Matt's botanical world at www.noburestaurants.com; Instagram @hoylematt

Notes
You can choose from a wide variety of fillings for your hand roll – mix and match what is in season or easily available. Some ideas include:

Pickles: I used yamagobo, which is a pickled burdock root that has a lovely earthy taste, adding texture to the roll, plus beansprouts given a quick sweet vinegar pickle. The pickled cabbages of all varieties – kimchi or sauerkraut, for example – work well with both texture and a depth of flavour.

Vegetables: Thin-sliced cucumbers and carrots form the base of the roll. These should be cut just shorter in length than the shortest edge of the nori. Kanpyo is a Japanese gourd (calabas) that has been dried and then braised in soy sauce. It adds a chew to the texture. Myoga is the flower bud from a Japanese ginger plant. These should be finely shredded and add a lovely fragrance. Regular root ginger, cut into fine slivers, could also be used.

Herbs: I used one piece of green shiso leaf per roll. I find the ones grown in Japan can easily be bitten through and so you can use them whole. I have some red shiso in my garden which has an excellent fragrance and flavour but is a much tougher leaf, so needs to be shredded. I also used peppermint and lemon verbena leaves from my garden to brighten up the flavours. Flower petals would also be an attractive addition.

Other leaves: Use your favourite salad leaves. Bibb lettuce (butterhead) works very well. If using crunchier varieties such as little gem or endive, cut them in half lengthwise.

Crunch: For crunch I have included my own blend of furikake. This is a mix of puffed rice, dry dulse (seaweed), lemon zest, sesame seeds, soba cha (buckwheat) and dried miso. This mixture is normally used as a condiment for plain rice.

1. To make the sauce, first grind the sesame seeds in a blender or with a pestle and mortar to form a paste. Gradually add the other ingredients and mix until harmonious.

2. To construct your hand roll, lay the sheet of nori along your left hand (right-handed people), lengthways and shiny side down. Arrange one piece each of the pickles, vegetables, herbs and other leaves on the left side of the nori.

3. Drizzle a tablespoon of sauce over the mixture.

4. With your right hand, roll the nori from the side closest to you as tightly as possible to make a cone.

5. Sprinkle furikake over the top to add crunch and eat straight away so that the seaweed is still crisp, contrasting with the textures inside.

Above The black wrappers used to create cones of temaki sushi (hand rolls) are made from nori, an edible seaweed species of the red algae genus *Pyropia* that has been consumed in Japan since at least the 8th century CE.

MY BOTANICAL WORLD . . .

'*Here in Harlem I have a small plot in a community garden. I grow herbs for teas (lemon verbena, lemon balm, pepper and spearmint) and herbs for cooking (oregano, lemon thyme, bay, savory, marjoram, cilantro [coriander] and parsley), which have done so well that, after drying, I have enough to keep me going until next summer. I'm also trying my hand at tomatoes, zucchinis [courgettes], cucumbers, raspberries and blueberries, while edible wildflowers have also made it back to our house occasionally.*'

Kitty Travers, La Grotta Ices, London

Bunch of herbs sheep's milk ice cream

Makes 1 litre (4½ cups) or 10 scoops

This recipe – a surprising and refreshing ice, delicious with a side of lightly sugared, sliced stone fruits – employs the use of a simple milk base, thickened with natural vegetable starch so as not to interfere with the pure flavour of the soft fresh herbs. Sheep's milk is a lot higher in protein and fat than cow's milk, meaning you only need add a little cream to this recipe. The protein adds body to the ice cream, improving the texture without the need for fillers like dry milk powder (a staple of most commercial ices). It is delicious in the summer when the sheep are grass-fed, and should taste sweet and clean – a perfect match for a bunch of freshly cut garden herbs.

130g (scant ¾ cup) unrefined sugar
15g (1 tsp) tapioca or cornstarch
550ml (2⅓ cups) sheep's milk
100ml (scant ½ cup) double cream
1 heaped tsp glucose syrup (optional, see note)
40g mix of freshly picked and roughly chopped garden herbs and flowers, such as mint, hyssop, verbena, rose petals, wild fennel, edible honeysuckle flowers (avoid berries as some are toxic), fig leaves or elderflower

Note
This ice is best eaten within 1 week – the recipe contains no egg yolk and little cream so it freezes quite hard and can become icy otherwise. The glucose syrup is optional but helps to keep it scoopable.

Prepare the milk ice

1. Fill a sink with iced water and set a timer to 15 minutes. Have a clean bowl ready with a fine mesh sieve over it.

2. In a bowl, whisk 2 tablespoons of the sugar into the tapioca or cornstarch.

3. Heat the rest of the sugar with the milk and cream; stir often using a whisk or silicone spatula to prevent scorching.

4. Once the liquid is hot and steaming, pour it into the bowl containing the starch. Whisk constantly to combine without lumps forming.

5. Return the mix to the pan, cook over a low heat, whisking constantly, just until it starts to simmer.

6. Remove the pan from the heat, stir in the chopped herbs and flowers, then cover the pan tightly with cling film and place it in the sink of iced water to cool. Start the timer.

7. After exactly 15 minutes, remove the pan and pour the mix through the sieve. Squeeze hard to extract as much flavour as possible from the herbs. You should see a tint of pale acid green seep into the mix with the last squeezes. Discard the remaining herbs.

8. Return the custard to the sink to cool completely before refrigerating, covered, overnight.

MAKE THE CONNECTION
Find out more about Kitty's botanical world at www.lagrottaices.tumblr.com; @lagrottaices

Make the ice cream

1. The following day, liquidise the ice-cream base with a stick blender for 1 minute to emulsify completely; this will also help liquefy the mix.

2. Pour the mix into an ice-cream machine. Churn according to the machine's instructions (20–25 minutes) or until frozen and the texture of whipped cream.

3. Scrape the milk ice into a suitable lidded container. Before sealing, top the milk ice mix with a piece of waxed paper to limit exposure to air, then cover and freeze until ready to serve.

Above The edible sweet nectar found in the tubular flowers of honeysuckle (*Lonicera*) give this beautiful flower its fitting name. Pull the style from the base of a flower and the stigma will force a droplet of nectar through.

MY BOTANICAL WORLD . . .

'*I am obsessed with edible plants, especially fruits – and dazzled by nature – it stimulates the soul. My small attempt to "capture" it, as a painter might, is to turn it into ice cream – my second obsession – and try to contain all that flavour and perfume and colour in a smooth, creamy frozen scoop.*'

AUTUMN

Autumn is peak harvest time. Gluts of summer fruits or tender vegetables can be turned into chutneys, jams, jellies or pickles. Squashes and gourds ripen and expand to impressive sizes and colours, providing much meaty flesh for soups, stews and bakes. Trees are laden with pome fruits such as apples and pears – perfect for pies, crumbles and puddings – or enhanced with warming spices or the fruits of a forage, such as blackberries, nuts and seeds. Plus hopefully there may still be some sun, so leave space for some optimistic chilled desserts.

What to eat now
The autumn edit

Apple (*Malus pumila*)
The perfect snack, the ideal ingredient for puddings, and more than 7,500 cultivars to choose from.
Goes well with cinnamon, raisins, cheese

Blackberry (*Rubus*)
Pick your own, devour some, then turn what's left of your forage into autumnal pies, jams and crumbles.
Goes well with apple, plum, apricot

Damson (*Prunus domestica* subsp. *insititia*)
This lovely, tart, dark-blue-skinned subspecies of plum works best when stewed down for jam or pudding.
Goes well with gin, cheese, sage

Sweetcorn (*Zea mays*)
There's almost nothing better than hot, buttered, roasted cobs of maize. It's also perfect for salsa.
Goes well with jalapeño chillies, lime, nutmeg

Beetroot (*Beta vulgaris* cultivars)
Choose from deep red, yellow and stripy varieties; perfect for salads, soups and roast vegetable dishes.
Goes well with aniseed, chocolate, fennel

Sweet chestnut (*Castanea sativa*)
Cook these lovely shiny nuts first to reduce the high (stomach-irritating) tannic acid content, and to sweeten and warm – ideally over an open fire.
Goes well with chocolate, cream, Brussels sprouts

Pumpkin (*Cucurbita maxima*)
Pumpkin's generous orange flesh can be cooked up in numerous ways, as can the nutritious seeds.
Goes well with rosemary, cumin, maple syrup

Pear (*Pyrus* genus)
Pears impart a sweet yet delicately acidic flavour and granular texture to sweets, savouries and drinks.
Goes well with blue cheese, hazelnut, ginger

Quince (*Cydonia oblonga*)
The sole member of the *Cydonia* genus is high in pectin, which makes it ideal for use in jams, jellies and quince 'cheese'.
Goes well with Manchego cheese, vanilla, fennel

Chard (*Beta vulgaris* subsp. *vulgaris*)
The jewel-coloured leaf stalks and dark cabbage-like leaves provide a nutritious meal all of their own.
Goes well with garlic, coriander, citrus

Sensory notes of autumn
Colourful • Aromatic • Comforting • Sweet and sour

Opposite Five species of gourd (*Curcurbita*) are grown worldwide, producing numerous cultivars including butternut squash (*Curcubita moschata*) and pumpkin (*Cucurbita maxima*), first domesticated for its flesh.

Try four things
The autumn edit

Rosehip syrup
Makes up to 840ml (3½ cups)

900g ripe rosehips
5 litres water
560g (2¾ cups) sugar

Rosehip syrup, often made from the hips of dog roses (*Rosa canina*), is renowned for its high vitamin C content, and a daily dose of this was advised during World War II when fresh fruit was in low supply. Rosehip syrup can also be made into a delicious mango-meets-lychee-tasting cordial or cocktail, or drizzled over desserts or breakfasts.

1. The UK's Ministry of Food produced a wartime recipe that provides a good base from which you can develop your own perfect concoction. It involves mincing hips in a coarse mincer (or food processor) and adding 1.7 litres (7 cups) of boiling water. You then bring it to the boil again in a pan, turn off the heat and set it aside for 15 minutes.

2. Pour the syrup into a muslin cloth or a jelly bag set over a bowl and allow to drip until the bulk of the liquid has come through. Return the hips to the pan, add the remaining water, stir and allow to stand for 10 minutes. Pour it through the cloth or bag once more. If you see any remaining hairs in either liquid, strain again, as these are irritants.

3. You are now ready to boil down your juice until it measures about 840ml (3½ cups), before adding the sugar and boiling for a further 5 minutes. Pour into hot sterile bottles and seal immediately.

4. Your syrup will last several weeks once opened, several months unopened. For higher vitamin C content you can make your syrup raw, layering hips and sugar, but the process takes several weeks.

Blackberry jam
Makes 2 (450g/16oz) mason or jam jars

350g (2¾ cups) blackberries
350g (1¾ cups) preserving sugar
Finely grated zest and juice of 1 lemon

Blackberries have been foraged and enjoyed for thousands of years, as the shiny red-to-black drupelets of late summer and early autumn are hard to resist. In cooking they translate particularly well to jams, jellies, pies and crumbles and pair well with apples.

1. To preserve a freshly picked glut, soak the fruit in salted water for a couple of hours to destroy any bugs, then rinse in cold water and combine with the zest and juice of 1 lemon in a stainless-steel pan on a low heat, stirring until softened.

2. Press gently with a masher to release more juice but avoid pulping the fruit. Add the sugar and stir for 10 minutes until it has all dissolved, then gently increase to a boil for around 7 minutes, using a sugar thermometer to check for a setting point of 105°C/221°F/gas mark ¼.

3. Remove from the heat, skim off any scum and leave for 5 minutes. Stir well, and test for setting point by placing a spoonful of the jam onto a chilled plate and leaving it in the fridge for a few minutes. When tipped, the jam should crinkle if it is ready – if not, boil for a further 2 minutes, then test again.

4. When ready, pour hot jam into warm sterilised jars (cleaned and heated in an oven at 180°C/350°F/gas mark 4. Add waxed discs and lids and store in a cool, dark place. Refrigerate once opened.

Hazelnut butter
Makes 1 (450g/16oz) mason or jam jar

250g (2 cups) hazelnuts
Salt

Hazelnuts begin to ripen when the leaves on the trees start changing colour in the autumn. Once the papery outer covering starts pulling back, the nuts are edible. However, it's quite hard to find enough ripe ones on a tree to make a batch of butter as the squirrels are unlikely to pass them by. In this case, you can try to grow your own and protect them from predators, or simply buy a bag of nuts – this delicious, nutritious spread is worth it.

1. To make your butter, roast the hazelnuts at 150°C/300°F/gas mark 2 for 10 minutes to add flavour and help release the oils. Place roasted nuts in a food processor and blitz until smooth, stopping at intervals to scrape the sides with a spatula and give your processor a rest. This can take up to 10–12 minutes.

2. Add salt to taste, if required, or chopped nuts for texture, before storing in a sterilised glass jar.

3. You can also substitute hazelnuts for cashews, peanuts, almonds or sunflower seeds. Or you can add additional ingredients towards the end of the mix, such as turmeric, maple syrup, cocoa or cacao powder. Then spread liberally on toast or drop dollops into bowls.

Poached pears
Serves 4

4 medium pears
1 litre (4 cups) water
300g (1½ cups) caster sugar
1 vanilla pod
5 cloves
1 cinnamon stick

'There are only ten minutes in the life of a pear when it is perfect to eat,' observed the American poet and nature lover Ralph Waldo Emerson – unless you poach them, that is. Not only does this extend the deadline, it enhances that soft, comforting pear sweetness that Emerson was talking about.

1. To prepare yours, choose nicely shaped, smooth-textured and not quite ripe pears to avoid disintegrating under heat. Peel the pears, leaving the stalk intact, and trim a thin slice off the bottom so they can stand up when serving. You can also carefully cut the cores out, which can help the fruit cook through.

2. Combine the water, caster sugar, vanilla pod, cloves and the cinnamon stick in a small deep pan and bring to a simmer. Add the pears, ensuring that they are all fully covered in liquid. Gently poach at a simmer for 1–2 hours until tender.

3. Remove from the heat, allow to cool and steep (overnight in the fridge gives a deeper flavour).

4. Remove the pears and simmer the liquid for 30 minutes to reduce to a sticky glaze. Pour over to serve.

BOTANICAL RECIPES
The autumn edit

MY BOTANICAL WORLD . . .
'Fennel is one of my favourite ingredients and particularly shines in late summer and autumn when the golden yellow umbels start to set seed. Pick one straight from the plant and savour the fragrant yet spicy taste and aroma – the inspiration for numerous sweet and savoury dishes.'

Tom Harris, The Marksman Pub and Dining Room, London

Orange, fennel seed and almond cake

Makes 1 cake

This wonderfully sticky citrus and nut cake can trace its origins back to the almond and orange groves of Spain and the Middle East. The addition of toasted fennel seeds and Noilly Prat vermouth adds floral yet warming spicy notes, while Seville or blood oranges, or even lemons, can be used into winter for a more tart, marmalade-like flavour.

2 oranges
6 eggs
220g (generous 1 cup) sugar
220g (2 cups) ground almonds
5g (1 tsp) fennel seeds, toasted
50g (⅓ cup) flour
50ml (¼ cup) olive oil
15ml (1 tbsp) Noilly Prat vermouth (Anise liquor)
Crème fraîche and a drizzle of honey, to serve

1. Boil the oranges whole in two changes of water until very soft – enough to push a finger through without resistance. You will need to simmer them for 1½–2 hours.

2. Once cool, purée the oranges in a blender and pass through a sieve to remove any pips.

3. In a mixer or bowl, whisk the eggs and sugar until light and fluffy and tripled in size.

4. Fold in the ground almonds and flour with the toasted fennel seeds, keeping as much air in the mix as possible. Then fold in the orange paste made in step 2, olive oil and the Noilly Prat.

5. Pour into a buttered and lined cake tin and bake at 180°C/350°F/gas mark 4 for approximately 25 minutes or until a skewer comes out clean.

6. Serve warm with a dollop of crème fraîche and a drizzle of honey.

Opposite Blood oranges, a natural mutation of sweet orange *Citrus* × *sinensis*, have a dark red flesh made by the antioxidant pigment anthocyanin, and a raspberry-citrus flavour. Try Tarocco, Moro and Sanguinello cultivars.

MAKE THE CONNECTION
Find out more about Tom's botanical world at www.marksmanpublichouse.com; @marksman_pub

Skye Gyngell, Spring Restaurant, London   *Serves 6*

Raw cabbage, fennel and pecorino salad

This simple salad is perfect for the early weeks of autumn when the sun is still slightly warm and the season's beautiful ingredients are beginning to trickle in. I like to use the ruffley Savoy cabbage because I love its texture and sweet, mild taste – look for tender young cabbages with vibrant leaves. This salad works best if dressed an hour or so before serving. It's lovely as a starter, with a few slices of the finest Parma ham served alongside.

1 Savoy cabbage
1 fennel bulb
150g (5¼ oz) pecorino, finely sliced
A small bunch of flat-leaf parsley, leaves only, finely chopped
120ml (½ cup) extra virgin olive oil
3 tsp good-quality red wine vinegar
Sea salt and freshly ground pepper

1 Wash the cabbage under cool running water, separating the leaves as you do so by snapping them off at the base. Pat each leaf dry, using a clean tea towel, then slice into fine ribbons and place in a bowl.

2 Remove the fibrous outer layer of the fennel, and then slice the bulb in half lengthways. Cut each half into very fine slices and add them to the cabbage ribbons in the bowl.

3 Add the pecorino slices and chopped parsley, then season the salad with salt and plenty of pepper. Drizzle over the olive oil and wine vinegar and toss really well to combine.

4 Leave the salad to sit for an hour or so, to allow the vinegar to soften and wilt the leaves slightly, before serving.

From *Spring* by Skye Gyngell (Quadrille Publishing Ltd, 2015)

MAKE THE CONNECTION
Find out more about Skye's botanical world at www.springrestaurant.co.uk; @skyegyngell

Opposite Fennel 'bulb' is actually the inflated leaf base of cultivated Florence fennel (*Foeniculum vulgare Azoricum* Group). This type of fennel is also one of the three main herbs used to make absinthe.

MY BOTANICAL WORLD...

'*Apart from being in fashion, nature has literally planned for our health and wellbeing. The fact that citrus is a winter fruit and it's full of vitamin C, and iron-rich foods like kale also grow in winter, shows how important it is to eat with the seasons. Even if it's just basil, rocket or some flowers for a salad, it's easy to grow some produce at home. And herbs give you an amazing connection to the world.*'

Gill Meller, author of Gather *and cookery teacher at River Cottage, Dorset*

Roast parsnips with blackberries, honey chicory and rye flakes

Serves 4 as a lunch or starter

The last of autumn's blackberries are ripe, salacious and full of flavour. Once all the berries are picked, the bramble retires back into the hedgerow, unnoticed and still for winter. It's easy to forget then how productive this thorny weed really is. In this gorgeous, colourful salad, I tumble the berries through a tray of roasted early winter parsnips, where they burst in the hot honey and olive oil, and are cut through by the bitterness of red chicory. A scattering of rye flakes adds a nutty crunch, which I really love. This is a real River Cottage favourite.

4 parsnips, quartered lengthways and cored
3 tbsp extra virgin olive oil
2 heaped tbsp rye flakes
2 tsp Dijon mustard
1 tbsp runny honey
1 tbsp cider vinegar
2 or 3 thyme sprigs
2 or 3 rosemary sprigs
About 100g (generous ⅔ cup) blackberries
1 firm red chicory, leaves separated
Salt and freshly ground black pepper

1 Preheat the oven to 180°C/350°F/gas mark 4.

2 Place the parsnip quarters in a large roasting tray. Combine the olive oil, rye flakes, Dijon mustard, honey and cider vinegar in a bowl and season generously with salt and pepper. Mix well, then pour the mixture over the parsnips and tumble them to coat thoroughly. Tear over the thyme and rosemary, then place the tray in the oven and roast for 40–45 minutes, turning once or twice with a spatula, until the parsnips are tender in the middle and crisp and caramelised on the outside.

3 Remove the parsnips from the oven, then scatter the blackberries and chicory leaves over them. Set aside to cool for 5–10 minutes, then serve as a warm salad with fresh bread, or as an accompaniment to good sausages, duck or pork chops.

Recipe from *Gather: Simple, Seasonal Recipes* from Gill Meller, Head Chef at River Cottage (Quadrille Publishing Ltd, 2016)

MAKE THE CONNECTION
Find out more about Gill's botanical world at www.gillmeller.com; @gill.meller

Opposite Award-winning photographer Andrew Montgomery perfectly captures the sweet yet tart flavours, jewel-like colours and earthy crunchiness of Gill Meller's perfectly autumnal dish.

MY BOTANICAL WORLD . . .
'*I've always liked the word "gather". It feels hopeful, natural and very human. It's a word that embodies many of the simple things we do every day. As people, we gather in one way or another all the time. It's what we do.*'

Adele Nozedar, Brecon Beacons Foraging, Wales

Elderberry flu remedy

Makes about 3 litres (12 cups)

An ancient tree, the elder. You might think that its name reflects its ancient wisdom; this isn't so. 'Elder' is derived from an Anglo-Saxon name, 'aeld', meaning 'fire'. But that's another story. And there's a use for elderberries which is equally ancient, but more valid. For millennia, we've used elderberries to boost our immune systems and help us fight infection. Until the plant was put through rigorous clinical trials in the last ten years or so, we didn't know exactly how this worked. Now we have evidence that our ancestors were right, and pharmaceutical companies sell elder-inspired remedies over the counter. You can buy these if you really want to, but if you prefer, you can step into the footsteps of generations before you and forage for the berries yourself. That way, you also get the advantage of taking a lovely walk on a late summer or early autumn day, when the berries are ripe.

2kg (4½ lb) elderberries
Sugar (jaggery or demerara) – 200g (1 cup) for every 500ml (2 cups) liquid produced
Juice of 1 lemon
5 cloves
1 cinnamon stick
1 lump of ginger
1 star anise
1–2 cardamom pods

Note
Your remedy can be taken out and used as needed and will last indefinitely. You can also pour it over your breakfast porridge, over ice cream, or even make it into a sorbet.

1 Pick a couple of kilos (pounds) of elderberries on a dry sunny day. Wash and drain them and strip the berries from the stalks. This won't take long.

2 Put the berries in a pan and just cover with water. Bring to the boil and simmer until they are soft. This will take about half an hour.

3 Strain through a fine wire-meshed sieve until the dripping stops. Then press the pulp a little to extract every last drop of liquid. Measure the liquid and return it to the pan.

4 For every 500ml (2 cups) of liquid, add 200g (1 cup) sugar (preferably jaggery or demerara), along with the lemon juice, cloves, cinnamon, ginger, star anise and a cardamom pod or two. Oh, and don't throw away the pulp!

5 Return the liquid to the heat and simmer for about 20 minutes. Then put the lid on the pan and let the mixture sit overnight so that all the flavours get to know one another.

6 Next day, strain the mixture again and let drip as before. Don't throw away the spices – they can still be used for something else.

7 Preserve by pouring the liquid into empty water bottles and popping in the freezer, leaving a space at the top for the liquid to expand into.

Recipe from *The Hedgerow Handbook: Recipes, Remedies and Rituals* (Square Peg, 2012)

MAKE THE CONNECTION
Find out more about Adele's botanical world at www.breconbeaconsforaging.com; @HEDGEROWGURU

Opposite Elderberries are the fruit of the elder or *Sambucus* genus of plants, and are edible when ripe and cooked. Before the fruit come lacy white flowers, ideal for sweet, fragrant cordial or pretty floral fritters.

MY BOTANICAL WORLD...

'*I'm an author and forager, passionate about the relationship between people and plants, and the stories shared by both. I grow flowers, vegetables and weeds in my garden in the Brecon Beacons mountains of Wales. I believe that human beings are an equal component within the natural world; not bigger, not smaller, and certainly no more relevant than the tiniest bug. Sometimes we think we're in charge. We're not. Realising this can be a huge relief.*'

WINTER

Winter brings shorter, colder days but also ample opportunity for feasting. Root vegetables such as carrots, parsnips and turnips provide the base for many a hearty stew or soup. Cabbages, spinach, chard and kale keep on going through plummeting temperatures and frost to provide shiny, nutritious sides, even more delicious with beans and nuts. While citrus fruits such as lemons and oranges bring a zing to grey days along with the sweetness of raisins, sultanas and dates. There may be fewer ingredients, but cooking with them is somehow just as inspiring.

What to eat now
The winter edit

Brussels sprouts (*Brassica oleracea* cultivar)
Employ ingredients such as cheese, vinegar, fruit, nuts or spices to perk up these lovely cabbage family buds.
Goes well with mustard, paprika, miso

Celeriac (*Apium graveolens* var. *rapaceum*)
This knobbly root vegetable might look ugly but it imparts a lovely celery flavour to dishes or as a side.
Goes well with cheese, chicory, cashew

Kale (*Brassica oleracea* cultivar)
Frost-loving kale has been making a comeback in recent years – try it wilted, in smoothies or roasted as crisps.
Goes well with parsley, squash, garlic

Blood orange (*Citrus* x *sinensis*)
Use the deep-red flesh and raspberry flavours of the blood orange for marmalade, baking or winter salads.
Goes well with bulb fennel, ginger, cardamom

Broccoli (*Brassica oleracea* cultivars)
One of the most commonly eaten flowers, available in large-headed, sprouting or purple varieties
Goes well with lemon, almonds, Parmesan

Cabbage (*Brassica oleracea* cultivars)
Cabbage is wonderful raw, cooked or pickled with a choice of Savoy, red, white and green leaves to try.
Goes well with apple, sultanas, clementine

Date (*Phoenix dactylifera*)
Naturally sweet, potassium-rich dates are lovely fresh, dried, stuffed, or in desserts, sauces or stews.
Goes well with ginger, pistachio, sesame seeds

Leek (*Allium ampeloprasum* cultivar)
Often used to flavour stocks, these mild onion-like leaves are also delicious raw, fried or boiled.
Goes well with chervil, thyme, pine nuts

Parsnip (*Pastinaca sativa*)
This pale, carrot-like taproot has an earthy but sweet flavour that works perfectly in roasts, soups or stews.
Goes well with honey, walnut, apple

Turnip (*Brassica rapa* subsp. *rapa*)
Don't discard the leafy green turnip tops – they work beautifully with gently sautéed root slices.
Goes well with sage, sweet potato, anise

> **Sensory notes of winter**
> Warming • Spicy • Citrus • Rooted

Opposite There are several cultivar groups of cabbage including crimp-leaved Savoy, loose-headed spring greens, slightly pointed green cabbage, red or purple cabbage, and smooth pale-green 'white' cabbage.

Try four things
The winter edit

Kimchi
Makes 1.25 litres (5 cups)

1 Chinese/napa cabbage (about 700g)
75g (¼ cup) sea salt or kosher salt
1.5–2 litres (6–8 cups) distilled water
5 cloves garlic, crushed
2.5cm (1 in) piece root ginger, grated
1 tsp golden caster sugar or 1 tbsp shredded apple
½ nori sheet or 3 tbsp fish sauce (optional)
3 tbsp gochugara (Korean chilli flakes/paste)
2 daikon radishes, finely sliced
4 spring onions, finely shredded

Kimchi is a Korean staple of highly seasoned, salted and fermented cabbage, and is eaten with almost every savoury dish. The most common cabbage used is the napa cabbage (*Brassica rapa* subsp. *pekinensis*), a cool-season annual, but you can substitute this for green cabbage or pak choi if a napa is not to hand.

1. Simply slice the cabbage into 2.5cm (1 in) strips and submerge them in a salt and distilled water mix. Cover with a plate (to fit the interior of your bowl) and weigh down for at least an hour, or ideally overnight.

2. Make the kimchi sauce by blending the garlic, ginger, sugar or shredded apple, nori or fish sauce (if using) and gochugara in a small bowl.

3. Rinse and dry the cabbage, transfer to a large bowl and toss with the kimchi sauce along with the radishes and spring onions.

4. Pack the mixture into one large 1.5 litre sterile jar or a couple of 750ml jars, seal tightly and leave to ferment in the fridge for 3 days. Refrigerated, the kimchi should keep for 3 weeks.

Kale crisps
Makes 1 bowl

200g kale (2 cups), ideally black kale, cavolo nero
1 tsp flaky salt
½ tsp smoked paprika
Zest of 1 lemon
Olive oil

Kale crisps are all the rage these days, proposed as a healthy alternative to deep-fat-fried chips or crisps. Done well, they can be just as delicious and add a nice seasonal touch, especially if you grow your own leaves. Black kale (cavolo nero) is best for this recipe as it doesn't frazzle as easily under heat as curly kale.

1. Simply heat the oven to 120°C/250°F/gas mark ½. Tear the kale's leafy sections away from the stems and wash well before drying in a salad spinner or with paper towels.

2. Grind the salt, paprika and lemon zest together in a pestle and mortar to a fine powder (you can substitute other spices such as chilli, cumin or garlic powder, if you like).

3. Coat the kale in olive oil, massaging into all the undulations, before adding the seasoning. Toss and then spread the leaf strips out in a single layer on baking trays lined with baking parchment. Bake for around 30 minutes, turning at the 15-minute mark.

4. Turn off the oven and loosen the leaves with a spatula. Taste, then add a touch more seasoning if required. Return to the oven for a further 15 minutes to crisp and gently cool. Serve as a side or nutritious snack.

Cranberry jelly

Makes 500ml (2¼ cups)

450g (4½ cups) cranberries
340ml (1⅓ cups) water
450g (2¼ cups) caster sugar

This seasonal Thanksgiving/Christmas favourite is also lovely year round served with roasts or leftovers. Indeed, cranberries were already a staple food of indigenous peoples when European settlers arrived in the 'New World' to which the large-fruited variety (*Vaccinium macrocarpon*) is native.

1. First prepare the cranberries by discarding any discoloured or shrivelled ones. Wash the berries and place in a saucepan with 340ml (1⅓ cups) water. Bring to the boil over a medium heat, then simmer until the cranberries are tender. The mixture can then be pressed through a sieve into a bowl using a spoon.

2. Return the resulting smooth, juicy pulp to the pan, bring to the boil and add the sugar. Allow to simmer for 10 minutes and then test for the setting point (see page 218 for blackberry jam).

3. When it's ready, pour into warm, sterilised jars.

4. Once opened, the jelly should keep in the fridge for up to three weeks. Unopened, you can store it in a dark larder for up to one year.

Blood orange marmalade

Makes 1 litre (4½ cups)

4–5 large blood oranges
1 large unwaxed lemon
750ml (3 cups) water
700g (3½ cups) granulated sugar

The amazing, dark-red flesh of blood oranges also carries a distinctive raspberry flavour alongside citrus notes. The most flavourful is the Moro blood orange, which makes lovely bitter marmalade. Seville or regular sweet oranges can also be used, and you can add spices such as ginger for even more intense, warming flavours.

1. First, top and tail the oranges and the lemon, then cut them in half lengthways and halve again into long wedges. Slice each quarter crosswise into very thin slices. Place the fruit (and seeds enclosed in a bundle of muslin) into a large bowl and cover with water. Place a plate on top, weigh down and refrigerate overnight.

2. Place the fruit and water (minus the seed bundle) in a saucepan and bring to the boil over a medium heat before simmering for around 40 minutes. When the rind is tender, add the sugar and bring back to the boil until the temperature rises to 105°C/220°F/gas mark ¼ on a jam thermometer. It should take 25–30 minutes. Check for setting (see page 218 for blackberry jam) before pouring into warm, sterilised jars.

3. Once opened, the marmalade should keep in the fridge for up to 3 weeks. Unopened, you can store in a dark larder for up to 1 year.

BOTANICAL RECIPES

The winter edit

MY BOTANICAL WORLD...

'*The inherent oddness of the knobbly yellow quince belies a pectin-rich, spongy flesh that gives off a sweet, delicate fragrance as it ripens and goes a rich ruby red when cooked (it's not great when raw). It's a complex flavour, but in a good way – one that seems destined to be shared, to lift the spirits or indeed a host of dishes, or to usher in the festive season.*'

Tom Harris, The Marksman Pub and Dining Room, London

Serves 1 as main

Red chicory, pickled quince, goat's curd and walnut salad

Bitter, crimson-veined radicchio – also known as red endive or red chicory – sweet pickled quince, waxy walnuts and tangy, salty goat's curd combine beautifully to create a crunchy yet mellow dish, perfect as a light yet full-bodied main or an uplifting, jewel-coloured side dish.

For the pickled quince
1 quince, peeled, cored and cut into quarters
500g (2½ cups) sugar
500ml (2 cups) red wine vinegar
1 slice of orange peel
1 cinnamon stick

For the walnut emulsion
100g (1 cup) toasted walnuts
2 eggs, boiled for 6 minutes and peeled
50ml (¼ cup) sherry vinegar
400ml (1¾ cups) grapeseed oil
20ml lemon juice
100ml (scant ½ cup) walnut oil
1–2 tsp salt

For the salad
1 radicchio (red chicory)
500g goat's curd, crumbled
1 toasted walnut

For the dressing
20ml sherry vinegar
20ml pickled quince syrup (produced while making the Pickled Quince)
100ml (scant ½ cup) olive oil
Salt and freshly ground black pepper

1. Place the quince quarters, sugar, vinegar, peel and cinnamon stick in a pan and simmer on a low heat until the syrup has reduced and the quince has gone a deep ruby red. For best results, simmer for 20 minutes, turn off the heat and leave to cool, then repeat this process five or six times. Set quince quarters and the reduced syrup to one side. (Quince pickle is also delicious with cheese, cured meats and ice cream, so make a big batch.)

2. Next, add the toasted walnuts, boiled eggs and vinegar to a blender and blend to a paste. Continue blending while slowly pouring in the grapeseed oil to form an emulsion. Add the lemon juice and salt, and a little water if the consistency is too thick.

3. Finally, quarter the radicchio and grill under a hot grill or on a hot griddle pan until charred on each side.

4. Prepare the dressing by combining the sherry vinegar, pickled quince syrup, olive oil and pour over the radicchio while still warm, adding salt and pepper to taste.

5. To serve, spread the walnut emulsion on the plate, cover with the grilled, dressed radicchio quarters, place some of the sliced quarters of pickled quince across the plate, crumble goat's curd over the top and finish with grated toasted walnut.

Opposite Radicchio is a cultivated type of leaf chicory (*Cichorium intybus*) with distinctive white-veined red leaves. Try tight-headed Chioggia or endive-like Treviso, grilled in olive oil to remove the bitter edge.

MAKE THE CONNECTION
Find out more about Tom's botanical world at www.marksmanpublichouse.com; @marksman_pub

Paola Gavin, vegetarian cook and author, London

Serves 3 to 4

Swiss chard with cannellini beans and chilli

This simple peasant dish from Sardinia makes a very good side dish or light main course with some wholemeal toast on the side. If you like, you can use canned cannellini beans, but rinse them well under cold water before use to remove salt.

450g (1 lb) Swiss chard
4 tbsp extra virgin olive oil
1 medium onion, finely chopped
2 garlic cloves, finely chopped
½–1 hot red chilli pepper, to taste, cored, deseeded and finely chopped
350g (2 cups) cooked and drained cannellini beans
Salt and freshly ground black pepper

1 Wash the chard and cut off the stalks. You can save them for a soup or stew. Cut the leaves into thin strips. Cook in a covered saucepan with a few tablespoons of water for 10 minutes or until tender. Drain and chop coarsely.

2 Heat the olive oil in a large frying pan and add the onion, garlic and chilli. Cook over a moderate heat for a few minutes or until the onion is softened. Add the Swiss chard and cannellini beans and season with salt and black pepper. Simmer for a further 8 to 10 minutes to blend the flavours. Serve hot.

Paola Gavin is the author of *Hazana: Jewish Vegetarian Cooking* (Quadrille Publishing Ltd., 2017)

MAKE THE CONNECTION
Find out more about Paola's botanical world in *Hazana: Jewish Vegetarian Cooking* (Quadrille Publishing, Ltd, 2017)

Opposite Swiss chard (*Beta vulgaris* subsp. *vulgaris*) is a cultivated descendant of wild sea beet (*Beta vulgaris* subsp. *maritima*). Grow cultivars with bright red or yellow leaf stalks as garden ornamentals as well as nutritious vegetables.

MY BOTANICAL WORLD...

'*I have been a vegetarian and a food writer for more than thirty years and am passionate about food and health. Fresh fruits and vegetables are the mainstay of my diet – especially dark green leafy vegetables. One of my favourites is Swiss chard. It is packed full of vitamins, minerals and antioxidants that boost immunity and may help prevent macular degeneration, heart disease and some cancers. It is also very versatile. It can be used in soups, stews, omelettes, rice dishes, as stuffing for pasta and savoury pastries or simply cooked and dressed with extra virgin olive oil, lemon juice and garlic.*'

MY BOTANICAL WORLD...

'*Living a natural life, alongside nature, means stepping outside of the four fixed seasons and constantly observing. It doesn't mean we can't talk of summer, autumn, winter and spring — these are rich terms, full of appropriate meaning — but the key is not being stuck to them on a calendar. Instead, use them as descriptors of a particular time in a particular place, and understand that sometimes none of them will fully apply — there are many times of transition, too.*'

Above Grow your own white turnips (*Brassica rapa* subsp. *rapa*) for their bulbous, vitamin-C-rich taproots but also for their mustardy top leaves.

Matt and Lentil Purbrick, Grown and Gathered, Victoria, Australia

Sweet and spicy turnips

Serves 4

There is a common misconception that all turnips are good for is soup. But they are, in fact, delicious when partnered with the right flavours, and the tops are a fantastic cooking green, a bit like spicy broccoli. Turnips are always more delicious after a frost, because once they get frozen in the ground they become sweeter in flavour.

When boiling or blanching vegetables, always liberally salt the water; it is a wonderful way to bring out their natural flavours and enhance their digestibility, while also ensuring their nutrients don't leach out into the water. Don't worry, it won't make things taste salty, just seasoned. The salt blocks the water-soluble vitamins from crossing the cell walls of the vegetables. And their stunning colour is testament to the nutrient density that is retained.

600g (1⅓ lb) small–medium turnips with tops
Extra virgin olive oil
Unrefined salt

For the reduction
45g (½ cup) sultanas
45g (¼ cup) prunes, pitted
1 tsp unrefined sugar (e.g. rapadura)
1 dried chilli
Juice of ½ lemon
Juice of ½ orange
1 clove
1 star anise

To serve
3 tbsp toasted almonds, roughly chopped
3 tbsp toasted pumpkin seeds
A handful of sultanas

Note
If you can't get turnips with tops for this recipe, use kale as a substitute for the tops. And if you can't get turnips themselves (or swede, turnip's cousin), try the dish with kale and another root vegetable like carrots.

1. Preheat the oven to 180°C/350°F/gas mark 4. Cut any medium-sized turnips in half and leave any smaller ones whole. Add them to a saucepan of boiling salted water (salty like the sea!) and boil for 8–10 minutes until just tender.

2. Drain, pat dry, place on a baking tray and bake in the oven for 5 minutes. Coat the turnips in oil, season with a pinch of salt and roast for 20–30 minutes until soft and golden.

3. While the turnips are cooking, make the reduction. Place the sultanas, prunes, sugar, chilli, lemon juice, orange juice and 125ml (½ cup) of water in a blender and process until smooth.

4. Tip the mixture into a heavy-based saucepan and add the clove and star anise. Bring to the boil then reduce the heat to low. Cook, uncovered, until the mixture thickens (about 20 minutes).

5. About 5 minutes before the reduction is ready, strip the leaves from the stalks of the turnip tops. Cut the leaves coarsely and slice the stalks finely.

6. Heat a big splash of oil in a frying pan over a medium heat. Add the turnip tops and sauté until wilted. Add a splash of water and season, then cover and cook until the tops have softened.

7. To serve, add the cooked turnips and tops to a large bowl and spoon over the reduction. Sprinkle the almonds, pumpkin seeds and sultanas on top and finish with a splash of oil.

From *Grown and Gathered* by Matt and Lentil Purbrick, published by Plum (2016)

MAKE THE CONNECTION
Find out more about Matt and Lentil's botanical world at www.grownandgathered.com.au; @grownandgathered

Pascal Baudar, Urban Outdoor Skills, Los Angeles

Native 'power food' – the energy bar

Makes 1 bar

This power bar is really not complicated to make and can be made with foraged ingredients or commercial chia seeds, pine nuts and other seeds such as sesame, fennel and flax, as well as dates or even dried figs. Local native peoples used to create a similar type of power food with ground dates, berries, nuts and seeds. In the North, where the colder climate dictated a much higher fat content, pemmican (the original survival food) was made using fat, animal protein and berries. The recipes would vary immensely based on the regional ingredients such as acorns, pinyon pine nuts, dates, wild seeds and berries (elderberries, Manzanita, currants and so forth) that would be mixed together in a thick mush, while in the North, pemmican was made using available meat and fat from such animals as elk, buffalo, bear, moose, deer and the local berries.

6 large dates (such as Medjool), pitted
80g (½ cup) chia seeds
1 tsp pine nuts, ground
40g (¼ cup) mixed edible seeds
2 tbsp (30ml) olive or other oil
Dehydrated berries, such as currants, blueberries or elderberries (optional)

1. Using a stone pestle and mortar or blender, crush and mix all the ingredients into a thick paste.

2. Using clean hands, roll the paste into balls or form into a rectangular shape similar to commercial energy bars. It's a bit of a messy and sticky business, but you don't have to do a perfect job shaping the thing.

3. Place in the refrigerator for a few hours or overnight. After chilling, the texture will be harder and you can shape it again more easily.

4. Wrap in baking parchment, and voilà! Your power bar should keep for many weeks in the fridge.

Recipe from *The New Wildcrafted Cuisine: Exploring the Exotic Gastronomy of Local Terroir* by Pascal Baudar (Chelsea Green Publishing Co, 2016)

MAKE THE CONNECTION
Find out more about Pascal's botanical world at www.urbanoutdoorskills.com; @pascalbaudar

Opposite The sweet edible fruit of the drought-tolerant date palm (*Phoenix dactylifera*) has been a staple food of the Middle East and the Indus Valley for millennia. It is one of a few crops that grow in the desert.

MY BOTANICAL WORLD . . .
'There is a very tangible pleasure to be found in wandering through the local wilderness, foraging seasonal wild edibles, and creating from scratch an incredible dish that could not be found anywhere else in the world. You get a feeling of independence from the regular food system, a sense of freedom and choice.'

T. 4. N.º 2.

BOTANICAL DRINKS & BOWLS

The botanical world has inspired numerous beverages, from tea, coffee, tisanes, smoothies, juices and cordials to wine, beer, spirits and cocktails. As civilisations evolved, humans found new ways with which to create drinks from the derivatives of plants. Some were produced for medicinal, religious or social reasons, others for the sheer joy of creating new flavours. While 'botanical drinks' might seem like a new trend, they're actually an extension of thousands of years of thirst-quenching experimentation, first founded in survival – worth raising a toast to.

In the mix
Botanical cocktails

Elderflower (*Sambucus nigra*)
A classic cordial ingredient that goes beautifully with sparkling water, gin, vodka, Prosecco or Champagne.
Goes well with gooseberry, rhubarb, lemon

Hibiscus (*Hibiscus rosa-sinensis*)
Lovely as a ruby-red, fruity-tart exotic tea, equally delicious in cocktails and as an edible flower garnish.
Goes well with peach, blueberry, raspberry

Lavender (*Lavandula*)
Lavender-infused syrup adds a delicate, flowery taste of summer to cocktails, lemonade or soda.
Goes well with honey, black pepper, rosemary

Mint (*Mentha*)
Add a bespoke twist to Mojitos, Mint Juleps or Mai Tais with chocolate, apple, pineapple or orange mint.
Goes well with cucumber, tea, strawberry

Pomegranate (*Punica granatum*)
Traditionally, the red layer of a Tequila Sunrise (via grenadine, see page 242) and lovely as a fresh juice or garnish, too.
Goes well with ginger, orange, lime

Ginger (*Zingiber officinale*)
Use fresh ginger or ginger beer to add spice, warmth and depth to vodka, rum or bourbon-based cocktails.
Goes well with lime, agave, grapefruit

Lemon verbena (*Aloysia citrodora*)
Bruise a leaf and muddle into citrus cocktails. You get a powerful hit of lemon but without the acidity.
Goes well with mint, lime, scented geranium

Lime (*Citrus*)
The sour component and garnish of many traditional cocktails, including the Margarita and the Gimlet.
Goes well with ginger, mint, chocolate

Orange (*Citrus* x *sinensis*)
Add orange bitters for depth, spice and complexity, or fresh orange juice for a fruity cocktail or mocktail.
Goes well with ginger, almond, pomegranate

Rose (*Rosa*)
Rose syrup can bring an alluring note of Turkish delight to cocktails. Prettify with candied petals.
Goes well with pistachio, raspberry, cardamom

> **Sensory notes for botanical cocktails**
> Sweet • Dry • Zesty • Aromatic

Opposite Use lime, a hybrid citrus fruit, to add a sour yet sweet note to cordials, cocktails and juices. Mint (*Mentha*) is another versatile drink ingredient, lovely as fresh mint tea or muddled with sugar syrup in mojitos.

Try four things
Botanical cocktails

Elderflower cordial
Makes 25–30 servings

30 elderflower heads
1kg (5 cups) caster sugar
1.5 litres (6 cups) boiling water
3 unwaxed lemons
2 unwaxed oranges
2 rounded tbsp (50g) citric acid

Made with the flowers of the European elderberry (*Sambucus nigra L.*), this popular summer drink – perfect with still or fizzy water or as a flavouring for cocktails or desserts – can be traced back to Roman times and is still popular across Europe. The plant itself has a much wider spread, being native to Europe, Africa and southwestern Asia, while American elderberry (*Sambucus canadensis*) can be found throughout North America.

1. To prepare, first collect some elderflowers (in the early midsummer months). Avoid polluted roadsides and choose a dry, sunny day, keeping flowers upright to retain the uniquely fragrant pollen.

2. Remove any bugs, trim the stalks and wash the flower heads.

3. Place the sugar in a large pan or bowl and cover with the boiling water to dissolve. Finely zest the lemons and oranges into the mix, then slice the fruit and add the slices along with the citric acid and flower heads. Cover the bowl with a clean cloth and leave to steep for 48 hours.

4. Strain the liquid through a clean, fine muslin cloth into a clean bowl or large jug. The cordial can then be funnelled into sterilised bottles and stored in a cool, dark place for several weeks. Once opened, store in the fridge.

Grenadine
Makes 500ml (2 cups)

4 pomegranates (120ml/½ cup juice per fruit)
400g (2 cups) sugar
1–2 dashes orange flower water

Deep-red, tart-sweet grenadine is a classic cocktail ingredient, most commonly known for the crimson layer of a Tequila Sunrise (tequila, orange juice and grenadine) or the original Sea Breeze (grenadine and gin). The fruit responsible for such deliciousness is the pomegranate (*Punica granatum*). It's easy to buy a bottle of grenadine, but making your own is simple, fun and likely to contain significantly less sugar.

1. You'll need about 500ml (2 cups) of pomegranate juice, which can be store-bought or obtained by deseeding and juicing fresh fruit (in season autumn–winter). To avoid too much mess, cut the top and bottom off the fruit, then gently cut the skin and prise into 6 sections, using the pith as a guide. Use your fingers to pry away the seeds into a bowl of cold water – any pith will float to the top. The juicy arils can then be slowly pulsed in a blender, a cup at a time, to avoid crushing the seeds. Push the results through a fine sieve with a spoon to collect the juice in a bowl below.

2. Combine 500ml (2 cups) of the juice in a pan with the sugar. Bring to a slow boil, stirring constantly until the sugar is dissolved. Reduce the heat, cover and simmer for 10–15 minutes.

3. Allow to cool, then pour into a sterilised bottle. Add orange flower water to taste (not too much), seal the bottle and shake to combine. Store in the fridge and use within 2 weeks.

Sloe gin
Makes 1.5 litres (6½ cups)

500g (3 cups) ripe sloes
250g (1¼ cups) golden caster sugar
1 litre bottle gin

Sloe gin is made from the fruit of the blackthorn (*Prunus spinosa*), native to Europe, Asia and northwest Africa but naturalised in North America and New Zealand, so foraging potential is pretty widespread. It's traditional to pick ripe sloes after the first frost of winter, but check for ripeness by testing if fruit can be squished easily between finger and thumb.

1. Pick enough to half-fill a tall, wide-necked jar (a 2-litre Kilner jar is ideal), washing and pricking drupes with a cocktail stick before use.

2. Add the sugar and gin, shake well, seal and store in a cool, dark place. Turn every day for 2 weeks, then once a week for 2–3 months. Made sufficiently slowly, the sugar and alcohol has time to extract the taste and pink colour of the sloe's flesh but also the unique almond-like flavour of the drupe's stone.

3. The liquor should then be strained through a muslin square, carefully decanted into sterilised containers and left to stand for another week. Fun cocktails include a forager's fizz of sloe gin and Prosecco or concoctions using sprigs of rosemary, cinnamon, cloves, orange or vanilla. Sloe gin also gets better over time, so consider making it one year ahead.

Orange bitters
Makes 625ml (2½ cups)

4 Seville oranges
2 whole cloves
2 whole coriander seeds
1 whole allspice
1 pod cardamom
500ml (2 cups) neutral-grain alcohol
    (vodka or Everclear)

Bitters, typically made with botanical ingredients such as spices, roots, fruit peel and aromatic herbs, can add depth and complexity to a cocktail, working something like black pepper in food. Orange bitters made using Seville oranges (*Citrus x aurantium*) were incredibly popular at the turn of the twentieth century and a key ingredient in cocktails such as the original Martini (gin, sweet vermouth, orange bitters), the Old Fashioned (bourbon, Angostura bitters, orange bitters, a sugar cube) and The Manhattan (whiskey, sweet vermouth, Angostura bitters, orange bitters). They're also experiencing something of a renaissance today, along with lavender and caraway versions.

1. To try your hand at orange bitters, carefully remove the peel from 4 oranges (discarding any pith) and dehydrate in a low oven at 105°C/225°F/gas mark ¼ for 1–2 hours until curling slightly.

2. Place the alcohol, oranges and spices in a sterilised glass bottle. Seal and store in a cool dry place for around 3 weeks to 1 month, agitating periodically. Taste occasionally by placing a few drops in a glass of water.

3. When you're happy with the flavour, strain through a coffee filter into small, sterilised pipette bottles.

Alan McQuillan, The Bloomsbury Distillery, London

Mandarin shrub

Makes about 150ml ($^2/_3$ cup)

The evolution of our western palates in the recent decade has presented a renewed interest in complex flavours for everyone to enjoy and explore. Complexity comes from a variety of areas when it comes to distillation (think Sáke, think Shochu, think the gin craze) but the most interesting avenue of exploration in bars and restaurants is currently coming from the flavours we get through fermentation and pickling.

The impact on spirits and cocktail culture is already apparent in London (where The Bloomsbury Distillery is based), with some of the city's venues fermenting and serving their own hedgerow brews, serving unfiltered wines and finding ways to incorporate herbs, botanicals and spices into their serves to create cocktails that offer something more.

Shrubs – often referred to as drinking vinegars – have a long and storied evolution through the last 300 years of our food and drink culture, but at their very essence they are an acidic beverage (alcoholic or non-alcoholic) incorporating fruit, vinegar and sugar. Shrubs can be used as cordial to be served over ice with soda water or they can be incorporated into cocktail-making as a substitute for citrus elements like lemon juice or similar. There are two main ways to make them, hot and cold:

Hot

1 Heat sugar and vinegar in equal parts on a medium heat for 10 minutes until dissolved.

2 Add fruit and/or herbs and continue to simmer to infuse flavours for a further 15 minutes.

3 Cover and let cool.

4 Strain through muslin or cheesecloth.

5 Seal in a clean glass jar or bottle and refrigerate for three days.

6 Filter again and add more vinegar or sugar to taste.

Cold

1 Combine fruit and vinegar in equal volume in a large, clean glass jar with sealable lid. Add a handful of any desired herb.

2 Shake 100 times.

3 Let stand for one week and shake every day.

4 Strain through muslin or cheesecloth into a clean jar.

5 Add sugar to taste (start with a quarter of the amount of sugar to the total volume of liquid; i.e. 250ml sugar if you have 1 litre of liquid) and shake until completely dissolved. Keep refrigerated.

These two methods will give interesting results for any number of combinations and should be a fun way to experiment. Try the following recipe to get started.

MAKE THE CONNECTION
Find out more about Alan's botanical world at www.bloomsburydistillery.co.uk; @wc1distiller

For a Mandarin shrub

Peels of 10 mandarin fruit

3 sprigs of thyme

480g (2¼ cups) sugar (refined white sugar is a good start but feel free to experiment with unrefined sugars and plant syrups)

150ml (⅔ cup) apple cider vinegar (try to use a vinegar with flavour – spirit vinegars will not add the complexity that makes a shrub interesting to drink)

1. Mix the fruit, herbs and sugar in a steel, lidded pot for 2 days to macerate.

2. Add the vinegar and gently heat through until the sugar dissolves. Do not boil.

3. Strain the mixture into a clean jar or bottle and add additional vinegar or sugar to taste.

4. Keep refrigerated and allow 2 days for the flavour to settle.

5. To serve, mix 25ml of Beefeater Gin (or similar), 30ml Mandarin Shrub and 15ml fresh lemon juice over ice in an 'Old-fashioned' glass. Top up with soda water and add a sprig of thyme. Or try the Mandarin Shrub over ice with soda water and garnish with a slice of orange for an immensely satisfying, long, non-alcoholic serve.

Above Sweet oranges (*Citrus × sinensis*), the spherical fruit of which are illustrated in this 19th-century Japanese woodblock print, are a hybrid of the oblate-fruited Mandarin orange (*Citrus reticulata*) and pomelo (*Citrus maxima*).

BOTANICAL RECIPES
Botanical drinks and bowls

MY BOTANICAL WORLD . . .

'*Plants and their properties, their smells and flavours, are complex universes, revealing culture and terroir that can transport you in an instant. Whether it's tea from a mountain at the top of Sri Lanka, to peat bog from the west of Ireland, plants and nature continue to inspire and drive my inquisitive approach to sharing experience through spirits.*'

Amy Zavatto, wine, spirits and food writer, New York City

Serves 1

DIY gin (Martini)

While gin certainly is its own unique animal in the booze kingdom, it starts out like any neutral spirit: clean, clear and unadulterated. Basically, vodka. What makes gin unique, of course, is all those delicious botanicals infused into it – from the basic to the out-and-out funky. Many new spirit producers are drawn to that unique stamp that can be put on gin; it inspired me, too.

The real way to make gin is to allow the rising vapours that evaporate during distillation to seep through a special, perforated basket that holds the roots, herbs and fruits of choice. But seeing as I don't have a still handy, a workaday ball jar and a little time did just fine. I used botanicals that were dried (juniper berries from my friend Ray Dowd's yard in West Hampton Beach) and fresh rosemary (from my windowsill), and citrus (from, well, the store, since winter is citrus season). I encourage you heartily to come up with your own ideas, too, working in small batches until you create your own signature blend.

Note
The recipe for DIY gin makes a 1-litre (32-oz) bottle or mason jar. Make ahead of time and keep on hand for when 'gin-spiration' hits you.

DIY gin
2½ tbsp dried juniper berries
1 tbsp (15g) coriander
4 cardamom pods, cracked with the thick part of a butcher's knife blade
2 generous peels of lime
2 generous peels of orange or lemon
2 rosemary sprigs
710ml (3 cups) vodka

1 Drop your botanicals and fruit peels into a 1-litre (32-oz) mason jar. Pour in the vodka and allow to sit in a cool, dry place for 24 hours.

2 Strain out the botanicals using a fine mesh sieve. Repeat until all traces are removed.

3 Pour the gin back into the clean mason jar. Keep indefinitely in the sealed jar in a cool, dry place.

DIY gin martini
75ml (⅓ cup) DIY gin
1 tbsp (15ml) dry vermouth
Orange or lemon peel, or sprig of rosemary, to garnish

1 Fill a mixing glass half-full with ice cubes. Pour in the DIY gin and vermouth and stir for 25–30 seconds. Strain slowly into a cocktail glass, and garnish with orange or lemon peel or a sprig of rosemary.

Recipe from *Forager's Cocktails: Botanical Mixology with Fresh, Natural Ingredients* by Amy Zavatto (HarperCollinsPublishers UK/ Sterling Publishing USA, 2015)

MAKE THE CONNECTION
Find out more about Amy's botanical world at www.amyzavatto.com; @amyeats

Above Adding a sprig of fresh rosemary (*Rosmarinus officinalis*) to a gin cocktail delivers a fresh, herbal scent with every sip, plus the woody stem doubles up as a stirring device.

MY BOTANICAL WORLD . . .

'Putting fresh ingredients in your cocktails isn't just a fun weekend experiment, it also makes your drinks better. From the days of the Carthusian monks, reaping a multitude of herbs from their monastic grounds and putting them into tinctures and liquors (the original wild cocktail concocters) for countenance-curing endeavours, the idea of preserving harvestables with spirits is an old trick that has become new again.'

In the mix
Juices, smoothies and bowls

Almond (*Prunus dulcis* syn. *Prunus amygdalus*)
Almond milk is a good dairy-free base for smoothies or bowls, while almond butter provides nuttiness.
Goes well with apricot, banana, orange

Avocado (*Persea americana*)
Take advantage of the avocado for its creamy smoothness, beneficial fatty acids and distinct taste.
Goes well with banana, kale, strawberry

Blueberry (*Vaccinium corymbosum*)
Antioxidant-rich blueberries also impart a rich purple colour. Great combined with other berries.
Goes well with mango, banana, summer berries

Chia (*Salvia hispanica*)
A popular addition to smoothies, juices and bowls, rich in fibre, protein, calcium and magnesium.
Goes well with raspberry, coconut, pomegranate

Apricot (*Prunus armeniaca*)
Use apricots in place of bananas for fibre, body and potassium, or just enjoy the zesty taste on its own.
Goes well with pineapple, strawberry, lemon

Banana (*Musa*)
A classic start-the-day smoothie component, great with berries, milk, yogurt, honey, oatmeal or nuts.
Goes well with honey, strawberry, peach

Carrot (*Daucus carota* subsp. *sativus*)
The ultimate juicing ingredient, rich in immunity-boosting beta carotene, vitamins and minerals.
Goes well with apple, ginger, coriander

Coconut (*Cocos nucifera*)
Get your fix of beneficial coconut via creamy coconut milk, chewy coconut flakes or coconut oil.
Goes well with banana, turmeric, pineapple

Oats (*Avena sativa*)
Add a handful of filling oats directly to smoothies, or eliminate texture by grinding in a food processor first.
Goes well with peanut, honey, raspberry

Spinach (*Spinacia oleracea*)
This leafy superfood is packed with vitamins and iron. Juicing it is one of the healthiest ways to consume it.
Goes well with celery, kale, apple

Sensory notes for juices, smoothies and bowls
Smooth • Uplifting • Fruity • Filling

Opposite Harness the bounty of fresh seasonal fruit and vegetables to create nutritious juices, smoothies and bowls. Keep nuts and seeds at hand to deliver additional vitamins, minerals, healthy fats and a protein boost.

Try four things
Juices, smoothies and bowls

Green machine
Serves 2

100g (2 cups) leafy greens, such as spinach, kale, Swiss chard or romaine lettuce
500ml (2 cups) liquid base, such as coconut water, coconut milk or almond milk
500g (3 cups) ripe fruits such as apples, avocado, mango, pineapple, banana or berries

Green smoothies typically consist of raw, leafy ingredients, such as spinach, kale or Swiss chard, paired with fruit and a liquidising element, such as coconut water or almond milk. Eating your greens has never been so popular, probably due to the sweetening effect of fruits such as mango, banana or berries, and the fact that the green components are blended out of sight – in all but colour, that is. The green smoothie has thus become synonymous with good health, containing immune-boosting doses of vitamin- and mineral-rich ingredients that might not otherwise be embraced.

1. To make, first blend the leafy greens and the liquid base. Then add fruits and blend again.

2. You can also boost with seeds (chia, hemp or flax), oils or butters (coconut, almond or cashew), protein powder or spices (turmeric, cinnamon or ginger root). Have fun experimenting with ingredients and amounts to find your perfect green creation.

Berry almond smoothie
Serves 2

125g (1 cup) frozen raspberries
150g (1 cup) frozen strawberries
100g (1 cup) frozen blueberries
2 medium bananas (roughly chopped)
250ml (1 cup) unsweetened almond milk
Honey or agave, to taste (optional)

Berry smoothies can absolutely be made with fresh or just-picked fruits, but the frozen kind adds instant chill and is still packed with antioxidants. If you find the taste is a little tart after blending, add a drop of honey or agave nectar.

1. Place the frozen raspberries, strawberries and blueberries with the chopped bananas and almond milk in a blender, ensuring there is a gap at the top for movement. Blend until smooth.

2. Add honey or agave to taste and blend gently to combine. Pour into a glass and drink while cold.

You could also make your own almond milk.
1. Simply take 135g (1 cup) raw almonds and cover with an inch of water. Leave overnight or for up to two days in the fridge.

2. Drain and rinse the plumped-up almonds thoroughly under cool running water. Then cover with 500ml (2 cups) water in a blender, pulse several times to break up the nuts and then blast at high speed for 2 minutes. Strain through a muslin square or nut bag to extract the smooth white milk – you should get about 500ml (2 cups) in total.

3. Sweeten with honey or agave, if required. The mealy remnants can also be used on oatmeal, in smoothies, in muffins or baked on a tray covered with baking parchment in a low oven (100°C/212°F/gas mark ¼) for 2–3 hours until dry.

Boost juice
Serves 2

6 carrots
4 apples
2.5cm (1 in) piece root ginger

This classic raw juice recipe combines the sweetness of carrot and apple with the spice of ginger to create a drink that feels warm, cleansing and energising. Indeed, carrots are a rich source of skin-nourishing, immune-boosting vitamin A and potassium, apples are rich in antioxidants and fibre, while ginger is said to be anti-inflammatory and aid digestion. The resulting tangy orange colour also provides a boost that can be especially cheering on a colder, autumnal day.
Lime juice, lemon juice and celery also go well, as does beetroot for a deep red version. Experiment with combinations until you get the perfect taste for you.

1. To make your daily dose, push ingredients through a juicer to extract the liquid. Or you can peel and layer them in a blender with some water and strain the juice out through a muslin cloth or nut bag afterwards. The juice will just be a little more diluted but should taste the same.

Chia breakfast bowl
Serves 2

4 tbsp chia seeds
500ml (2 cups) almond or coconut milk
1 tbsp honey or maple syrup
75g (¾ cup) blueberries
3 fresh apricots, pitted
Handful of toasted coconut flakes

Make your breakfast the night before with a pre-made bowl of grains, seeds, oats, plant-based milk and fruits. A lovely approach is to design your bowl around seasonal ingredients – raspberries, strawberries and drupes in summer; pomegranates or apples in autumn; oranges in winter; or rhubarb in spring, for example. Not only does this bowl taste great, apricots are a good source of vitamin A and C, while blueberries are the king of antioxidant foods. The relatively less well-known chia seed hails from the *Salvia hispanica* plant, a member of the mint family (Lamiaceae). Chia is rich in fibre, protein and omega 3 fatty acids and can help with fullness – the ideal ingredient with which to start the day.

1. Combine the chia seeds with the almond or coconut milk and stir well. Add the syrup or honey.

2. Place in the fridge and leave overnight or for at least 3 hours so that the chia seeds soften and plump up.

3. When you're ready to eat, layer up with blueberries, chopped apricots and coconut flakes.

MY BOTANICAL WORLD...

'*When we're making recipes at the shop it can get quite complicated, with sometimes 50 different ingredients for one batch. The batches are always changing, which is how we keep our relationship with the botanicals alive. We actively encourage tapping into your inner artist and making a recipe your own. In early winter, for example, the wild rosehips are ripe for picking. Cacao butter is a great delivery mechanism for their immune-boosting goodness and tastes amazing too.*'

Charisse Baker, East London Juice Co., London

Rawow chocolate butter

Makes about 600g (2 cups)

If you love store-bought chocolate spread but aren't into refined sugar, vegetable oil, milk solids, emulsifiers and 'flavours', this is for you.

This chocolate nut formula is raw, which makes all the difference. All the natural goodness is kept, and this butter will keep you moving until the very last second. For extra bedroom – or anywhere – energy, add a spoonful of maca root powder (or to taste) along with the other ingredients in step 4. Enjoy it as a spread on your toast, crackers or favourite food base. Melt into a sauce and pour onto ice cream, pudding or a partner. Or add to smoothies, milkshakes or cakes – additions multiply the good times.

70g (4½ tbsp) chopped cacao butter (raw, organic and Fairtrade if possible)
100g (scant ½ cup) virgin coconut oil
250g (2½ cups) dry activated almonds (or nuts of choice)
¼ tsp Himalayan or natural blue salt
140g (¾ cup) coconut sugar
140g (1 cup) cacao powder

Note
Activating the nuts ensures maximum nutritional value and digestibility, but if you are making a last-minute batch, you can skip this step and use untreated nuts. In this case, you may not need to add all of the oil mixture, as the nuts will release their own oils.

Phase 1: Nut preparation

1. Cover 250g (2½ cups) almonds (or other favourite nut, such as hazelnut or walnut) with filtered water and a large pinch of salt. Soak for 1 hour, then rinse the nuts clean. Repeat, soaking a second time for around 12 hours.

2. Spread the activated almonds (or chosen nuts) evenly in a dehydrator or oven heated to 40°C/104°F until dried to original weight prior to activation phase – in this case 250g. This may take 12 hours to 1 day. The almonds are now ready to use in the following recipe.

Phase 2: Making the butter

1. Melt the butter and coconut oil gently in a double boiler, stirring continuously.

2. Blend the nuts together with the salt and three-quarters of the oil mixture. The quantity of ingredients should be adjusted to the size of the blender – the nuts should just cover the blades when placed in the blending jug.

3. Slowly blend until smooth, making sure not to heat above 40°C/104°F as you may lose some benefit of the raw state.

4. Blend together all the remaining ingredients until smooth.

5. Jar it and hide it from your friends at the back of the refrigerator for your own enjoyment for up to 6 weeks.

6. Don't forget to wipe the inside of the jug clean using a slice of soft bread, or just add milk and blend further to create a delicious instant milkshake.

Opposite Cocoa beans are the fermented, dried seeds of the cacao tree (*Theobroma cacao*), from which nutritious pale yellow cocoa butter can be made. The seeds are found within fruit growing directly on the tree trunks.

MAKE THE CONNECTION
Find out more about Charisse's botanical world at www.eastlondonjuice.com; @eastlondonjuiceco

Rawia Baitalmal Edwards, The Secret Sage

Mango coconut raw oat porridge

Makes one breakfast bowl

This filling, naturally sweet and nutritious porridge is the ideal way to kickstart a busy day, not least because you can prepare it the night before. The spices are a wonderful way to support the body, especially turmeric – a powerhouse spice from the ginger family known for its anti-inflammatory, carminative and potent antioxidant properties. Turmeric can be wonderful for gut health and digestive comfort, as can flaxseed, which is full of fibre, rich in omega 3 oil and great at absorbing any residual liquid. Regular helpings of flaxseed can also help improve skin and hair and help balance hormones. All in all, a delicious, guilt-free celebration of the botanical world.

For the porridge
100g (1 cup) rolled oats
1 ripe banana, mashed
2 tsp ground flaxseed (crushed gold and brown, or seeds – find your preference)
½–1 tsp turmeric
½ tsp grated nutmeg
½ tsp ground cinnamon
Pinch of salt
250ml (1 cup) coconut milk (if you prefer less richness you can put half/half with water)

For the dairy-free mango cardamom lassi
5–6 crushed cardamom pods (or fewer to taste)
1 ripe mango
1 tsp good-quality honey, maple syrup or your preferred natural sweetener
1 x 400ml tin full-fat coconut milk

For the breakfast bowl
1 ripe banana, sliced
Toppings of your choice

Phase 1: The night before

1. First make your overnight raw rolled oat porridge. Place the oats, mashed banana, flax, spices, salt and coconut milk in a bowl. Evenly mix, then cover and leave to soak overnight.

2. Next prepare your dairy-free mango cardamom lassi. Place all the ingredients in a blender and blend until smooth. Place in the fridge overnight to set.

Phase 2: The morning after

1. To assemble your breakfast bowl, first line a ceramic or glass breakfast bowl with sliced banana. Secure the bananas in place with a large tablespoon of overnight rolled oat porridge followed by a layer of mango lassi. Repeat, layering oat mix and lassi until you have the desired portion, finishing with a lassi layer.

2. Sprinkle with toppings such as crushed cardamom, crushed pistachio, desiccated coconut, fresh or freeze-dried raspberries, chunks of mango, soaked goji berries, almond flakes or hulled hemp and savour every mouthful, making time to digest as well as eat.

MAKE THE CONNECTION
Find out more about Rawia's botanical world at www.thesecretsage.com; @thesecretsage

Opposite Common oats (*Avena sativa*) were domesticated around 3000 years ago, in the wetter, colder conditions of Europe. Not only are they a comforting breakfast food, oats are also great for skin too (see page 264).

MY BOTANICAL WORLD...

'*Plants are amazing organisms, capable of transforming sunlight into food energy. Imagine what that power can do for our bodies? A respectful, more mindful approach to the food we consume helps foster these connections and a healthy future for humans and the Earth. Plants are my "happy" – out in nature or in the kitchen, experimenting with all the wonderful and nutritious ingredients that the botanical world provides.*'

'Everything on the Earth has a purpose, every disease an herb to cure it, and every person a mission. This is the Indian theory of existence.'

Hum–Ishu–Ma or Mourning Dove, born Christine Quintasket,
Native American author (1884–1936)

CHAPTER FIVE

BOTANICAL REMEDIES

NATURE'S APOTHECARY

Explore the healing and transformative power of nature for health, wellbeing, beauty and home use by immersing yourself in the wonders of the Plant Kingdom. Setting up a home apothecary is a lovely way to begin that journey, and to add an element of plant-based peace to your home and daily routine.

There are an increasing number of modern herbals being produced on the medicinal, therapeutic and rejuvenating benefits of plants – many of them outlined in the pages to follow, along with inspiring remedies from the wonderful herbalists and aromatherapists who have taken time to create and share them. But there's no better way to acquaint yourself with such knowledge and wisdom than to immerse yourself in nature. Take a walk, get to know the plants in your local vicinity or garden and learn to identify the species that are well known for their medicinal properties – under the tutelage of an expert if you can (see Further Reading, page 392).

Choose a specimen that intuitively appeals to you – something simple, such as a herb that is known to promote calm (lavender), sleep (chamomile) or better digestion (peppermint or fennel), for example. Then gather your herb – with care to the wild (always get permission if required, avoid protected specimens, don't pick too much and give thanks to the plant in question) – or from a reputable vendor that bears sustainability in mind.

Dry herbs if necessary, laying them out on drying racks, hanging them in small bunches, placing them in the oven on low heat or in a dehydrator. Once dried, store in an appropriate glass jar, label with common name, botanical name (see page 124) and the date and location where they were collected, and store in a cool, dark place to protect sensitive ingredients. You are now ready to try your hand at preparing a few basic remedies – an infusion (see page 271) or aromatic massage oil (see page 271 or 284) perhaps.

Once you've mastered a remedy or become familiar with a herb, move on to the next one. Soon you'll find that your home apothecary has grown organically to include numerous jars of herbs or shelves of successful preparations for you or loved ones. You'll also find that you become familiar with commonly used apothecary-worthy ingredients, too, such as plant-based essential oils, carrier oils, salts, butters, waxes and clays – plus any vital rules of use.

Notice how your knowledge of plants increases accordingly. There are, for instance, optimum times to pick leaves, flowers, stems and roots, which requires a greater awareness of the seasons, plant morphology (see page 92) and plant anatomy (see page 88). Buds and flowers are best harvested just as they are opening, leaves often before a plant is in full bloom (although there are exceptions), and roots or bulbs in autumn or spring when the energy of the plant remains below ground. Some 'magical' potions may also require picking herbs in tune with the moon, or as part of an intention-setting ritual.

Most of all, learn to trust your instincts and enjoy the unfurling of your wisdom as you become more familiar with aromas, flavours, various medicinal properties and indeed your own senses through plant identification, remedy preparation, aromatherapy or massage. Remedy your way to better health and wellbeing, and enhanced beauty and a more restful home will most likely follow.

Opposite The 1224 Arabic edition of *De Materia Medica* by Greek physician and botanist Pedanius Dioscorides (c. 40–90 CE) shows two physicians harvesting a medicinal plant.

الصنفُ الرابعُ الذي يقال له الملوس بوسُن فانَّ له ورقًا شبيهًا
بورق الاخرُ غيرَ انَّه قريبُ الشكل من ورق بقلة الحمقا الاندا ادقُّ
منه واندا اسغندان وله قضبان اربعة او خمسة مخرجها من اصل

واحدٍ طولُها نحوٌ من شبرٍ وثمارٌ ململمة من لبن وله رأسٌ شبيهةٌ برأسِ
الشبت وثمرانه موضوعٌ في رؤوسٍ وجمة هذا النبات
سقلا مع اسفل الثمن ولذلك سُمّي الپوسقويون ومعنا

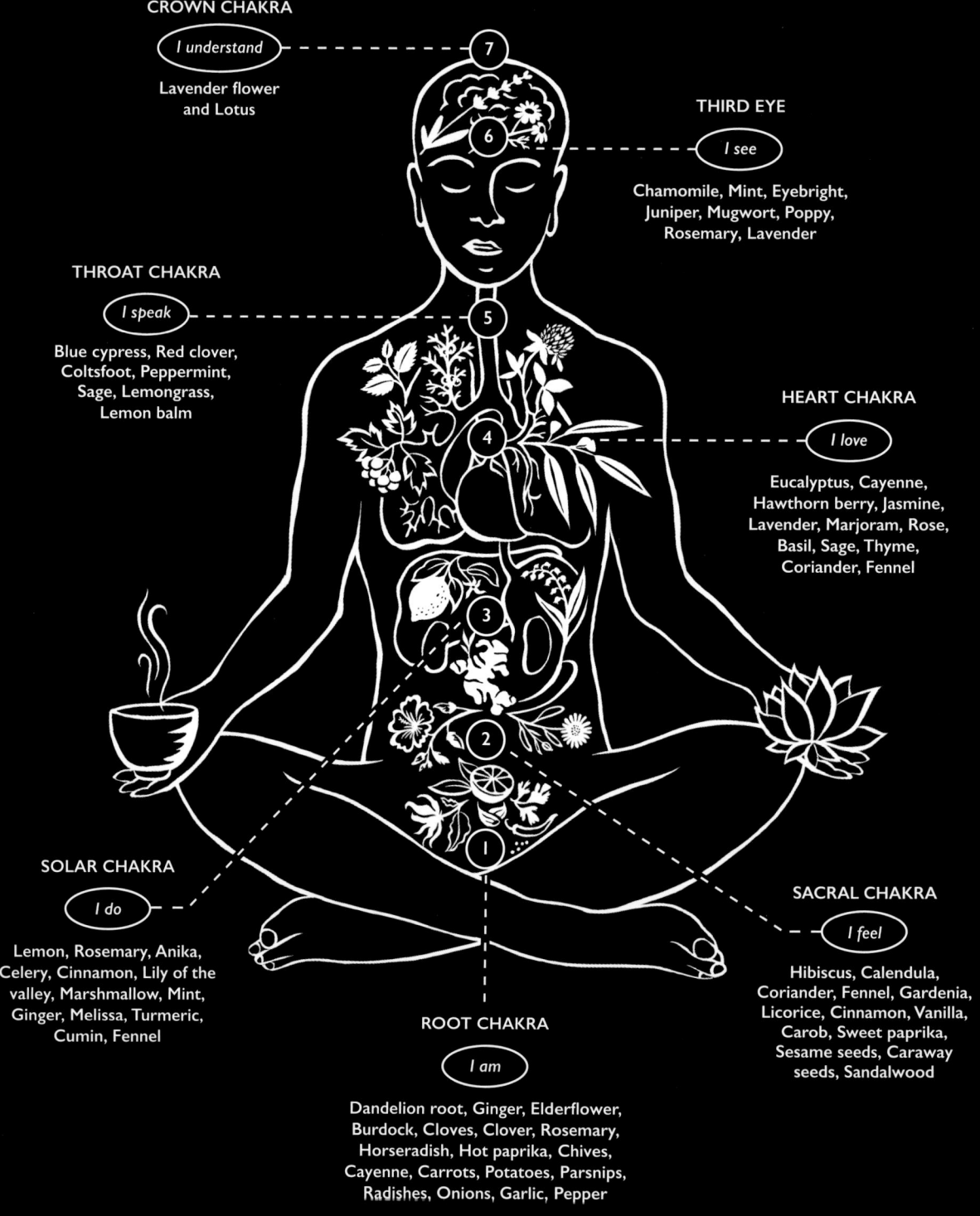

THE HEALING POWER OF PLANTS

Of the many connections between plants and people, one of the most ancient and vital is the use of plants to heal, leading to millennia of remedies for health, wellbeing and beauty, and paving the way for much of modern medicine.

At its most ancient, the medicinal properties of plants are believed to have been harnessed by early Prehistoric hominids using local specimens or those accessed by a nomadic hunter-gatherer lifestyle. Around three million years later, Kew's second *State of the World's Plants* report (2017) stated that some 28,187 known plant species are recorded as being of medicinal use. Of these, 4,478 specimens are cited in a medicinal regulatory publication such as a modern pharmacopoeia, a significant enough amount to beg the much-debated question: how did humans discover all these wonderful properties?

The most obvious explanation is that humans must have discovered such medicinal herbs and the most therapeutic preparations of them empirically, through millennia of accidental discovery, observation and trial and error. The same premise would have applied to which plants could be eaten as foods. Indeed, many well known medicinal herbs such as peppermint and rosemary have culinary properties, too.

The other route to the potential healing power of plants stems from the notion that people received this knowledge directly from plants. Such received wisdom was the result of communing with plant spirits, often via a traditional or spiritual healer such as a medicine person or shaman. It's an abstract idea, but not without foundation. Of the 4,478 medicinally regulated plants in Kew's report, many have curative qualities first rooted in the transcendental art rather the science of plant-based healing. At the same time, those who work with plants or their derivatives – through gardening, cooking or, most relevantly, herbalism – will most probably report a certain amount of intuition where their usage is concerned, along with a boost in wellbeing from communing with nature.

Consider, too, the way in which plants affect us on a cellular level or in relation to certain body parts, organs or systems, resulting in physiological and emotional healing. This surely deserves room for a bit of magical thinking – the ultimate proof of just how connected plants and people are.

Whichever school of thought you subscribe to, plants are undeniably linked to human survival, longevity and quality of life, from disease-busting drugs such as quinine (from *Cinchona officinalis*), painkillers such as morphine (from *Papaver somniferum*) and circulation-boosters such as ginger (*Zingiber officinale*) to multi-tasking herbs such as lavender (*Lavandula spp.*) – relaxant, skin tonic and antibacterial among its talents.

Modern medicine is vital to the treatment of many life-threatening diseases or conditions, but time-honoured practices such as herbalism, aromatherapy, Chinese herbalism and Ayurveda also have their place, not least by empowering people to self-heal (for minor ailments at least), prevent rather than cure, potentially gather materials directly from nature and, importantly, treat body and mind holistically for optimal health, wellbeing and happiness – chakras be open, all praise to the botanical world …

Opposite According to Ancient Indian wisdom there are seven spiritual energy centres in the human body – the chakras. One way to support them is with therapeutic herbs.

MAKING PREPARATIONS

One way to create a home apothecary is to gather together key tools plus some basic ingredients, including herbs, oils, waxes, butters, clays, salts, vinegars and honey. Another, more intuitive, way is to get out into your garden or the countryside and become more familiar with your botanical world. Take the route that works for you, savouring those moments where the real alchemy takes place: the satisfaction of plant identification, the creativity of preparing remedies, the power to enhance the human experience, the joy in communing with nature.

The home apothecary
Basic ingredients

Herbs
Medicinal herbs form the backbone of an apothecary, gathered sustainably from nature or sourced from a responsible supplier – read on to find herbs suitable for health, wellbeing, beauty and the home.

Essential oils
Essential oils are also key to many herbal remedies and, of course, aromatherapy. *The Complete Aromatherapy and Essential Oils Sourcebook* (2018) by leading British aromatherapist Julia Lawless, is an ideal introduction to their properties and uses.

Carrier oils
Ideal for massage, 'carrying' essential oils onto skin or hair or whipping up into moisturising 'butter', experiment with grapeseed, sweet almond, avocado, coconut, argan, castor, macadamia nut, rosehip, sesame, olive, wheatgerm or black seed oils.

Clays
Use clay as a base for masks to help absorb impurities and soothe and heal skin. Choose from bentonite, fuller's earth, kaolin, French green or Moroccan rhassoul clay, depending on skin type or condition.

Vinegar
Apple cider vinegar can help aid digestion, improve immunity, soothe skin and increase mineral absorption. It can also be used to make alcohol-free tinctures. Neutralising white vinegar is ideal for home cleaning.

Butters
Plant-based butters can be melted, whipped or stirred into lotions, lip balms, salves, soaps and other bodycare recipes. Choose from commonly used shea or cocoa butter, or antioxidant mango.

Waxes
Beeswax is a natural emollient and can help soften the skin. It is also useful as a thickener in recipes and is perfect for making scented candles. Vegan alternatives include soy and candelilla wax.

Salts
Dead sea salt, Epsom salts and Himalayan pink salt are all packed with soothing and detoxifying minerals. Gather a selection of these beneficial salts for making salt scrubs, salt soaks, bath tea and bath bombs.

Flower waters
Flower waters (or hydrosols) are produced by distilling leaves, fruits, flowers or other plant materials and are useful for perfumes, skincare and cleaning recipes. Try lavender, rose, peppermint or chamomile.

Raw honey
Raw honey is antibacterial, antifungal, packed with antioxidants, vitamins and minerals and, when sourced locally, can help with pollen allergies. Use it in pastilles, elixirs, oxymels, teas or infusions, or apply topically to skin.

Opposite The most important element of a home apothecary is to really know your medicinal plants. Setting a space up to enable the gathering of this wisdom, and reflect who you are, can be a wonderful part of this journey.

Alcohol
Alcohol is a convenient and effective menstruum (solvent) for creating herbal tinctures. In most cases, a 50 per cent (100-proof) vodka is ideal. Brandy, high-proof grain alcohol or vinegar can also be used.

Vegetable glycerin
This colourless, odourless, viscous sugar alcohol is emollient, sweet-tasting and ideal for skin, hair, nails and lips. It is usually derived from the somewhat controversial palm oil, or palm-free soy, rapeseed or coconut oil.

Oatmeal
Oats are naturally anti-inflammatory and can be added to simple face masks or tied into muslin squares to help cleanse, soothe and heal skin conditions. Nutritious oats can also aid digestion.

Lemons
Naturally alkalising, packed with vitamin C and wonderfully scented, lemons can help energise, boost immunity, cleanse, tone, soothe and heal. A bowl of lemons – organic, preferably – will always be useful.

Activated charcoal
Activated charcoal is extremely porous, making it useful for drawing toxins from your body or the home, or to whiten teeth. Source a brand that's made from charred bamboo or coconut shells rather than fossil fuels.

Bicarbonate of soda
Sodium bicarbonate is a white, gently abrasive, alkaline powder that reacts with acids to produce carbon dioxide. Use it in home cleaning remedies, toothpaste or neutralising mouthwashes and deodorants.

Sugar
Granulated sugar is ideal for creating gently exfoliating skin scrubs. It's also handy for creating restorative syrups from hips, haws and berries. Source raw, organic cane sugar whenever possible.

Castile soap
This non-toxic, grease-grabbing, olive-oil-based hard or liquid soap is named after the Castile region in Spain, where it was historically produced using olive and laurel oil. It's ideal for skin, hair and use in the home.

Witch hazel
Native American peoples used a distillate of naturally astringent witch hazel (*Hamamelis virginiana*) to treat wounds, bruises, sores and swellings. Use for soothing and healing toners, misters, sanitisers and liniments.

Aloe vera
Antiviral, antibacterial, mineral-rich Aloe vera has been used for millennia, to soothe skin, cleanse hair, improve digestion and heal burns. Source organic, as pure as you can get, gel or juice, or harvest your own.

Opposite Find a warm, dry place to dry aromatic and healing herbs and flowers such as thyme (*Thymus*). Ensure that the diameter of your herb bundle is not too large to allow for good airflow and discourage mildew.

Pflanzen. LXIV.   Plantes. LXIV.

Fig. 1.
Fig. 2.

The home apothecary
Key tools

Expert guidance
Learn from and be inspired by reputable experts via books such as *Rosemary Gladstar's Medicinal Herbs: A Beginner's Guide* (2012) – or consider joining an accredited course. Medicinal herbs can be powerful substances.

Awareness
Be plant aware by picking or sourcing herbs sustainably, and observing legal codes of conduct where necessary. An increased awareness of basic human anatomy can also help in the application of remedies.

Intuition
Trust your intuition. As your practical knowledge of botanicals and remedies grows, so should the courage to tap into the wild wisdom that lies within. Don't be afraid to commune with and give thanks to nature.

Storage containers
Collect a variety of containers for storing herbs and dispensing remedies, such as glass mason jars, dark-glass or amber 'medicinal' bottles, tins and salve jars. It's fun to scour thrift shops for some of these.

Lids and droppers
Some preparations require storage vessels to be airtight – lever-armed, rubber-sealed preserving jars are ideal. Plastic lids can be easily cleaned, while pipette lids are also handy for tinctures and oils.

Bowls, pots and pans
When making herbal remedies, choose bowls, pots or pans made from non-reactive and non-toxic materials such as glass, stainless-steel, ceramic, marble, wood or cast-iron. Never use aluminium or plastic.

Measuring and mixing utensils
Utensils such as cups, funnels, spoons and scoops should also be made out of non-reactive and non-toxic materials. Glass measuring jars and funnels are handy, as are stainless-steel and wooden spoons.

Fine-mesh strainer
This is useful when straining infusions. Choose one that incorporates a funnel for easy decanting. Squares of cheesecloth, muslin or hemp fabric are also ideal for straining.

Muslin cloth
Muslin can be indispensable in the home apothecary and adds a nice traditional touch. Use large squares to strain liquids and smaller ones to squeeze lemons through, or fashion into tea or bath sponge bags.

Grater
Keep a large grater on hand that you use only for grating waxes or soaps – these materials can be quite hard to clean off. Set aside a smaller grater or zester for grating lemons or other food items.

Grinder
Some herbalists choose to grind up their herbs with a traditional pestle and mortar, others with a small food processor or coffee grinder (again, if you are using a coffee blender, keep one for this purpose only to prevent the coffee flavour transferring). A stick blender can also be useful for formulations such as body butter.

Opposite The shea tree (*Vitellaria paradox*; fig. 1), native to Africa, shown here alongside the drumstick tree (*Moringa oleifera*; fig. 2) produces a pulpy, plum-like fruit that houses the oil-rich seed from which shea butter is extracted.

Capsule machine
This handy machine can turn dried herbs or other matter into uniform capsules – ideal for preparations that don't taste so good. You can also fill capsules (gelatin or vegetarian) by hand.

Herbal inventory
It's a good idea to keep an inventory of ingredients and preparations in your apothecary to keep track of collection and use-by dates or items that need replenishing. A notebook or index cards work well.

Tea infuser
A mesh tea-infusion ball makes it easy to prepare loose leaf tea, one cup at a time. Individual portions of dried herbs can also be placed in muslin bags. For multiple cups, try a glass teapot infuser in one.

Scissors and secateurs
Set aside scissors and secateurs to use just for cutting herbs. Keep them clean to avoid cross-contamination and sharp to avoid unnecessary damage to plants that you harvest or forage from.

Double boiler
This is simply a pot that can be placed over a pan of boiling water to melt the substance in it above, often with pouring spouts at either end. You could also just use a glass bowl that fits safely over a pan.

Labels and pens
Get into the habit of labelling your herbs and preparations from the start. This makes it easier to find the appropriate herbs, oils or other ingredients and successfully reproduce winning formulas.

Small kitchen scale
Sometimes recipes call for more specific weights or measurements – not all 'handfuls' of herbs weigh the same – in which case a small digital kitchen scale can be useful.

Basket
Keep a basket handy for moments when you have the impulse or need to go and forage for your own ingredients in your garden or, with permission, in the wild (see page 161). Let nature inspire you.

Herb dryer
Ensure dried herbs are fully dehydrated – they should crumble easily and be crispy to touch. Dry yours with a drying screen (DIY it with a wooden frame and mesh), a dehydrator, or hang them in bunches in a cool dry place, upside down.

Opposite Sweet-smelling coconut oil is extracted from the meat of mature coconuts harvested from the coconut palm (*Cocos nucifera*). Use on hair or skin to deliver essential fatty acids, which can help nourish and moisturise.

Try four things
Making preparations

Herbal tincture
Makes enough to fill 1 (450-g/16-oz) mason or jam jar

Chosen herb or herbs (pesticide free and cleaned)
40–50 per cent (80–100-proof) alcohol, such as vodka, gin, brandy or rum, or vegetable glycerin or apple cider vinegar – enough to cover herbs in chosen jar

A tincture is a very concentrated liquid extract of beneficial herbs, usually made with an alcohol solvent, although non-alcohol preparations can be made using vegetable glycerin or apple cider vinegar. It's a quick and easy way to get the health benefits of herbs into your system via a dropperful or two in warm water, tea or juice (while you can drink tinctures straight, they don't tend to taste great), and once made they can last for years. Vinegar preparations don't last as long (one year, possibly) but where appropriate can be incorporated into meals (salad dressing, for example). Likewise, food-grade vegetable glycerin is sweet-tasting and can be easier to prescribe to children or sneak into food (this type of tincture should last for 2–3 years, while alcohol-based tinctures can last for 5–10 years or more if stored correctly).

1. To make your tincture, simply chop the selected herbs finely and place in a sterilised glass jar with a tight-fitting lid. Pour the alcohol, vinegar or glycerin over to cover, leaving a 5–7.5cm (2–3 in) margin or liquid at the top.

2. Seal the jar and place in a warm, sunny spot for 4–6 weeks to allow the herbs to macerate.

3. Shake daily to help the mixing process along. When ready, strain the liquid through a strainer or muslin cloth into another clean, sterilised glass jar with a tight-fitting lid.

4. Now store in a cool, dark spot and use when required serving with a teaspoon or pipette to dispense the correct amount.

Herbal infusion (steeping)
Makes around 1 litre (4½ cups)

4–6 tbsp dried or 6–8 tbsp fresh leaves, flowers or buds, or select berries, seeds and other aromatic plant parts
1 litre (4½ cups) boiling water

Leaves, flowers and buds are generally steeped in boiling water, rather than cooked, so as not to destroy important medicinal components such as enzymes, vitamins and essential oils. The end product is called a herbal infusion. The longer such components are steeped, the stronger the infusion becomes. This is not always advised, however, as extra time can bring out unwanted flavours.

1. To create an infusion, select dried or freshly picked herbs and place into a 1-litre (32-oz) glass mason jar or centre area of a similar-sized straining teapot. Cover with boiling water and leave to steep for 30–40 minutes.

2. If prepared in a jar, strain through a fine-mesh sieve or muslin square into a clean jar, then serve by the cup. Or remove herbs from the central portion of the teapot and pour as desired. The used herbs can be placed on the compost pile.

3. Your infusion can be kept for 1–2 days and reheated if not used all at once. If it tastes stale or flat or bubbles form, make a new pot.

Herbal decoction (simmering)

Makes around 1 litre (4½ cups)

4–6 tbsp dried or 6–8 tbsp fresh fibrous or woody plant parts, such as roots, bark, twiggy parts and some seeds and nuts

In order to extract the less forthcoming beneficial constituents of roots, bark, twiggy elements and some seeds and nuts, slow simmering in lightly boiling water is required. This end product is called a decoction.

1. Simply place the selected fresh or dried plant parts in a non-reactive pan with 1 litre (4½ cups) cold water and, on a low heat, bring to a slow simmer. Cover and continue simmering for 25–45 minutes.

2. Strain into a clean jar using a fine-mesh sieve or muslin cloth and serve by the cup. For a stronger decoction, simmer the herbs for 20–30 minutes, then set the jar aside to infuse further overnight. This may affect the size of the serving, as more concentrated decoctions will potentially be stronger in flavour and prescription.

Note
Infusions and decoctions can also be made without boiling the water, by warming the infusing mixture in a glass jar in direct sunlight for several hours – a lovely way to while away the time outside using pure solar power.

Infused oil

Makes enough to fill 1 (450-g/16-oz) mason or jam jar

Chosen herbs (ideally as dry as possible)
Extra virgin or high-quality olive oil or other vegetable oil, such as jojoba or coconut oil

Infused oils can be used as remedies or as therapeutic bath or massage oils, depending on the key herbal ingredients and the oil chosen to be infused. Olive oil is a good choice as it's stable, doesn't become rancid easily and brings beneficial qualities of its own, but a good-quality carrier oil such as jojoba or coconut oil may be more pleasant for massage or skincare. Mix and match herbs and oil to see what suits you, choosing edible and food-grade ingredients for anything that will be ingested.

1. The quick method is to chop the herbs and place them in a double boiler (or suitably fitting glass bowl) on top of a pan of water. Cover the herbs with 2.5–5cm (1–2 in) of your preferred oil, heating until the oil reaches a slow simmer. Keep simmering for 30–60 minutes, checking frequently to ensure that the oil doesn't overheat.

2. When it's golden or golden-green and smells beautifully fragrant or aromatic, remove from the heat. Strain into the sterilised jar through a fine-mesh strainer or muslin cloth, if required.

3. Use when cooled.

Note
Infused oil can also be made by covering herbs with 2.5–5cm (1–2 in) of oil in a sterilised (450-g/16-oz), tightly-sealed mason or jam jar and leaving in a warm, sunny spot for 2 weeks. For best results, replenish the herbs after that time and leave for another 2 weeks.

HEALTH

Humans have been using medicinal plants for millennia. Some botanical remedies are relatively well known and widely used, such as peppermint for digestion, lavender for relaxation, or lemon and honey to help ease the symptoms of cold or flu. Others are beginning to find their way back into the mainstream as more people embrace the benefits of holistic healing practices. From therapeutic teas and salves to immune-boosting syrups and vinegars there's a wealth of plant-based wisdom to tap into, benefit from and hopefully share with others.

Select ingredients
The health edit

Onion (*Allium cepa*)
A mainstay of the household remedy chest since ancient times, onion stimulates the circulation, to help the body 'weep' out toxins and infection. It can be taken orally, applied raw or as a poultice.

Turmeric (*Curcuma longa*)
In Ayurveda, turmeric is revered as a symbol of prosperity, a cleanser and an immunity booster for all systems of the body. Use as a spice or oil infusion for digestive issues, colds and joint pain.

Peppermint (*Mentha* x *piperita*)
Peppermint is most often used as a herbal leaf tea or as a refreshing essential oil. In its appropriate form, peppermint is used for digestive issues, as an expectorant, and to refresh and cool skin.

Chamomile (*Chamaemelum nobile*)
Roman (*Chamaemelum nobile*) or German (*Matricaria chamomilla*) chamomile flowers can be used as tea to help reduce anxiety or digestive upset. Essential oils of both can help soothe skin.

Nettle (*Urtica dioica*)
Nettle leaves can be used as an iron- and vitamin-rich culinary ingredient, taken as a mildly diuretic infusion or rubbed onto stiff or arthritic joints or insect bites to counteract pain.

Echinacea (*Echinacea purpurea*)
Echinacea can help build the immune system, and fight off disease and infection. Harvest leaves, buds, flowers and roots over a growing season for a restorative whole-plant tincture.

Garlic (*Allium sativum*)
Naturally antiseptic and antibiotic, garlic is an exceptional respiratory disinfectant that can be taken raw, pickled, in milk or as an oil infusion.

Yarrow (*Achillea millefolium*)
Historically, antiseptic, anti-inflammatory and astringent yarrow was used to treat soldiers' wounds. It's still useful to help stem bleeding or swelling, or as a tea to help reduce fever.

Elder (*Sambucus nigra*)
Elderflower ointment has long been a remedy for chilblains and chapped hands, while tea can be brewed for hayfever and sinusitis. Try elderberry syrup or remedy (page 226) to help fight colds.

Ginger (*Zingiber officinale*)
Use warming, stimulating ginger for circulation, to help ease sickness and for menstrual cramps. Prepare edible preparations using the fresh rhizome; topical ones with essential oil.

Fennel (*Foeniculum vulgare*)
Fennel is a popular culinary herb that, as a tea, can help to soothe digestive issues or lift the spirits. As an essential oil it can be stimulating, balancing, restorative and cleansing.

Eucalyptus (*Eucalyptus globulus*)
The Australian blue gum eucalyptus is commonly used to produce the essential oil, a powerful antiseptic and renowned decongestant that's used mainly for coughs and colds.

Licorice (*Glycyrrhiza glabra*)
The glycyrrhizic acid in licorice is fifty times sweeter than sugar – eat the root straight for a natural candy. Licorice can also help soothe inflamed throat or bowels, or balance hormones.

Ginseng (*Panax ginseng/Eleutherococcus senticosus*)
Widely used in the Far East, ginseng is said to replenish vital energy, strengthen the immune system and increase concentration. Use in powdered form to make a tonic wine.

Opposite Plant a patch of German (*Chamaemelum nobile*) or Roman (*Matricaria chamomilla*) chamomile in your garden for the sweetest daisy blooms, which can be harvested and dried throughout summer to make tea.

Try four things
The health edit

Herbal poultice/compress
Makes 1 poultice/compress

Mashed or grated fresh herbs or
  dried powdered herbs
Cotton flannel or cloth

In *Medicinal Herbs* (2012), Rosemary Gladstar describes one of her earliest teachers, the great English herbalist Juliette de Baïracli Levy, as being especially fond of the 'laying on of leaves', as she called poultices and compresses. Both can work to draw out impurities from the skin, increase circulation and in some cases reduce inflammation.

1. Add enough boiling water to mashed or grated fresh or powdered herbs to make a paste or pulp.

2. Apply the paste directly to the skin or fold into a cotton flannel or bandage and place in the appropriate spot – on a bruise, wound, insect bite or sore muscles, for example – for up to an hour. Placing a hot water bottle or heat pad on top can help keep the heat in.

Note
A less potent delivery of herbs can be made via a compress, which is a linen or muslin cloth soaked in a double- or triple-strength herbal infusion or decoction. The compress can be held in place by a bandage or wrap for up to 30–45 minutes. A hot compress can help draw blood to the skin's surface and therefore draw out impurities. A cold compress, either cooled in the fridge or with ice cubes, can help reduce inflammation and swelling. This can help with ailments such as sunburn, swollen glands and mastitis.

Herbal inhalation/facial steam
Makes 1 treatment

Herbal infusion or decoction – a handful of fresh
  or dried herbs or 2–3 drops of essential oil
Bowl of steaming water
Bath towel

Warm, moist air can relieve many respiratory problems and allow the healing power of plants and other products to enter the bloodstream through the lungs.

1. To prepare an inhalation, half-fill a big bowl with steaming water, then add your chosen herbal infusion or decoction, or 2–3 drops of essential oil.

2. Place your face about 15cm (6 in) above the bowl and inhale the steam. It can be easier to achieve a healing herb vapour by draping a towel over your head to create a tent. Around 5–7 minutes at a time should be enough to ease congestion.

Warm and moist steam can also help promote sleep, so this is an ideal remedy to action before bedtime. Try thyme, sage, rosemary, ginger, orange peel and eucalyptus to help combat symptoms of colds and bronchial infections such as coughs, congested noses or sinuses, and sore throats. Or use calendula, Roman chamomile, rose, lavender, citrus or peppermint for a facial steam that can help open the pores – to remove impurities and allow herbal properties to infiltrate skin – and stimulate circulation. In this case, follow with a natural cleanser, scrub or mask, with a splash of cold water or toner at the end to close the pores again.

Herbal syrup

Makes around 1 litre (4½ cups)

1 litre (4½ cups) herbal tea or decoction (see page 271)
350–700g (1–2 cups) honey or other sweetener, such as maple syrup, vegetable glycerin or brown sugar for every 500ml (2 cups) of concentrated herbal tea or decoction

Syrups are essentially cooked-down and concentrated infusions or decoctions with added sweetener, such as sugar or honey. Sweetening some herbs can often make them infinitely more palatable – especially useful for children, the elderly or reluctant medicine-takers – while helping to preserve their beneficial qualities and use-by date at the same time.

1. First simmer 1 litre (4½ cups) of a herbal decoction or infusion down over a low heat to around half the original volume. Measure to check you have around 500ml (2 cups).

2. Strain any herbs from the pot (if relevant). Then add 350g (1 cup) of honey or another sweetener – up to 700g (2 cups) if you prefer a sweeter syrup.

3. Warm the mixture over a low heat, stirring well until the honey or sweetener combines with the concentrate. Decant into a 1-litre (32-oz) sterilised bottle and then store in the refrigerator for up to 3 months.

Herbal salve/ointment

Makes around 10 portions of salve, in 10 (28-g/1-oz) glass jars or non-reactive tins

250ml (1 cup) herbal oil (see note)
50g (¼ cup) shaved beeswax

Another way to apply herbs is via an emollient salve or ointment, a combination of medicinal herbs, vegetable oil and beeswax.

1. Measure out the oil – for every cup of infused oil you will need 50g (¼ cup) shaved beeswax.

2. Heat the shaved beeswax in a pan over a very low heat, stirring occasionally until it has melted. Pour the oil over the top and melt together, stirring gently.

3. When mixed, test the consistency of your salve by placing a spoonful in the freezer for 1–2 minutes or fridge for 10–15 minutes to cool. If you want a harder salve, add more beeswax, for a softer one, add more oil.

4. When the consistency is ideal, remove from the heat and pour into the glass jars or non-reactive tins.

5. Store in a cool, dark place. Use salve as required to soothe or heal dry or sore skin, ease aches and pains, or for massage. For the latter, coconut oil is particularly good as it melts easily on contact with warm hands or skin.

Note
If you don't have a pre-made infused oil to hand, make one by placing your chosen herbs in a double boiler under 2.5–5cm (1–2 in) of olive oil (or other vegetable oil) and bringing to a gentle simmer on a low heat until a golden or golden-green oil is produced. This can take 30–60 minutes. Strain out the herbs.

BOTANICAL REMEDIES
The health edit

MY BOTANICAL WORLD . . .

'*It seems that as one begins to study herbs, the plant's essence infuses one's entire life with joy. People become happier, healthier, more in balance and in tune with their inner dreams. The beauty of the herbs work their gentle magic on the heart of the user.*'

Rosemary Gladstar, Herbalist, Sage Mountain Retreat Center and Native Plant Preserve, Vermont

Fire cider vinegar

Makes 1 litre (4½ cups)

This is my favourite herbal vinegar, first concocted in the kitchen at the California School of Herbal Studies, where I taught between 1978 and 1987. I was constantly experimenting and concocting medicinal herbs into a variety of recipes; those that turned out well were shared freely with my students, our community, and later as I travelled around. The idea was to bring medicinal herbalism back into people's kitchens, as part of their food and as a way of being, not just for medicinal purposes. The original formula contained garlic, onions, horseradish root, ginger root, hot peppers, sometimes turmeric, and often echinacea; all powerful immune-enhancers that help ward off infections, colds, flu and bronchial congestion. It's an amazingly effective remedy for staying healthy in the winter and keeping colds and flu at bay and is delicious. Use it as a salad dressing, but be sure to save some for medicinal purposes.

1 medium onion, chopped
4–5 cloves garlic, coarsely chopped
3–4 tbsp freshly grated ginger root
3–4 tbsp freshly grated horseradish root
Up to 1 litre (4½ cups) apple cider vinegar (preferably unpasteurised)
Honey, to taste
Cayenne powder, to taste

1. Combine the onion, garlic, ginger, and horseradish in a widemouthed glass 1-litre (32-oz) jar and add enough warmed apple cider vinegar to cover them. (Warming the vinegar allows it to more actively draw the properties out of the herbs.)

2. Place in a warm spot (near a sunny window is fine) and let sit for 3–4 weeks.

3. Strain, then discard the spent herbs. Now the fun part: Add honey and cayenne to taste. The finished product should taste lively, hot, pungent, and sweet.

4. Take 1–2 tablespoons at the first sign of a cold, and repeat the dose every 3–4 hours until symptoms subside. To avoid garlic breath, chew a few parsley sprigs or anise, fennel, or dill seeds.

From *Rosemary Gladstar's Medicinal Herbs: A Beginner's Guide* by Rosemary Gladstar (Storey Publishing, 2012). Also see *Rosemary Gladstar's Herbal Recipes for Vibrant Health: 175 Teas, Tonics, Oils, Salves, Tinctures, and Other Natural Remedies for the Entire Family* (Storey Publishing, 2008).

Opposite The flower- and leaf-bearing rhizome of common ginger (*Zingiber officinale*) not only allows the plant to spread, it contains volatile oils including gingerol which deliver its warming aroma, flavour and therapeutic properties.

MAKE THE CONNECTION
Find out more about Rosemary's botanical world at www.sagemountain.com; @sagemountainherbal

Dr JJ Pursell, Fettle Botanic, Oregon

Immune support syrup

Makes 6–8 (225-g/8-oz) bottles

When winter is around the corner it's time to make a batch of syrup loaded with vitamin C and immune-boosting herbs. Each fall I wander the Pacific Northwest woods and collect the elderberries for this recipe and am reminded of the roots upon which my herbal passions have grown from.

1 tbsp (½oz) reishi mushrooms
1 tbsp (½oz) ashwagandha
1 tbsp (½oz) rosehips
1 tbsp (½oz) elderberries
2 litres (8 cups) distilled water
1.4kg (4 cups) honey
   or 800g (4 cups) cane sugar

1. Put all herbs and the elderberries into a stock pot and add the distilled water.

2. Cover and bring to a boil. Reduce the heat, set the lid ajar and reduce by half.

3. Strain three times.

4. Return to clean stock pot and add the honey or cane sugar. Slowly stir until completely dissolved.

5. Allow to cool and then transfer into the bottles.

Find more remedies for optimum health and wellbeing in *The Herbal Apothecary: 100 Medicinal Herbs and How to Use Them* (Timber Press, 2015) and *The Woman's Herbal Apothecary: 100 Natural Remedies for Healing, Hormone Balance, Beauty and Longevity, and Creating Calm* (Fair Winds Press, 2018) by Dr JJ Pursell.

MAKE THE CONNECTION
Find out more about Dr JJ's botanical world at www.fettlebotanic.com; @fettlebotanic

Opposite *Sambucus nigra*, engraved and hand-coloured by H. Weddell from John Stephenson and James Morss Churchill's *Medical Botany: or Illustrations and descriptions of the medicinal plants*, 1831. See page 227 for a photograph of elderberries.

MY BOTANICAL WORLD...
'Herbs and their healing potential began at a young age as I'd walk around my dad's country farm and gardens. Their colors, textures and scents inspired me to seek out others who knew unique and different things about what was growing around me. I began to meet plant pathologists, genetic professors who cross-bred plants, master gardeners and then herbalists. Learning about how I could use herbs for my health was a pivotal point in my life. It was like a light switch was clicked on, and I began to immerse myself in the study of health and herbs.'

WELLBEING

Perhaps the most powerful benefit of plants is their ability to promote wellbeing, emotional healing, connectedness and spirituality. Plants feature prominently in many rituals, ceremonies and offerings by way of energy-clearing smudge sticks, meditative incense, anointing oils, floral garlands, blessings or spells. While those seeking to treat anxiety, stress, insomnia or hormonal imbalance are more frequently turning to the power of herbs, essential oils, flower remedies, or teas to curb, ease or even cure symptoms of life in the modern world.

Select ingredients
The wellbeing edit

Sage (*Salvia officinalis*)
Heal home spaces or yourself with a bundled smudge stick of herbs. Sage is thought to dispel negative energy, followed by sweet grass for healing and purification, or lavender for calm.

Spikenard (*Nardostachys jatamansi*)
This Valerian family member was used by the Ancient Egyptians as a luxurious aphrodisiac, and also to anoint tombs. Healing to skin and the nervous system, use for peace and assurance.

Cedarwood (*Cedrus atlantica*)
One of the first oils to be used in ancient medicine and ritual, cedarwood is associated with spiritual sight and development and often used as a grounding temple incense. Use for meditation.

Galbanum (*Ferula galbaniflua*)
Distilled from the resin of a tall, umbelliferous Middle Eastern herb, galbanum is commonly combined with frankincense to make holy incense. Use for the heart and crown chakras.

Palo santo (*Bursera graveolens*)
Palo santo meaning 'holy stick' is thought to ground energy and promote creativity for those open to its magic. Burn as incense or use the heavenly-scented essential oil in massage or baths.

Vetiver (*Vetiveria zizanioides*)
Known for its grounding and centring properties, this grassy plant is often sown to help prevent soil erosion. Harness the tranquil, aphrodisiac and earthy essential oil in massages, baths and scents.

St John's wort (*Hypericum perforatum*)
The leaves and flowers of St John's wort have anti-depressant qualities. Make a restorative solar oil infusion for skin or for a massage oil or salve.

Linden (*Tilia x europaea*)
Linden trees produce sweet-scented lime flowers in spring that can be infused into a calming, soothing tea or gentle herbal bath. A 'family herb', it is said to open the emotional heart.

Vervain (*Verbena officinalis*)
This unprepossessing wayside plant with small, unscented lilac flowers provides a surprisingly nourishing tonic that can help with exhaustion, insomnia or paranoia, or to aid flexibility of view.

Myrrh (*Commiphora myrrha*)
The resin or gum known as myrrh has been used for thousands of years to heal and embalm. Deliver its soothing, antiseptic, balancing, fear-calming actions through massage or incense.

Bergamot (*Citrus bergamia*)
This small lemon-coloured orange is the ingredient in Earl Grey tea and eau de Cologne. Joyous and uplifting, it is a powerful anti-depressant that can help boost mood or help combat stress.

Juniper (*Juniperus communis*)
Used in Ancient Greece and Egypt to combat disease, juniper is physically and emotionally cleansing. Use the essential oil to help treat skin and hair problems and relieve nervous tension.

Saffron (*Crocus sativus*)
The sultry orange crocus pistils known as saffron make a spiritually awakening if regally expensive premenstrual tea. It's also said to awaken energy centres, and to attract love, beauty and money.

Hawthorn (*Crataegus monogyna*)
Hawthorn has been used for centuries as a natural remedy for the heart, while legend has it that the tree itself is sacred. Use berries for syrup, or blooms and branches for rituals.

Opposite Mike Hill's manipulated plant mandala pays homage to an ancient tradition of arranging flowers and petals in repetitive form, both mimicking cyclical designs found in nature and thus instilling a sense of wellbeing.

Try four things
The wellbeing edit

Smudge stick
Makes 1 smudge stick

Selection of cleansing, calming or uplifting herbs, such as sage, rosemary, lavender, sweet grass, lemon balm, mugwort, peppermint or juniper
Cotton or hemp twine
A feather

Smudge sticks – tied bundles of sacred, aromatic, often resinous herbs such as white sage (*Salvia apiana*), sweet grass (*Hierochloe odorata*) and red cedar (*Juniperus virginiana*) – have been ritualistically burned for thousands of years by some Native American and First Nations peoples to clear out negative energy and connect with the earth and ancestors. Herbs would be chosen and blessed for their purifying or energising properties and traditionally gathered at certain times of the day, month or year. Today, 'smudging' has been appropriated into other belief systems, with the understanding that the ritual burning of scented or magical herbs is something that many cultures and religions have always done.

1. To make a version, select and give thanks for your chosen herbs (a mix of sage, rosemary or lavender is cleansing, calming and uplifting), then cut them to 20cm (8 in) lengths. Allow to wilt overnight, then bundle together. Use twine to wrap from the base to the tip and then criss-cross down again, folding the leaves down at the tip. Tie at the base and trim the stalks.

2. Set to dry for 7–10 days.

3. When required, light the tip and use a feather to fan scented smoke over anything or anyone you wish to cleanse or bless.

Moon bath
Makes 1 (1 litre/32-oz) mason jar

2 tbsp Dead Sea salts
2 tbsp Himalayan bath salts
1 tbsp each dried herbs, such as calendula, lavender flowers, lemon balm, rose petals and jasmine flowers
Essential oils, such as lavender, jasmine and rose
500–750g (2–3 cups) Epsom salts

Full moon bathing is another lovely way to harness the power of plants and the energy of the moon itself – a roughly 29.5-day, tide-changing phenomenon that many people believe has the power to affect human and animal moods, behaviour and physiological changes plus plant growth and even events. Whether you believe this or not, the ritual of taking a monthly spiritual bath is undeniably a good one, especially if you factor in detoxifying herbs and salts.

1. To help cleanse body and soul, first prepare your herbal bath blend. Place the Dead Sea salts and Himalayan bath salts in the glass mason jar. Add the dried calendula, lavender, lemon balm, rose petals and jasmine flowers plus 3 drops of lavender, 2 drops of jasmine and 4 drops of rose absolute oils – or create a bespoke blend of herbs and oils.

2. Seal and shake gently, then leave overnight to infuse.

3. Measure 75g (¼ cup) of the infused salts into a muslin bag (or muslin square bundle) and add to a hot bath with the Epsom salts bath. Dry brush the skin for 2–3 minutes and then bathe for a further 20–30 minutes. Try using scented candles, ritual tea, meditative music and crystals to boost the lunar effect.

Ritual tea
Makes 1 (1 litre/32-oz) mason jar

Green or black loose leaf tea
Fresh jasmine flowers

In many cultures and belief systems the ancient art of drinking tea is as much a ritual as a thirst-quenching or pleasurable experience, a vehicle for both inner peace, health, wellbeing, inter-connectedness and happiness. Some tea preparations, such as green tea, an antioxidant-rich brew of unprocessed, unfermented and often handpicked tea leaves (*Camellia sinensis*), are also thought to have medicinal properties. While fragrant options such as jasmine tea, a green, white or black tea blended with jasmine blossoms (*Jasminum officinale*), use powerful olfactory notes to help soothe and destress.

1. To create your own ritual jasmine tea, pour 1 tablespoon (¼ cup) of green or black crushed tea leaves into a 1 litre (32-oz) mason jar. Cover with freshly picked jasmine flowers – evening picking is best, when the scent is at its strongest – and repeat. Finish with a layer of tea, weigh down and seal. Allow jasmine flowers to perfume the tea for at least 24 hours, ideally several weeks.

2. Remove the weight, extract the jasmine, if you like, and store the tea in a cool dark place.

3. To prepare a ritual brew, steep a heaped teaspoon of jasmine tea in a *gaiwan* (lidded tea cup) of just-boiled water for around 3 minutes, using the lid to artfully brush away tea leaves before drinking. Allow yourself time to savour the flavours and fragrance. Give yourself permission to relax.

Love potion
Makes 1 (1 litre/32-oz) mason jar

2–4 tbsp dried aphrodisiac herbs, such as damiana or rose petals
700g (2 cups) organic honey
1 vanilla pod or ginger (optional)

The Plant Kingdom is positively rife with natural aphrodisiacs – those chemical constituents that can help raise libido and increase sexual desire by relaxing, energising or stimulating the body or mind. Among these lusty herbs are anti-depressant damiana (*Turnera diffusa*) said to stimulate the sacral chakra; energising ginseng (*Panax ginseng*); circulation-boosting and warming ginger (*Zingiber officinale*); sensuous hibiscus (*Hibiscus rosa-sinensis*); libido-enhancing maca (*Lepidium meyenii*); nourishing oatstraw (*Avena sativa*); healing and loving rose (*Rosa rugosa*), said to open the heart chakra; hormone-regulating sarsaparilla (*Smilax ornata*); and exotic, heavily fragranced vanilla (*Vanilla planifolia*). A great way to deliver all this loving goodness is via a herbal honey 'love potion', which can be eaten on its own or added to cordials, desserts or other recipes.

1. Using a basic formula of 1–2 tablespoons of dried herbs per 350g (1 cup) of honey, fill the mason jar with dried damiana and dried rose buds or petals, then cover with organic honey – for added love or lust, add ginger or a vanilla pod.

2. Stir with a chopstick and then fill to the top with more honey. Wipe rim, seal tightly and leave to infuse for at least 5 days. Turn a few times to ensure herbs stay coated.

3. Strain the honey into a clean jar and use liberally.

Julia Lawless, Aromatherapist, Aqua Oleum, Stroud

Warming body oil

Makes 1 (30ml/1oz) bottle

During winter, it's easy for our skin to become undernourished. We are also much more likely to be affected by chills and viral infections such as colds and flu. Jojoba is an excellent nourishing carrier oil, being suitable for all skin types, while the combination of rosemary, lavender and marjoram essential oils is my favourite 'proven' blend for cold, achy joints.

Adding a few drops of spicy oils helps warm the entire system, plus they are among the best antiviral agents available. More specifically, ginger combats cold and dampness in the body, black pepper acts as an overall tonic and improves muscle tone, while cardamom is antimicrobial and brings instant heat to the metabolism.

Note
Warming the oil a little before application makes the blend even more effective and yummy.

20ml jojoba oil (or light coconut oil)
5ml avocado oil (to provide extra nourishment for the skin)
2 drops of black pepper essential oil
3 drops of marjoram essential oil
3 drops of rosemary essential oil
4 drops of lavender essential oil
2 drops of ginger essential oil
1 drop of cardamom essential oil

1 Blend the oils together in a glass beaker or bowl and mix well. Decant the blend into a dark-glass 30ml (1oz) bottle and seal.

2 To use, gently massage the oil into sore muscles and joints or apply as a warming preventative remedy and skincare treat.

See *The Complete Aromatherapy & Essential Oils Sourcebook* by Julia Lawless (HarperCollins Publishers, 2018, Sterling Publishing Co., Inc., 2017)

MAKE THE CONNECTION
Find out more about Julia's botanical world at www.aqua-oleum.co.uk; @aquaoleum

Opposite Elizabeth Blackwell's delicate study of black pepper (*Piper nigrum*) for *A Curious Herbal* (1737–1739) belies the hot, pungent spiciness of its round, berry-like fruits, which are steam-distilled to extract the essential oil.

BOTANICAL REMEDIES
The wellbeing edit

MY BOTANICAL WORLD . . .
'*I grew up beside the River Wye in rural Herefordshire. Wild flowers proliferated on its banks so I learned all their names, pressed them in diaries and marvelled at their beauty. I have always been a complete devotee of the natural world and still consider nature to be my perfect refuge. My summers as a child were spent in Finland, since my mother was Finnish, where I enjoyed a sense of wilderness. Later I became a passionate gardener, as well as training as a medical herbalist and aromatherapist. Today I am responsible for the formulation of aromatic products for Aqua Oleum Ltd., my family business.*'

Black Pepper.   *Piper nigrum.*

Eliz. Blackwell delin. sculp. et Pinx.

MY BOTANICAL WORLD...
'My botanical world consists of connecting with the plants wherever I go, from mugwort in the cracks of city sidewalks to the majestic redwood cathedral of northern California – I take time to send my love to the plants. Before I harvest any plant I offer it some tobacco and then connect to the plant spirit, asking if and how it would like to work with me. I ask these questions through song or meditation – whichever feels more appropriate for that specific plant. I began working with plant spirit medicine a decade ago and still have so much to learn. I'm grateful to be a student of Mother Earth in this way.'

Deborah Hanekamp, Mama Medicine, New York

Inner radiance ritual bath

Makes 1 ritual bath

In spiritual practices all over the world, bath rituals serve as a form of rebirth, cleansing, and healing. We unify with the sacred waters for detoxification, regeneration, and clarity. As we bathe, we're taking a moment to reflect, set intention, and wash away old habits and patterns that no longer serve our highest good, all while making our skin sing from the inside out. I was personally introduced to ritual bathing during my shamanic studies traveling back and forth from the Peruvian Amazon over the course of eight years. Working with plants, plant spirits, crystals, and mantras I came up with a recipe that would nourish the mind, body, and spirit in one sacred bath. Practise once a week as a special treat to honor yourself and your total beauty.

For the altar
A candle you love
Rose quartz crystal
A flower that represents beauty to you

For the bath
160g (2 cups) dried calendula flowers
Lemon balm tea
100g (1 cup) cacao powder
520g (2 cups) pink Himalayan salt
250ml (1 cup) red wine vinegar (recommended) or apple cider vinegar
160g (2 cups) fresh or dried pink rose petals
Raw local honey
White sage

Note
You can steep dried calendula flowers, lemon balm tea and cacao powder together in a tea prior to the bath, then add the brew to your tub. Or toss them all in as is for a pretty (albeit messy) bath.

Opposite Rose (*Rosa*), calendula (*Calendula officinalis*), sacred sage (*Salvia apiana*) and lemon balm (*Melissa officinalis*) come together with healing crystals, salts, cacao and honey to bring radiance back to the self.

Create a sacred space
Make a small altar near the tub using a candle, a rose quartz crystal, a flower, and any other personal power items that represent beauty to you.

Perform the ritual

1 Set up your altar and combine all the bath ingredients in your tub.

2 Cover yourself head-to-toe in raw local honey (yes, even your hair!), then light a piece of white sage with the flame from your altar candle and burn it around your entire body, even under the soles of your feet (lifting your feet to do so).

3 As you stand there naked, covered in honey, burn sage around you and take a moment to ask yourself if there are any outdated, untrue, or unkind beliefs you've held about yourself.

4 After letting the honey set into the skin for 7 minutes, step into the bath.

5 Practise a type of yogic breathing called *Kapalabhati* (shining skull breath) by exhaling from the nostrils, inhaling three-quarters of capacity and then taking at least eight quick, sharp, forceful exhales through the nose. Repeat three times. Then say this mantra at least three times: I am precious, I am worthy, I am beautiful, I am.

6 Close your eyes and soak for as long as you'd like in the powerful energy you've created.

7 When you're done with the bath, blow out the candle to signify the close of the ritual, and enjoy all its beautiful benefits.

MAKE THE CONNECTION
Find out more about Deborah's botanical world at www.mamamedicine.nyc; @mamamedicine

Nicola Cunningham, Stem Apothecary, London

Breathe in the forest salt bath and balm

Makes up to 4 bath treatments

The forest is a place close to my heart, a place where I can breathe so deeply. To recreate this breathe-easy aromatic experience at home, combine woody and uplifting oils with remineralising Epsom salts and add to the bath. Let all their medicinal goodness sink in while you lie back and let your mind wander through the woods.

Cypress (*Cupressus sempervirens*) is ideal for its immune-boosting and vasoconstrictive properties, helpful for easing conditions such as varicose veins. Pine (*Pinus sylvestris*) is a natural pain reliever, which can help clear congestion and is naturally uplifting. Lemongrass (*Cymbopogon citratus*) provides the perfect top note, a synergy of zest and earth and is also a tissue toner, eases muscles, and lifts and energises the nerves. While vetiver (*Vetiveria zizanioides*) has a musky earthy aroma that can only come from the ground and has centring and relaxing properties, which can help relieve anxiety and provide a protective forcefield for sensitive souls.

For cold spells or to help relieve symptoms of a chest infection, finish with a chest rub of frankincense (*Boswellia carteri*), eucalyptus (*Eucalyptus smithii*), rosemary (*Rosmarinus officinalis*) and wheatgerm oils to help loosen mucus, support and strengthen immunity and bring lasting woody warmth.

Notes
If you are pregnant, it is not recommended to use any essential oils other than mandarin. Please also note that pine essential oil can raise blood pressure with prolonged use. Do not use with clinically high blood pressure. In addition, rosemary essential oil should be avoided if you suffer from epilepsy. Consult your doctor or an aromatherapy practitioner before beginning if you have concerns.

MAKE THE CONNECTION
Find out more about Nicola's botanical world at www.stemapothecarystore.com; @stem_apothecary

Breathe in the forest salt bath
5 drops of cypress essential oil
3 drops of vetiver essential oil
8 drops of pine essential oil
4 drops of lemongrass essential oil
100g (scant ½ cup) Epsom salts

1. Add the oils to the salt and mix well in a bowl. Decant into a sterilised 100g (3.5oz) glass jar with a lid and store in a cool, dry place.

2. To best absorb the benefits use about 25g (1oz/a small handful) per bath and soak for 30 minutes.

Breathe in the forest chest balm
Makes 1 x 28-g (1-oz) jar or tin
1 tbsp beeswax beads
1 tbsp wheatgerm oil
5 drops of frankincense essential oil
5 drops of eucalyptus essential oil
2 drops of rosemary essential oil

1. Melt the beeswax in a bain marie, add the wheatgerm oil and stir well.

2. Remove from the heat, add the essential oils and stir well again.

3. Pour into a small, sterilised 28-g (1-oz) ointment jar or non-reactive tin and leave to cool.

4. Once cooled, seal with a lid and use in the morning or evening as required.

Opposite *Breathe in the Forest Bath* provides an aromatic portal to enchanting forest realms such as Hoh Rain Forest's verdant cathedral of ancient moss- and fern-covered deciduous and coniferous trees in Olympic National Park, USA.

MY BOTANICAL WORLD . . .
'*Everything smells for a reason. We must learn to read with our noses for the benefit of our health. Just like our tastebuds, our sense of smell will lead us to what our body does and doesn't need. A few years ago I couldn't bear the scent of clary sage, but as I mature I am beginning to like it. It will have something that my body can now use. Throughout life we have different needs and we can find the plants to help us at any stage in our life. Plants can be a forcefield if we use and respect them wisely. Here on Earth animals and plants rely on each other chemically and can nurture each other full circle.*'

BEAUTY

Plants have helped to smooth, perfume and pamper the body beautiful since the dawn of civilisation: the Ancient Greeks stained their lips with crushed mulberries; the Ancient Egyptians used vegetable oil emollients to protect their skin, essential oils for perfume and aloe to heal; in India, Ayurvedic beauty practices included infused-oil treatments, henna tattoos and spiritual cleansing; while the Romans used rose, frankincense and calendula to nourish, scent and soothe. All in all, chemical-free, beautifying plant power that still holds sway today.

Select ingredients
The beauty edit

Lavender (*Lavandula angustifolia*)
Lavender helps calm, soothe and balance and is a natural antiseptic. Use the essential oil to make relaxing spritzers, dried lavender buds to infuse massage oil, or scent as a natural perfume.

Frankincense (*Boswellia carteri*)
Frankincense or olibanum hails from the resin of the plant and is wonderfully calming and richly fragrant. Use the essential oil to help firm mature skin, soothe inflammation or for meditation.

Rose otto (*Rosa damascena*)
Rose essential oil has many healing qualities, including the ability to help soothe dry, sensitive, inflamed or allergy-prone skin. Or use petals or buds for scented waters, oils, infusions and toners.

Rose geranium (*Pelargonium graveolens*)
Rose geranium essential oil comes from the same plant as Geranium oil but has a rose-like scent. It has similar skin-soothing and uplifting properties.

Clary sage (*Salvia sclarea*)
Nutty-scented clary sage essential oil also has antidepressant and astringent actions that make it ideal for uplifting skin lotions. It's also good for muscle relaxing, balancing or aphrodisiac soaks.

Patchouli (*Pogostemon cablin*)
Patchouli essential oil has a musky, spicy aroma plus astringent, antiviral, antiseptic and antidepressant properties that can help soothe inflamed skin, dandruff, eczema and stress.

Tea tree (*Melaleuca alternifolia*)
Aboriginal Australian peoples have long harnessed tea tree for its antifungal, antibacterial and antiviral properties. Use the essential oil for inflamed skin, dandruff or dry scalps.

Marshmallow (*Althea officinalis*)
Create a tea to help soothe dry skin by soaking mucilaginous root or leaves overnight in cold water. Or combine marshmallow root powder and chamomile tea to help heal irritated skin.

Plantain (*Plantago major/ Plantago lanceolata*)
Harvest the broad-leaved or ribwort plantain from wayside or lawn weed. Finely chopped leaves in vegetable glycerin can make a healing skin lotion. It's also great for lips and hair.

Neroli (*Citrus aurantium*)
This exquisitely fragranced symbol of innocence and fertility comes from the blossoms of the bitter orange tree and can help to tone the skin, improve elasticity and prevent wrinkles.

Calendula (*Calendula officinalis*)
Pot marigold has been used for centuries for its healing and anti-inflammatory properties, via herbal teas, baths, oils and salves. It's particularly good for healing inflamed skin, and for children.

Thyme (*Thymus vulgaris*)
Thyme is a popular culinary herb but less well known for its antioxidant and antiseptic qualities. Use as a tea or as a fresh-scented essential oil, to help skin conditions such as acne and eczema.

Comfrey (*Symphytum officinale*)
Comfrey contains the naturally softening compound allantoin, while the mucilaginous root can help heal irritated skin. Or try a comfrey leaf infusion as a nourishing hair rinse.

Cucumber (*Cucumis sativus*)
Cucumber is widely used in folk medicine to reduce inflammation. It's also rich in vitamin C and good for external cooling and cleansing. Place circles over eyelids to help reduce puffiness.

Opposite The earliest roses (*Rosa*) are thought to have bloomed on Earth 35 million years ago, their petals providing beautifying scent and restorative healing to people from ancient civilisations to the present day.

Try four things
The beauty edit

Skin scrub
Makes 1 portion

Exfoliating ingredient(s), such as sea salt, pure cane sugar crystals, oatmeal, bicarbonate of soda (sodium bicarbonate), honey or kefir
Skin-friendly carrier or massage oil(s), such as coconut, jojoba, wheatgerm, grapeseed or almond oil (for use with salt or sugar exfoliants)
Skin-friendly essential oil(s), such as lavender, chamomile, orange or tea tree

Help promote a fresh-faced or body-beautiful glow by sloughing off dead skin cells with a gentle exfoliating scrub. Making your own scrub means you know exactly what you're putting on your skin, plus you can avoid synthetic exfoliating particles that are damaging to the environment. There are several exfoliating ingredients to experiment with, ideal for face or body. Sea salt is full of rejuvenating minerals and is great for dry skin, while sugar is a natural source of glycolic acid, which can help boost new cell production. In both cases, combine with your favourite massage or carrier oil – coconut, jojoba, wheatgerm, grapeseed or almond oil – and suitable skin-friendly essential oils such as lavender. For sensitive skin, naturally anti-inflammatory oatmeal can be a good choice, mixed with honey, kefir or yogurt. Or opt for a natural bicarbonate of soda exfoliation followed by a facial oil, balm or serum (see page 275).

1. Suggested starter recipes include 3–4 tablespoons of oil mixed with 2 tablespoons of brown sugar or 500g (2 cups) of coconut oil mixed with 250g (1 cup) Epsom salts and 8–10 drops of lavender essential oil.

2. Simply combine ingredients of your chosen scrub in a small non-reactive bowl, stir with hands and use as required. Or increase amounts proportionally to almost fill a 450-g (16-oz) lidded mason jar, shaking gently to blend.

Clay mask
Makes 1 mask

1 tsp bentonite or experiment with kaolin, French green, rhassoul or fuller's earth clay
½ tsp powdered calendula or chamomile flowers (optional)
1 tsp raw honey
1 tsp water, flower water or herbal infusion
2 drops of lavender essential oil (optional)

A clay face mask can help to cleanse, tone, moisturise and soothe your skin. Taking the time to apply a face mask and allow all the beneficial ingredients to do their thing can also help promote calm and relaxation. For the full spa experience, try making a mud mask using bentonite clay – a grey-cream, velvety, toxin-absorbing and naturally healing clay composed of volcanic ash.

1. Simply mix the bentonite powder or clay with powdered chamomile or calendula flowers, if using. Add the honey to form a thick paste, then enough water – or skin-soothing herbal infusion, decoction or flower water – to thin it out to an easily applicable texture. A few drops of lavender essential oil can also be soothing and add a lovely, uplifting fragrance to the mask.

2. Immediately apply to the face and neck in a circular motion, avoiding the eyes. Then relax for 10–15 minutes or until the mask has dried.

3. Wash off with a muslin cloth soaked in hot water and pat dry.

4. For best effect, experiment with different combinations of clays, herbs, flower waters, herbal infusions and essential oils to suit your skin type.

Floral/herbal hydrosol

Makes 1 hydrosol

1 part dried or fresh plant material, such as lavender flowers, rose petals or peppermint leaves
2–6 parts distilled water
Ice (preferably in sealed bags)

A hydrosol is the aromatic water that remains after botanical material has been steam-distilled. It contains the essence of the plant, including small quantities of essential oil. Popular hydrosols include sweet-scented rose, lavender and orange blossom, useful in aromatherapy, as perfume, or for facial toner, hair rinses, deodorant or air freshener.

1. To make a hydrosol, place a ceramic ramekin upside down in the centre of a large, glass-lidded pan. Surround the ramekin with the chosen plant material and distilled water – 1 part dried or fresh herbs/flowers to 2–6 parts distilled water – ensuring that the mix does not rise above the ramekin.

2. Stand a bowl on the ramekin, then place the lid on the pan, upside down so that the handle faces down into the bowl. Fill sealed bags half-full with ice and place on the lid.

3. Transfer the pan to the stove and heat on low. Replace the ice bags as each one melts, until the herbs or flowers look spent, or their fragrance is no longer emitted.

4. Allow the pan and contents to cool to room temperature. Then transfer the captured hydrosol into a sterilised jar or spray bottle.

5. Store in a cool, dark place and use as required. For additional scent and medicinal benefits, add a few drops of the relevant essential oil.

Body butter

Makes enough to fill 1 (450g/16oz) mason or jam jar

110g (½ cup) shea butter
100g (½ cup) coconut oil
120ml (½ cup) almond or jojoba oil, or a herbal infused oil such as calendula oil
110g (½ cup) cocoa butter or mango butter (optional)
10–30 drops of skin-friendly essential oil, such as lavender or chamomile (optional)

Rich, nourishing body butter relies on several key ingredients. Shea butter is rich in fatty acids and vitamins A and E, and can help stimulate the production of collagen. Coconut oil soaks deep into your skin, promoting healing and regeneration, while a suitable vegetable or herbal infused oil soothes, smooths and adds glide. To this core base you can add rich, chocolatey cocoa butter for added vitamin E, naturally emollient, sweet-scented mango butter for extra fatty acids, or 10–30 drops of your favourite skin-friendly essential oil.

1. To make, place all ingredients apart from the essential oils in a double boiler. Bring to a medium heat and stir the mixture until melted.

2. Remove from the heat and let cool. Then add the essential oils, if required, and transfer to the fridge to cool further for an hour. The blend should harden slightly but stay soft.

3. Now use a stick blender to whip the butter for 10 minutes or until fluffy. Then return to the fridge to set.

4. Store in a 450-g (16-oz) glass jar with a lid and use as required to slather face or body with moisture.

BOTANICAL REMEDIES

The beauty edit

MY BOTANICAL WORLD...

'There is something so tempting about harnessing stinging nettles into something soothing and restorative for the skin – I had to explore it. Nettles are full of vitamins A, C and E, antioxidants that can help protect the skin from bacteria, viruses and pollution. While not an actual sunscreen, vitamins C and E can also help to repair the skin from UV rays. Plus vitamin C helps the skin to make collagen, which keeps it youthful and plump.'

Nicola Cunningham, Stem Apothecary, London

Nettle facial toner and balancing facial oil

Makes 1 (60-ml/2-fl oz) bottle & 1 (15-ml/½-fl oz) jar

Nettle facial toner

To begin, you will need to find a patch of young, fresh nettles. Wearing gloves to avoid the sting, snip the top leaves – not the flowers or fruit – into a colander until it is loosely full. Shake your collection of leaves to release any creatures, then wash and lay out on kitchen paper or a cotton towel in a warm, dry place for a few days (this is also how you dry nettles for making tea). I prefer to work with slowly dried herbs when making skincare products both to avoid any creatures in the plant material, and also because drying removes the water so the vitamins and minerals can be harnessed more intensely in an infusion.

The next stage is to make an infusion using the leaves. A few tablespoons of dehydrated nettles in a mug of boiled, distilled water will give a gorgeous rich green colour. When it is cool, you are ready to make your facial toner.

50ml (¼ cup) cooled nettle infusion
10ml (2 tsp) witch hazel
¼ tsp bicarbonate of soda
½ tsp raw honey

Note

You can also try using rose or calendula petals to make an infusion toner. Harvest the flowers when they are just beginning to bloom. Rose reduces inflammation, tightens the skin's tissues and promotes feelings of optimism and love, and calendula encourages cell growth and repair and is gentle and soothing.

1. Mix all the ingredients together in a bowl, stirring briskly to work the sticky honey into the mix.

2. Transfer into a dark-glass, lidded bottle to be used as part of your facial routine. Your toner should stay fresh for around 8 days, or up to 2 weeks in the fridge.

Opposite The best time to harvest stinging nettles (*Urtica dioica*) is in the first few weeks after they come up in the spring, before flowering. Crops are usually prolific enough to pick the top few leaves and then move on.

Balancing facial oil

A facial oil is a sebum-balancing, carefully blended nutritious oil that is applied after toning the skin but before you moisturise. You can use it to target specific skin concerns such as wrinkles, acne, oily or dry patches, selecting beneficial ingredients for various skin types, and floral or woody notes to suit your aromatic preference. Make up one batch of the base oil recipe, then add 2 drops of one of the aromatic oils suggested below to suit your skin type.

For dull congested skin: Add 2 drops of geranium essential oil (*Pelargonium graveolens*), a sweet, floral, full-bodied mid note.
For acne prone or oily skin: Add 2 drops of cedarwood essential oil (*Cedrus atlantica*), a woody, clean mid note.
For wrinkles, dry and sensitive skin: Add 2 drops of rosewood essential oil (*Aniba roseadora*), a sweet, floral woody sent which is slightly spicy, giving a mid to base note.

5ml (1 tsp) cold-pressed rosehip oil
10ml (2 tsp) apricot kernel oil
1 vitamin E capsule or a dash of vitamin E oil

Note

See page 288 for notes regarding pregnancy.

1. Simply place all oils in a dark-glass, lidded 15-ml (½-oz) jar and shake gently to mix. Store in a cool, dark place.

2. To use, massage into the face using a gentle gliding movement to avoid stretching the skin. Work upwards to help lift facial muscles, acknowledging pressure points along the way to relieve tension.

MAKE THE CONNECTION
Find out more about Nicola's botanical world at www.stemapothecarystore.com; @stem_apothecary

Kim Walker and Vicky Chown, Handmade Apothecary, London

Melissa lip balm

Makes 2–3 (30ml/1oz) salve tins

This lip balm contains moisturising oils and antiviral St John's wort, lemon balm and eucalyptus. Eucalyptus may not be the first scent you think of when deciding on a lip balm flavour, but is full of antiviral agents that blitz the cold-sore virus. It also gives the lips a fresh and tingly feeling.

60g (¼ cup) shea butter
5g (1 tsp) beeswax
10ml (2 tsp) lemon-balm-infused oil (page 271)
10 drops of St John's wort tincture (page 270)
10 drops of eucalyptus essential oil

1. Melt the shea butter, beeswax and lemon-balm-infused oil together in a bain marie.

2. Remove from the heat and whisk in the St John's wort tincture and eucalyptus essential oil.

3. Pour into 2–3 30ml (1oz) small pots or lip balm tubes, seal, label and date.

4. To use, apply throughout the day, as and when required, as you would a normal lip balm.

Note
Shelf life is up to one year (store as you would a shop-bought lip balm).

From *The Handmade Apothecary: Healing Herbal Remedies* by Vicky Chown and Kim Walker (Kyle Books, 2017)

MAKE THE CONNECTION
Find out more about Kim and Vicky's botanical world at www.handmadeapothecary.co.uk; @handmade_apothecary

Opposite Crush the leaves of Lemon balm (*Melissa officinalis*) to release a delicious floral-lemony aroma produced by the volatile oils responsible for many of its medicinal properties including antiviral and carminative actions.

MY BOTANICAL WORLD...
'Vicky is from the city and I (Kim) hail from the countryside, but we stand united in the quest for something beyond the manmade, for peace through reconnection with the natural world. By collaborating as herbalists we are better placed to rediscover and share knowledge of how to optimally live with plants – for food, for medicine and for self-care. As educators, we can also help others to conserve and give back to nature, no matter where they come from or are going in life.'

MY BOTANICAL WORLD...

'This is the best DIY face mask we've made so far – we actually notice a difference. The propolis came from our bees, for which we have planted loads of flowers at our Field Station as a thank you. It's vital to our wellbeing to give back to nature.'

Above Give yourself over to the incredibly restorative power of the ocean, the primary life support system on Earth, where life evolved 3 billion years before life on land.

Charisse Baker, Field Station, Surrey

Ocean potion mask

Makes 1 mask

The tide is high and you're floating – put this rehydrating, remineralising, anti-inflammatory, soothing and repairing mineral magic on your face and you'll be soaking in goodness, exorcising the urban grit, wiping off the past and tickling the circulation to glow with inner and outer beauty.

Kombu (*Saccharina japonica*) is a marine species of brown algae in the Laminariaceae family (thus not technically part of the Plant Kingdom – see page 24), widely eaten in Japan and China and famous for its umami taste. It is packed full of minerals such as calcium and a rich source of iodine and vitamin B2.

Propolis – the true botanical component of the mix – is reputed to have been coined by Aristotle from the Greek words *pro* (before) and *polis* (city), meaning 'defender of the city', the substance being used to seal and protect bee larva chambers. It is collected by bees from tree buds, sap flowers and other plant sources to create a naturally antibacterial resinous mixture, which is soothing, as well as healing for skin.

180ml (¾ cup) cooled distilled water, plus extra for soaking
15g dried kombu seaweed, cut into small pieces (or forage for kelp at the seaside)
10 drops of propolis extract/tincture

Note
Avoid kombu if you suffer from thyroid problems due to high iodine levels, while those allergic to bees or bee products should avoid propolis.

1 Using the cooled distilled water, rinse the kombu and soak for 1 hour in a suitably-sized jar or bowl.

2 Strain and place the liquid to one side. Use the water from soaking the kombu in miso soups or as a stock for other culinary recipes.

3 Repeat, soaking the kombu for an additional 2 hours in a bowl or jar placed in the fridge.

4 Strain and place the soaked kombu, 180ml (¾ cup) distilled water and 10 drops of propolis in a blender. Blend until smooth.

5 Turn your phone off and the bath taps on. Relax and let the steam open up your pores. Then solicit a fellow mermaid or merman to apply your ocean potion mask to face and neck.

6 Sip a cool glass of water, put your feet up and enjoy the bath for 10–15 minutes as you let your mask work its magic.

7 Then rinse, pat dry with a towel and wait for 30 minutes before applying any further products.

8 Follow up with shea butter or a gently emollient plant oil.

9 Your ocean potion mask should keep in the fridge for a week, or place in the freezer to extend lifetime for up to 2 months.

MAKE THE CONNECTION
Find out more about Charisse's botanical world at www.eastlondonjuice.com; @field_station

HOME

Plants have also been used throughout history to clean and fragrance the home – and for the most part, they have done a perfectly good job. Sadly, in the modern quest for extreme, germ-free cleanliness, chemical-based formulas have risen to the fore, bringing unwanted toxins into the domestic setting and the environment at large. It's easier than you think to switch back to more natural solutions, however, once you know the power of a few simple grime-busting or freshening ingredients such as lemons, vinegar and a few basic essential oils.

Select Ingredients
The home edit

Lemongrass (*Cymbopogon citratus*)
The refreshing citrus smell of this tall aromatic grass can help repel flies or get rid of unwanted pet odours. The essential oil is also useful for skin.

Grapefruit (*Citrus* x *paradisi*)
Grapefruit essential oil has a fresh, tangy scent that enlivens the mind. Use in a mister, candles or home cleaners, or in body scrubs, soaps or massage oils. Grapefruit juice can also boost skin.

Ylang ylang (*Cananga odorata*)
Harness the uplifting yet sedative, aphrodisiac aroma of sweet yet spicy ylang ylang flowers in room misters or candles. The essential oil is also good for skin, circulation and the nervous system.

Pine (*Pinus sylvestris*)
The fresh scent and antibacterial actions of pine are long known. Use Scots pine essential oil for inhalations or pine-needle-infused vinegar to help make an invigorating home cleaner.

Sandalwood (*Santalum album*)
Sweet, woody, antiseptic and soothing, sandalwood has been used for millennia to heal and uplift. Use drops of the essential oil sparingly in the home or in massage oil for a mental boost.

Cardamom (*Elettaria cardamomum*)
The sweet, warming scent of cardamom brings joy and clarity to the mind and can help stimulate energy flow. Add cardamom oil to diffusers or pot pourri in winter for cosiness.

Rosemary (*Rosmarinus officinalis*)
Rosemary lifts the spirits, improves circulation, aids memory and is wonderfully uplifting in massage oils or air fresheners. Add essential oil to water and vinegar for a great natural cleaner.

Clove (*Syzygium aromaticum*)
This culinary and medicinal spice is deodorising and sweet, ideal for pot pourris, incense and air fresheners. Hang clove and orange pomanders to help repel insects, or use oil in natural cleaners.

Lemon (*Citrus limon*)
Antacid, antifungal, antiseptic, vitamin-C-rich lemons are well known for their therapeutic and culinary properties. Fresh or as essential oil, use for insect repellents, cleaners or home scents.

Jasmine (*Jasminum grandiflorum*)
It takes masses of flowers to produce a small amount of this warm, exotic essential oil, so use wisely for fragrance or in massage. Garlands of fresh flowers are a lovely way to enjoy the scent.

Mandarin (*Citrus reticulata*)
Delicate, fruity and floral, mandarin orange oil has a gentle healing action that can help with skin, digestive or circulatory issues. Diffuse in children's rooms to boost smiles and sleep.

Lemon balm (*Melissa officinalis*)
The lemony scent of melissa is easy to introduce via dried bouquets, or infuse into vinegar for a natural, mildly antiviral cleaner. Essential oil in a diffuser can also be restful and insect repellent.

Cinnamon (*Cinnamomum zeylanicum*)
In Ayurveda, the antiseptic, antibacterial, antifungal and astringent bark and leaves of cinnamon are used widely. Pair cinnamon leaf essential oil and bark for a warming pot pourri.

Lavandin (*Lavandula* x *intermedia*)
A more camphorous, sharp-smelling hybrid of true lavender and spike lavender, lavandin is strongly antibacterial, antiviral and antifungal and ideal for freshening your home or linen.

Opposite Jasmine (*Jasminus officinalis*) flowers have captured the imagination of poets and perfumers since ancient times. Plant or place pots under windows and let the soulful scent drift into your home.

Try four things
The home edit

Aromatic reed diffuser
Makes 1

Small glass or ceramic vessel with a narrow neck
Essential oils such as lavender, rosemary, peppermint, mandarin, cinnamon and sandalwood
Sweet almond oil (enough to quarter-fill a vessel)
Rattan reeds

Reed diffusers are great for scenting areas where you might not want to leave a candle burning unattended. They work by drawing single or blended essential oils, in a base of carrier oil or alcohol, through thin rattan reeds in order to diffuse the air with beneficial aromas. The beauty of making your own is that you can create bespoke blends, plus the DIY approach substantially keeps the cost down without skimping on style.

1. You will need a small glass or ceramic jar with a narrow opening. This can be purpose-bought or found in a charity shop.

2. Next, prepare the fragrant blend at a ratio of 25 drops of your chosen essential oil or oils – lavender, cinnamon, peppermint, rosemary, mandarin, sandalwood and sweet orange all work well – to 60ml (¼ cup) sweet almond oil. Make enough blend to fill the vessel at least a quarter of the way up. You can also add a teaspoon of 90 per cent (180-proof) vodka or perfumer's alcohol to the mix, which can help draw the oil up the multiple channels of the reed stems.

3. To finish, pour the oil blend into your diffuser jar and place the reeds in the neck so they point up. Turn the reeds once a week so that the oil permeates them fully. Replace the reeds when saturated and the oil blend when diminished.

Scented candle
Makes 1 candle

Mixture of soy wax and beeswax flakes or pellets at a 75:25 ratio (to almost fill your desired candle vessel)
Essential oils to suit mood or setting
Waxed and wired cotton wicks with metal tabs
Glass jar or suitable vessel

Store-bought scented candles are often made with synthetic fragrance and paraffin wax, which can emit harmful chemicals into the air. This can easily be avoided by making your own candles using plant wax such as soy or naturally glowing beeswax.

1. An easy way to make a candle is to mix soy and beeswax pellets or flakes at a 75:25 ratio. Pour the mix into the candle vessel (a glass jar or ceramic cup is ideal) until it reaches 2.5cm (1 in) from the top, then tip into a double boiler (or glass bowl above a pan of hot water) and repeat. Now turn on the heat and melt the wax until liquid.

2. Remove from the heat and allow to cool to around 140°C/275°F/gas mark 1, checking with a thermometer. You can now add your essential oils – experiment with citrus, floral and woody scents and blends to match your mood or setting, adding 40–50 drops of oil per candle.

3. Place the metal tab of the wick at the base of the candle vessel and hold the cotton part up while you pour the wax around it to desired depth. Hold the wick vertical by wedging it between two sticks or reeds until the wax has cooled and set.

4. Leave the candle to cure for 2 days, then light and enjoy the subtle scent and lovely, cosy glow.

All-purpose cleaner
Makes 2 (500-ml/16-oz) spray bottles

250ml (1 cup) white vinegar (325ml/1⅓ cups for a more active solution)
250ml (1 cup) water (175ml/¾ cup for a more active solution)
1 x 500-ml (16-oz) glass or plastic spray bottle
Up to 50 drops of naturally cleaning essential oils, such as lavender, lemon, peppermint, eucalyptus or tea tree

Making your own multi-purpose cleaner using a solution of white vinegar, water and naturally antibacterial, antiviral or antifungal essential oils is a great first step to removing unnecessary chemicals from your home. This is no old wives' tale – it really does work, thanks to the acetic acid content of vinegar, which helps combat germs and mould, loosens mineral deposits and dissolves soap scum, combined with naturally antibacterial, antiviral, antifungal aromatic essential oils, such as lavender, lemon, peppermint, eucalyptus and tea tree.

1. To make, pour equal parts vinegar and water into a 500ml (16-oz) spray bottle – a glass spray bottle is ideal or refill a clean plastic one. Up the vinegar content to a 2:1 ratio if you want more action.

2. Add 12–24 drops of essential oil per cup of vinegar-water (250ml) – so 24–48 drops combined of your chosen oils for a 500-ml (16-oz) bottle – or until you have the fragrance you desire. Shake gently, replace spray top and use around the home to clean surfaces, avoiding reactive granite, marble, latex grout and bleach.

This handy solution can also be used to clean glass or mirrors with a soft cloth, or as a deodorising air freshener. While for extra cleaning power in the bathtub, toilet or oven, for example, sprinkle a little baking powder on surfaces first to make a paste, then scrub away with a sponge.

Lemon magic
Various applications

1 lemon

If there's one plant that deserves an award for all-round natural 'green' cleaning, it has to be the lemon (*Citrus limon*), thanks to its naturally antiseptic and antibacterial properties, neutralising and reactive citric acid, and its amazing scent.

* One great trick is to collect discarded lemon (or other citrus) peels in a 1-litre (32-oz) mason jar. When the jar is half full, pour white vinegar over the top, seal and leave in a dark place for 2 weeks. Strain and use the liquid as a fresh-scented multi-purpose cleaner.

* To clean a microwave, combine the juice of 1 lemon with a cup of water in a suitably-sized bowl. Add the squeezed lemon halves, microwave on full power for 3 minutes and let stand for a further 5 minutes so that the citrus steam has a chance to loosen any grime. Then clean away any residue with a soft cloth.

* Use lemon and salt to clean copper pans, or use lemon halves or juice to wipe and add sparkle to shower doors, taps or non-brass-plated cutlery (finish all with a rinse of water).

* A cup of lemon juice in the bottom shelf of your dishwasher, run on a rinse cycle, will also freshen and clean your machine.

* If your whole room could do with a refresh, add the peel of several lemons to a pan of boiling water, and let those neutralising and aromatic properties permeate the air – a quick, easy and non-toxic way to energise your home and boost your mood at the same time.

Marlene Adelmann, the Herbal Academy, Boston, United States

Makes 900ml (32oz)

Herbal cleaning spray

Use this fragrant, naturally cleansing spray for cleaning surfaces in the kitchen and bathroom.

75g (1 cup) finely chopped fresh citrus peels (orange, lemon, and/or grapefruit), or 28g (¾ cup) dried mint family herbs (lavender, thyme, sage, rosemary, and/or lemon balm)
500ml (2 cups) white vinegar
500ml (2 cups) distilled water
20–40 drops of lavender, tea tree, and/or sweet orange essential oils (optional)

1. Combine the finely chopped fresh citrus peels or mint family herbs with white vinegar in a 1-litre (32-oz) canning jar.

2. Place a piece of wax paper between the lid and the jar to prevent corrosion of the metal lid. Seal the jar.

3. Let steep for 2–4 weeks in a dark place, shaking every couple of days. Top up with more vinegar if needed to ensure that herbs are covered.

4. Strain the herbs from the vinegar using a fine-mesh sieve or a clean coffee filter.

5. Combine the distilled water and 500ml (2 cups) of herbal infused vinegar into a 900ml (32-oz) spray bottle.

6. If desired, add 20–40 drops total of lavender, tea tree, and/or sweet orange essential oils.

7. Shake well before using. Spray the surface with the solution and wipe with a clean cloth.

MAKE THE CONNECTION
Find out more about Marlene's botanical world at www.theherbalacademy.com; @herbalacademy

Opposite Lemons (*Citrus limon*), such as this bijou cultivar illustrated by Walter Hood Fitch (see page 323) for *The Florist and Pomologist* periodical (1878–1884), are quite simply one of the most useful fruits on the planet.

MY BOTANICAL WORLD...
'*Plants are our nourishment, medicine, shelter, clothing, and inspiration for art; we are their caretakers and cultivators. Together, we are part of the beautiful and mysterious cycle of life. I am fortunate that I get to teach and share this love every day with my work at the Herbal Academy. The school is building a herbal community and collectively we have chosen to study the plant world, discovering plants on an intimate level – how they grow, their life cycles, how they protect themselves, and how they nurture and support us. Quite simply, people and plants are meant for each other.*'

W. H. Fitch, del.

*'Painting from nature is not copying the object;
it is realizing one's sensations.'*

Paul Cézanne
French artist and Post-Impressionist painter (1839–1906)

CHAPTER SIX

INSPIRED BY NATURE

A BLOOMING HISTORY

Flowers and plants are an obvious muse for their ephemeral beauty, realms of discovery, aptitude for symbolism and because they are subject to decay, but what's most apparent from millennia of art history is how wonderfully varied our view of the botanical world can be.

In 2017, the British artist Gavin Turk curated an exhibition at the Museum Van Loon, in Amsterdam, entitled *Turkish Tulips*, in which he invited a selection of artist friends and contacts to contribute tulip-inspired works – among them Sir Peter Blake's *Tulips after Van Brussel, Flowers in a Vase* (1995–1996), Michael Craig-Martin's *Tulips (after Mapplethorpe)* (2016), Matt Collishaw's *Tulip Mania* (2017), the late Rory McEwen's exquisite botanical watercolour on vellum *Old English Striped Tulip 'Sam Barlow'* (1974–1976), and Turk's own bronze box of *Turkish Tulips* (2017).

In this challenging exhibition, the artists reference the birth of capitalism as well as the beginning of academic learning in the University and Museum tradition taken for granted today. The artworks explore the history of trade routes between the Middle East and Europe through to the modern iconography of the tulip emoji.

For those with botanical leanings, it's an obvious coming together – all those tulip-inspired works hanging alongside each other in a space that has a floral history of its own, in a country with a renowned past and present flower trade. But on deeper reflection the historic depiction of flowers and the presentation of such works is surprisingly complex. 'Botanical art' and 'botanical illustration' happily sit side by side, but are not so often displayed next to 'flower paintings'. Indeed, the notion of a 'flower painting' genera itself isn't straightforward outside of nominal groups such as the 'Dutch Flower Painters' (see page 72), given that many artists see plants, flowers or nature as just one potential theme among many.

However, there are recurring motifs as artists, designers and craftspeople actively choose plants and flowers (such as the tulip) as their main muse or medium. In this case, plants are the most compelling or obvious way to both explore the world and engage others at the same time. While art history has for the most part explored such groups of works separately, it's also liberating to view them in closer proximity.

Thus, illustrator Katie Scott and digital artist Macoto Murayama are found within a section on Botanical Art & Illustration (see page 312) alongside Pierre-Joseph Redouté and Rosie Sanders; Rob Kesseler's zoomed-in shots of pollen rub shoulders with the surreal art-fashion photography of Viviane Sassen in Photography & Film (see page 332); Dutch flower painter Rachel Ruysch shares the pages of Painting & Drawing (see page 344) with the hyper-realistic cacti of Kwang-Ho Lee; Rebecca Louise Law's ephemeral floral installations and NILS-UDO's environmental art invite conversation in Sculpture & Installation (see page 356); and plant-inspired William Morris prints segue into online sensation Plants on Pink in Craft, Design & Style (see page 368).

There may not be room to include all the creative genii who were ever touched by a plant or flower but there is surely enough aesthetic and thought-provoking brilliance to pay homage to perhaps one of the most potent qualities of the botanical world – its unrelenting ability to inspire us.

Opposite *A Group of Tulips*, from Robert Thornton's *Temple of Flora* (1807). Engraved by Richard Earlom.

ART FORMS IN NATURE

A chapter entitled 'Inspired by Nature' must surely pay homage to the art of nature itself, from the curious phenomenon of symmetry to Fibonacci sequences in plant forms and flowerheads. Indeed, nature's mandalas are all around.

The landmark work of German biologist and artist Ernst Haeckel (see page 319) – *Art Forms in Nature*, published between 1899 and 1904 – is one of the best introductions to patterns in nature. In this work, Haeckel depicted numerous life forms including those from the Plant and Animal Kingdoms, displaying them in such a way that the graphic nature of their markings, textures, shapes or structures were brought to the fore, including one of the most fascinating of all: symmetry.

Symmetry, the 'quality of being made up of exactly similar parts facing each other or around an axis', is a curiously ubiquitous motif, one that is deeply ingrained in our sense of beauty. In flowers, there are various types of symmetry to hunt for.

Actinomorphic (also known as radially symmetrical or regular) flowers are star-shaped or radial, and can be divided into three or more identical sectors that relate to each other by rotation around a flower's centre. Buttercups (*Ranunculus*), bindweed (*Convolvulus*) and wild roses such as dog rose (*Rosa canina*) fall into this group. Zygomorphic flowers can only be divided by a single plane into two mirror-image halves and include orchids (Orchidaceae), sweet peas (*Lathyrus odoratus*) and snapdragons (*Antirrhinum*). While very few plant species are assymetrical – Valerian (*Valeriana officinalis*) and African arrowroot (*Canna indica*) are a couple of odd ones out.

In some cases, symmetry is not readily discernible. Daisies and dandelions, for example, are actually clusters of tiny (not necessarily actinomorphic) flowers arranged in a roughly radial symmetric inflorescence known as a pseudanthium. The so-called 'petals' of this type of flower are actually individual zygomorphic ray flowers, while the clusters of disc flowers in the centres are actinomorphic – many flowers for the price of one.

Early Greek philosophers such as Plato and Pythagoras spent many an hour trying to explain such orders in nature. As did an Italian mathematician known as Fibonacci (c. 1175–1250), who popularised the Hindu-Arabic numeral system in the Western world via his 1202 book *Liber Abaci* (*Book of Calculation*). He also took Europe by storm with his sequence of Fibonacci numbers, where every number after the first two is the sum of the two preceding ones: 0, 1, 1, 2, 3, 5, 8, 13, etc.

Such sequences are commonly found in plants, often revealing themselves as spirals. The phyllotaxis (arrangement of leaves on a plant stem) of spiral aloe (*Aloe polyphylla*) is one example; the centre of a sunflower (*Helianthus annuus*) probably the most well known. Once you notice nature's Fibonacci sequences it can actually be hard to stop seeing them everywhere, from flowerheads and pine cones to pineapples and the outlandishly fractal Romanesco broccoli (*Brassica oleracea*).

Our relationship with such natural phenomena has been to wonder at them, try to explain them and copy them for our own ends in the name of spiritual mandalas, art and biomimicry, resulting in some amazing visions and inventions. Actively look for the patterns in nature and it really does open up a whole new world.

Opposite The forms of plants and flowers are naturally pleasing to the human eye, and offer a wealth of inspiration, as illustrated by Lynn Hatzius' prismatic *Botanical World* mandala (see page 373).

BOTANICAL ART & ILLUSTRATION

Botanical art experts such as Wilfrid Blunt, William T. Stearn, Martyn Rix, Margaret Stevens and Dr Shirley Sherwood are understandably keen to highlight the difference between 'botanical illustration' and 'botanical art'. In short, the former is highly accurate as required for scientific purposes while the latter retains more scope for artistic expression (see page 79). The following collection of 'botanical artists' seeks to push those boundaries further, juxtaposing inspiring works, past and present, with suggestions for a more digitised future.

Pierre-Joseph Redouté
Belgian botanical artist
(1759–1840)

Pierre-Joseph Redouté began his journey towards the moniker of 'the Raphael of Flowers' as a teenage scenery painter in the theatres of late-eighteenth-century Paris. There he met the botanists Charles Louis L'Héritier de Brutelle and René Desfontaines, who encouraged him to switch to botanical illustration. The affluent L'Héritier also introduced him to key figures in the court of Versailles, including Marie Antoinette, who became his patron – the first of a long line of royal and aristocratic fans of Redouté's work. Stints at the Jardin du Roi, Paris and the Royal Botanic Gardens, Kew, resulted in astutely observed botanical renderings of plants and flowers, while Redouté's natural artistic flair and charm served to influence the exquisite composition and brushwork of his frequently reproduced master compendiums, *Les Liliacées* (1802–1816) and *Les Roses* (1817–1824).

> '*A great botanical artist must have a passion for flowers. You can set a good architectural draughtsman to draw a flower, and he will give you, if he thinks the subject worthy of real effort, a careful and precise study of the plant before him. But unless he loves what he is drawing, unless he knows the flower in all its moods, in all the stages of development, there will be something lacking in his work.*'
>
> William T. Stearn and Wilfrid Blunt, *The Art of Botanical Illustration* (ACC Art Books, 2015)

Georg Dionysius Ehret
German botanical artist
(1708–1770)

Georg Dionysius Ehret's happenstance journey from taking drawing lessons from his market-gardener father to becoming one of the greatest botanical artists of all time also reads as something of a Who's Who of eighteenth-century botanical breakthroughs. While the great Swedish botanist Carl Linnaeus is credited with formalising the modern system of binomial nomenclature (see page 123) in *Systema Naturae* (1735), it is Ehret's illustrated tabella of Linnaeus's *Methodus Plantarum Sexualis* (*Sexual System of Plants*) that made the idea infinitely more palatable. Despite not receiving a credit, the ever-progressive Ehret ran with his newfound knowledge, introducing reproductive details of flowers, fruit and seeds into his beautifully coloured, highly realistic, yet majestically rendered watercolour illustrations and engravings – helping to set the scene for the 'Golden Age of Botanical Art'.

Opposite *Iris pumila floribus caeruleis* from Redouté's collection *Les liliacées* (1802–1816). Find an example of *Les Roses* on page 78. **Right** Ehret depicts frangipani (genus *Plumeria*), a fragrant member of the dogbane family.

Katie Scott
Illustrator
(1988–)

Katie Scott's work appeals on so many levels, but perhaps the most vital of these is her ability to bring the intricate beauty and connectedness of life on Earth to people of all ages. In terms of plants, this is illustrated in the botanical feast that is *Botanicum* (Big Picture Press, 2016), part of a series of vintage-style 'Welcome to the Museum' themed books, of which Scott also illustrated *Animalium* (2014). With text by Professor Kathy Willis, Kew's Director of Science and author of *Plants: From Roots to Riches* (John Murray, 2014), *Botanicum* opens with an invitation to explore: 'This is no ordinary museum. Imagine you could wander through every field, wood, tropical rainforest and flower glade in the world. Think what it would be like if you could see the most beautiful, exotic and weird plants all at once. Have you ever wondered what you would see if you could stroll back in time, to the beginnings of Life on Earth? You can in the pages of *Botanicum*.'

Thus follows a journey through the Plant Kingdom in all its verdant and technicoloured glory, at once sitting comfortably with the lineage of scientific illustration while playing with fantasy and the aesthetics of contemporary design – a style that has also attracted numerous commercial clients, collaborations and a growing personal fanbase. While inspired by all sorts of old medical and natural history artwork, including weird and wonderful illustrations by the likes of Ernst Haeckel (page 319), Fritz Kahn and Albertus Seba, Scott veers away from simple reproductions or scientific renditions in favour of a more contemporary, imaginative visual language, with its own distinctive palette and space to play with reality and fantasy and draw on multiple influences. That said, her illustrations for *Botanicum* began with life drawings of specimens in the Royal Botanic Gardens, Kew, of which she was given free rein. By getting up close and personal with plants, Scott could observe firsthand their shapes, structures, growth patterns and colours before manipulating preparatory drawings into style back in the studio.

In 2017, Katie collaborated with Japanese flower artist Azuma Makoto (page 359) and London-based motion graphics designer James Paulley, to create *Story of Flowers* – a lush, wordless botanical animation that tells the story of the continuous, endless cycle of nature.

MAKE THE CONNECTION
Find out more about Katie's botanical world at www.katie-scott.com; @katiekatiescott

BOTANICAL ART & ILLUSTRATION

Below *Giant Waterlily* from Katie's richly illustrated *Botanicum* (Big Picture Press, 2016) **Opposite** There is tremendous variation among the members of the Cactaceae family, in another print from *Botanicum*.

My botanical world . . .

'*From an artistic perspective I'm drawn to plants as a seemingly endless resource for inspiration, there's so much more to study than can ever be seen in a lifetime. Even in London I regularly find new plants, forms and structures just in front gardens. You may think you know a particular species quite well, but you might never before have noticed what the seedpod looks like in the first stage of forming, or the underside of the leaf.*'

INSPIRED BY NATURE 315

Franklinia alatamaha (Franklin tree)

Eucalyptus pruinosa (silver-leaved box)

William Bartram
American naturalist/artist (1739–1823)

Third son of the great American botanist John Bartram, William Bartram was born and bred among the plants and flowers of Bartram's Gardens (now North America's oldest surviving botanic garden). Inspired by travels with his father and as commissioned by fellow Quaker and British botanist Dr John Fothergill, William organised an extensive plant-hunting expedition throughout the southeast of America (1773–1777), observing plants, animals and the lives of Native American peoples such as the Cherokee peoples. His flowery penmanship went on to influence writers and poets such as William Wordsworth and Samuel Taylor Coleridge, and extended to scientifically accurate yet romantically composed illustrations of native flora and fauna. First published in 1791, these insightful renderings are now known as *Bartram's Travels*.

Franz and Ferdinand Bauer
Austrian botanical artists (1758–1840 and 1760–1826)

Franz and Ferdinand Bauer first learned their trade from Dr Norbert Boccius, botanist and prior of the monastery at Feldsberg, who took them in upon the untimely death of their father, court painter to the Princes of Liechtenstein. They became acute observers of nature, propelled by the introduction to the Linnaean system of classification and microscopy by eminent Viennese botanist Nikolaus von Jacquin. Although strikingly similar in artistic style and attention to detail, the Bauer brothers followed their own paths. Franz was hired by Sir Joseph Banks as the first resident artist at the Royal Botanic Gardens, Kew, pioneering the use of microscopes for greater illustrative detail. Ferdinand, meanwhile, accepted a post as natural history artist on HMS *Investigator*, documenting many undescribed species found on the journey around Australia.

Cydonia oblonga (quince)

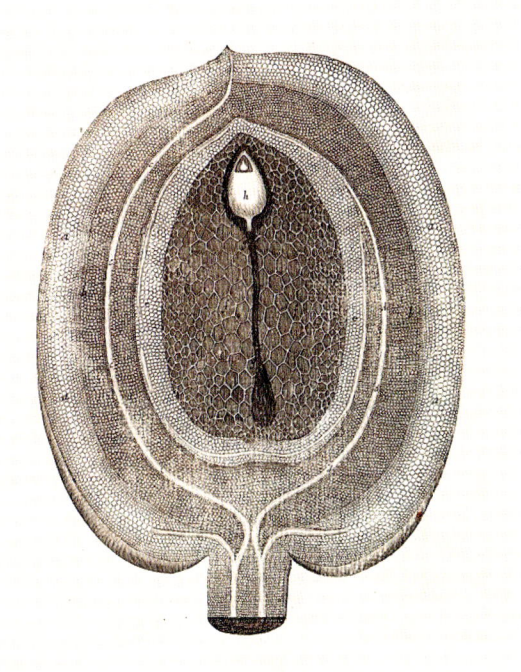

Slice of Young Fruit. From *Anatomy of Plants*, 1682.

Elizabeth Blackwell
Scottish botanical illustrator
(1707–1758)

Best known as the artist and engraver for the plates of *A Curious Herbal* (1737–1739), what's less well known are the curious circumstances in which this New World field guide was created. Following a secret marriage to her somewhat shady cousin Alexander Blackwell, Elizabeth soon found herself destitute when Alexander's medical practice was exposed as illegitimate and he was sent to debtor's prison. To make ends meet she determined to illustrate a new herbal commissioned to describe exotic species of the New World; botanically-trained Alexander would write the descriptions of plants from his cell. Elizabeth duly moved near to the Chelsea Physic Garden (established in 1673) where she could draw and hand-colour the New World plants that grew there from life. The resulting book included 500 such meticulous yet expressive illustrations, and was enough of a success to secure – for a while – Alexander's release.

Nehemiah Grew
English plant anatomist and physiologist
(1641–1712)

Nehemiah Grew's four-part, 82-plate *Anatomy of Plants* (1682) revealed for the first time the inner structure, function and 'mechanical way' of plants. The work earned him the title 'Father of Plant Anatomy', establishing – along with fellow microscope aficionado Marcello Malpighi – the observational basis for botany for the next 100 years. Using comparative analysis, he observed differences in pollen size and shape between species and various components of seeds. He was also the first to observe sap circulating through tissue, the complicated folding of expanded leaves within a bud, and note that the stamen was the male organ of a flowering plant. While Nehemiah's illustrations certainly enhance his philosophies and findings, it's also clear that he found great satisfaction in their engineering – each finely drawn graphic sings in its own right.
The picture above, from *Anatomy of Plants*, shows a magnified slice of a young fruit.

II. Papaver flor. miniato plens.   I. Papaver multiplex album oris rubicundis.   III. Papaver flore plens argentei color.

Basilius Besler
Bavarian apothecary and botanist
(1561–1629)

Hortus Eystettensis (1613), Basilius Besler's masterpiece codex, took no less than 16 years to complete but when finally published it changed the face of botanical art overnight. The emphasis had previously been on medicinal and culinary herbs, with illustrations often serving as appendages to the written text. Commissioned by the Bishop of Eichstätt in Bavaria to produce a record of his garden, Besler's presentation included 1,084 species of garden flower, herb, vegetable and exotic plant, depicted at life size where possible. Given the large-scale format, the inclusion of 367 intricately-detailed copper plates and the option of a luxury hand-coloured version, Besler could afford to demand a high price for his work and thus finally earn some well-deserved prosperity. His seasonal showpiece also prospered, as one of the most influential, stunningly beautiful botanical artworks of all time.

> 'What is the difference between botanical art and botanical illustration? In art, the finished painting is the whole object of the artist, and it has no further purpose than to be admired. A botanical illustration has scientific purpose, to illustrate a book or act as a record of a plant species or plant part. The illustration should have a generality that ignores the imperfections of the individual specimen, and so can represent a species or particular form of a species.'
>
> Martyn Rix, *The Golden Age of Botanical Art* (Andre Deutsch Ltd., 2012)

Ernst Haeckel
German biologist, naturalist and artist
(1834–1919)

Nearly 300 years after Basilius Besler presented his gamechanging *Hortus Eystettensis*, German-born biologist, naturalist, artist, philosopher and doctor Ernst Haeckel presented a similarly groundbreaking view of the world via *Kunstformen der Natur* (1904), or *Art Forms in Nature*, as it is now commonly known. But while Besler scaled up, Haeckel zoomed in, exploring the minutiae of nature through the all-seeing lens of a microscope. He also illustrated not just from the front but through gaps and holes in organisms so as to create 3D images, unrivalled for their graphic precision, meticulous attention to detail and emphasis on the essential symmetries and order of nature. It's clear that Haeckel could find the biological beauty in almost anything and by doing so show just how connected nature's numerous art forms indeed were.

Opposite Three varieties of poppy (genus *Papaver*) from Besler's *Hortus Eystettensis*. **Right** Haeckel's *Orchidae* depicts the dramatic orchid family.

Jess Shepherd / Inky Leaves
British botanical artist and curator
(1984–)

Take a leaf, study it closely for the duration of its life cycle, zoom in with a microscope to observe the smallest detail, before heading back into the field to record the soundtrack of the leaf's natural habitat. Then weave these components together into a realistic yet emotionally charged painting, blown up to incredible proportions, pushing the boundaries of page and quite possibly existence. Welcome to *Leafscape* by Jess Shepherd (also known as JR Shepherd), a series of images that illustrate the 'personality' of plants and flowers yet tread the boards of scientific illustration at the same time. While Shepherd originally trained as a botanist, gaining an MSc in plant taxonomy and working in the University of Plymouth's herbarium for a number of years, art was always there in the background. Growing up in a museum environment where her mother worked as a ceramicist, she developed a curatorial view of the world. This was further cemented by a later stint at the Shirley Sherwood Gallery at the Royal Botanic Gardens, Kew, where she developed her vision: to enrich current perceptions of botanical art, its applications and how it sits within the larger scope of the visual arts. *Leafscape*, presented as an exhibition and artist's publication in 2017, certainly achieves this aim; each larger-than-life watercolour painting forces the viewer to look closer both at the work itself and the Plant Kingdom at large.

Inspired by the process and the response, Shepherd is currently working on a new series of investigative blue flower paintings that will take her around the world. While *Leafscape* kept her largely studio-bound, observing the plant forms of her garden, the *Blueflower* project requires a peripatetic lifestyle across landscapes as far afield and diverse as the Himalayas (in search of the blue poppy), Thailand (to find the blue vanda) and France (to pay her respects to the cornflower, or Bleuet de France). Blue, as Shepherd explains, is a rare colour in nature, yet it is a shade that explores the full spectrum of emotions and deepens, fades or glows with the given light. Inspired by the Early German Romanticist writer Novalis, who sent his protagonist in search of a blue flower in his unfinished novel *Heinrich von Ofterdingen* and by the lustrous, light-filled works of artists such as Rory McEwen, the quest for Shepherd's manifested 'cure for plant blindness' continues.

My botanical world . . .

'*When I paint plants I try to become them. There is a great deal of observation involved even before my brushes touch the paper. I have to understand the personality of a plant through an entire life cycle in order to paint it properly and go beyond depicting the mere physical.*'

MAKE THE CONNECTION
Find out more about Jess's botanical world at www.inkyleaves.com; @inkyleaves

Opposite *Castor Oil Plant* (Ricinus communis) *leaf*, 2016. Watercolour on paper, 76 × 56cm (30 × 22 in).

Protea and Golden-breasted Cuckoo, of South Africa.

Rosa Caucasea

Marianne North
Engish biologist and botanical artist (1830–1890)

To get a true sense of the spirit of Victorian botanical artist Marianne North, pay a visit to the Marianne North Gallery located in the Royal Botanic Gardens, Kew. There, you'll be met by 833 wall-to-wall paintings of tropical and exotic plants and flowers made by the intrepid artist while travelling around the world, often to areas she was strongly persuaded not to visit. Although North did not receive formal training in illustration, it is obvious that she possessed a natural talent that allowed her to sketch rapidly in pen and ink on heavy paper and then complete the story with a vivid palette of oil colours. Unlike many Victorian flower painters, she preferred to paint plants in their natural setting, going to great lengths to meet her brief. Spot the depictions of tea and coffee over her gallery doorways – a rebellious action taken when she was denied the pleasure of serving drinks to visitors.

John Lindley
English botanist, illustrator and orchidologist (1799–1865)

There is much for which to thank John Lindley: recommending that Kew Gardens should become a centre of botanical science complete with herbarium and library; posthumously lending his name and vast collection of books to the Royal Horticultural Society's Lindley Library; publishing and editing various books and periodicals including his expert tome *The Genera and Species of Orchidaceous Plants* (1830) and horticulture periodical *The Gardeners' Chronicle* (founded in 1841 and later absorbed by *Horticulture Week*); and organising the first annual flower show in Britain. Lindley was also a talented artist, illustrating his own monographs such as *Rosarum monographia; or, a botanical history of roses* (1820). Lindley's polymathy, artistic talent and love of roses later manifested itself in the luminous flower paintings of Scottish artist and musician Rory McEwen (1932–1982), Lindley's great-grandson.

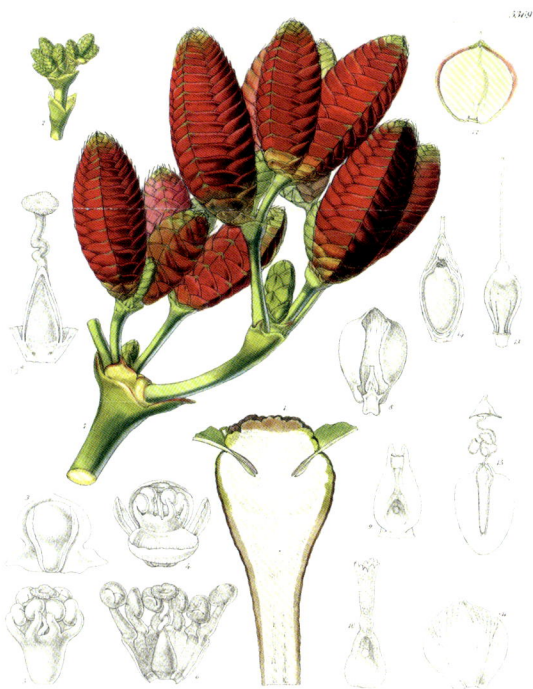

Welwitschia mirabilis (welwitschia)

Metrosideros excelsa (New Zealand Christmas tree)

Walter Hood Fitch
Scottish botanical artist
(1817–1892)

Leaf through almost any illustrated botanical or horticultural publication of significance published in Britain from the 1830s to the 1880s and you'll find a work by Walter Hood Fitch – not difficult considering that he illustrated more than 2,700 plants for *Curtis's Botanical Magazine* (founded 1787) and published more than 10,000 illustrations in total. As well as prodigious talent, Fitch also benefited from a close working relationship with William Hooker, editor of *Curtis's Botanical Magazine* and Director of the Royal Botanic Gardens, Kew. When Hooker required an illustrated description of the botanical behemoth 'Queen of the Waterlilies' – *Victoria regia* (later renamed *Victoria amazonica*), discovered in 1837 – he turned to Fitch, who duly completed a portfolio of plates illustrating all stages of the flower's epic unfolding. The Queen was impressed, further cementing Fitch's reputation and helping to secure him a well-earned pension.

Sydney Parkinson
Scottish botanical illustrator and artist
(1745–1771)

The name Sydney Parkinson will forever be synonymous with that of Joseph Banks, James Cook and the globe-trotting HMS *Endeavour*. Commissioned by Banks to travel with him on Cook's significant first voyage to the Pacific in 1768, Parkinson made nearly a thousand drawings of plants and animals, trooping on even as swarms of flies ate his paint or he was squished in a small cabin with specimens. To keep up with the pace of 'discovery', Parkinson also adapted his style of painting, completing sketches but only partially colouring them in so other artists could complete them or create new watercolours – entirely necessary as herbarium specimens would have faded during carriage. The completed works are known collectively as *Banks's Florilegium*, first published in full colour between 1980 and 1990 (as a limited edition) some 200 years after Banks first commissioned them.

Jung Koch and Quentell
Botanical wall chart producers
(Late 1800s)

There's something about a black background that can persuade one to stand back and look objectively and zoom in to study the beautifully presented detail. This is the essence of a Jung Koch and Quentell botanical wall chart, first produced by a team of three visionaries in the late nineteenth century: painter and biology professor Gottlieb von Koch, college director Dr Friedrich Quentell and teacher Heinrich Jung. Now considered as objets d'art, the charts were originally made devoid of text so that they could be studied in any language. Some popular posters include illustrations of horse chestnut, fern, primrose and hazelnut on the signature black background, but the team also produced zoological studies including the metamorphosis of a butterfly and the workings of a gorilla. Hitting new heights of classroom use and renewed popularity in the late 1950s, the charts are still produced by educational publisher Hagemann.

Arnold and Carolina Dodel-Port
Botanical chart producers
(Late 1800s)

The Dodel-Port Atlas was created by Swiss botanist Dr Arnold Dodel-Port (1843–1908) and his illustrator wife Carolina Dodel-Port between 1878 and 1883, and contains 42 charts of detailed, wall-worthy botanical schematics. Indeed, the Dodel-Ports' representations of everything from single-celled organisms and green algae to the workings of the dodder plant parasite or the pollination of quince flowers are beautiful enough to exist as works of art in their own right. However, the main raison d'être of these wonderful renderings was as visual aids to be used at 'every level of botanical teaching and in every branch of botanical knowledge', the Dodel-Port belief being that natural, scientifically reliable wall charts were 'more enlightening than the spoken word'. Find out more and pore over images in Anna Laurent's wonderful book *The Botanical Wall Chart* (Ilex, 2016).

'Europe was enjoying a golden age of scientific discovery; naturalists were exploring the globe and there was a clamouring for knowledge of the natural world. A pedagogical curiosity was no longer limited to elite salons and research; education was now considered a right afforded to all, in classrooms across the continent. And thus the botanical wall chart was born: a synthesis of art, science, and education.'

Anna Laurent, *The Botanical Wall Chart* (Ilex, 2016)

Left This Jung Koch and Quentell wall chart depicts the arum lily, with its pistil, stamen and pollen. **Opposite** *Ulothrix*, a genus of green algae, hails from an ancient branch of the Plant Kingdom.

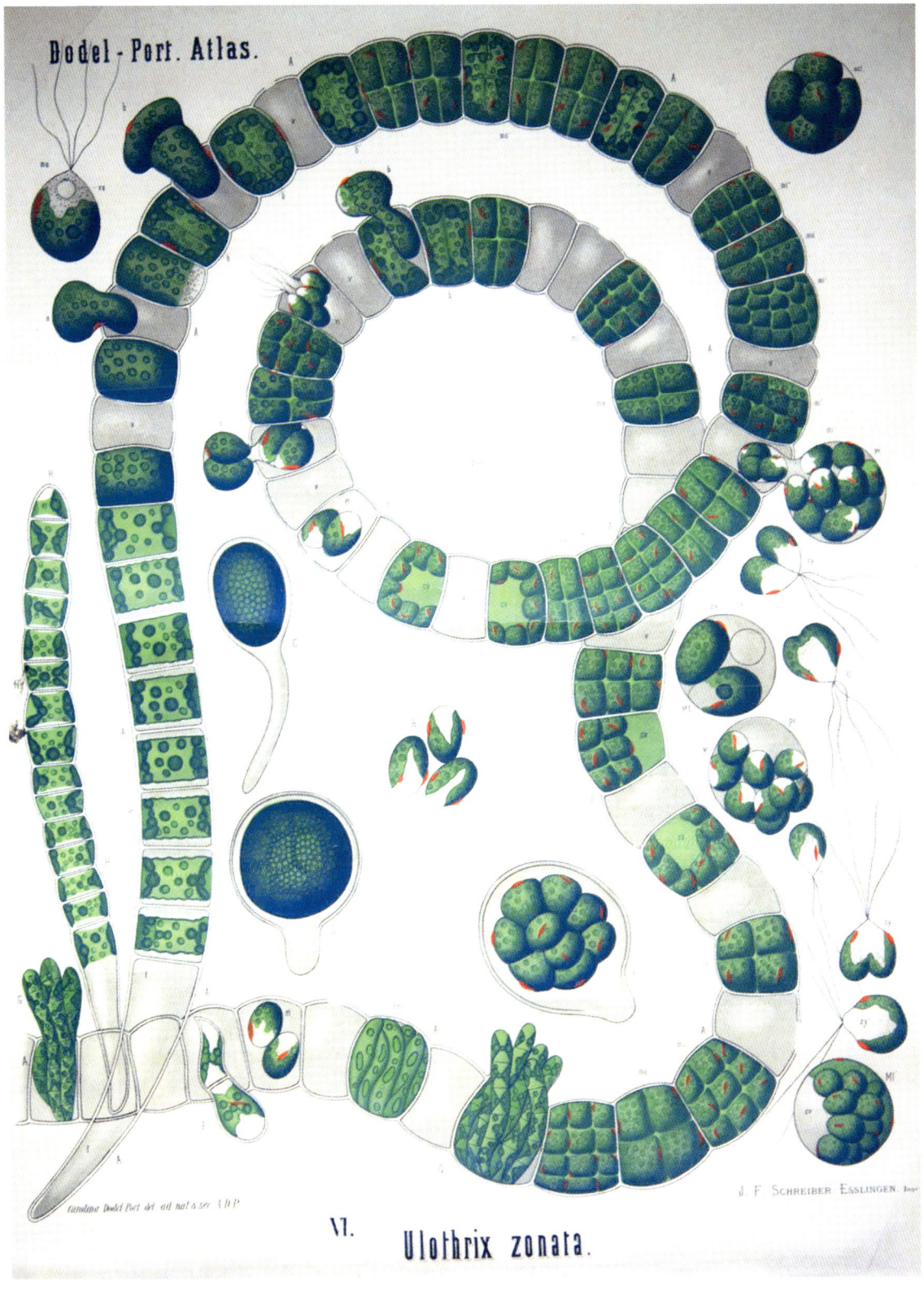

Rachel Pedder-Smith
Botanical artist
(1975–)

Rachel Pedder-Smith's route into botanical art was via an MA and PhD in Natural History Illustration at the Royal College of Art, London. Four Gold Medals from the Royal Horticultural Society, three solo shows and numerous group exhibitions later, her paintings are now included in many notable collections including those at the Alisa and Isaac Sutton Collection, the Hunt Institute for Botanical Documentation in Pittsburgh, USA, and the Lindley Library in London. However, the work that really propelled Pedder-Smith to stardom was her 2012 masterpiece *Herbarium Specimen Painting*. Created as part of her PhD – examining how material culture theory can be applied to natural history specimens – the 5-m (18-ft) long watercolour painting took 766 days to complete and depicts one specimen for each of 506 flowering plant families painted in the order of a contemporary DNA-based classification system. To create the mammoth work, Pedder-Smith studied pressed, dried plant specimens in the herbarium of the Royal Botanic Gardens, Kew. She then painted scientifically accurate illustrative renditions, in unique and striking compositions. So vivid are the details and translucently fragile the rendition, it's hard to believe that the 700 or more gnarled pods, splitting husks, curling fronds and pastel-shaded flowers are not actual specimens. A groundbreaking celebration of the Plant Kingdom, the work is also something of a hymn to Kew's labyrinthine herbarium, representing, as Pedder-Smith attests, 'at least one specimen from every year since it was founded in 1853' as well as 'the people who have come through its doors, the collections made by eminent scientists, and the stories that go with them'. Indeed, Kew's herbarium holds more than 7.5 million dried plant specimens and many handwritten fieldnotes from esteemed collectors such as Charles Darwin and David Livingstone. Pedder-Smith first visited the herbarium in 2001 as an MA student and quickly became obsessed, working from the desiccated, sculptural forms she found within. From the primitive Amborella (see page 23) to the Angiosperm Phylogeny Group II (see page 40) through to the remnants of an olive branch, cannabis leaf and piece of cinchona bark, among other species, Pedder-Smith's epic work perfectly captures the kaleidoscopic beauty of nature and the many connections and stories within.

MAKE THE CONNECTION
Find out more about Rachel's botanical world at www.rachelpeddersmith.com; @poddersmith

My botanical world...

'*My parents brought me up to love and respect nature. I have always produced artwork focusing on the natural world, particularly botanical. Herbarium specimens are the perfect subject matter for me, the aesthetic beauty and the ingenious design of the botanical world carefully pressed, displayed, catalogued and stored for eternity.*'

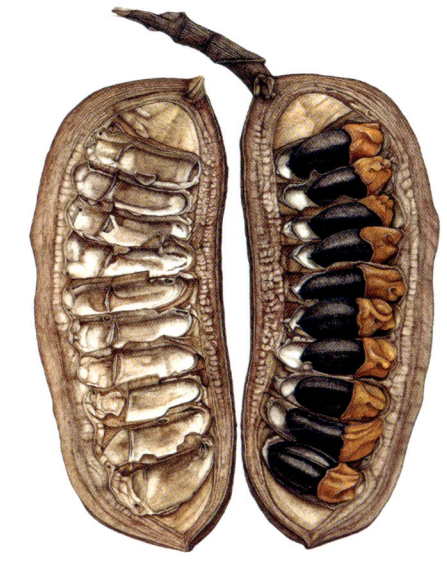

Top *Afzelia quanzensis*, 2010. Watercolour on paper. 47 × 47cm (18½ × 18½ in). **Below** *Bean painting, specimens from the Leguminosae family*, 2004. 59 × 81cm (23 × 32 in). **Opposite** Rachel's pattern *Floral Eve* (based on the painting, *Herbarium Specimen Painting*) featured on a Vivienne Westwood dress, spring/summer 2013 collection.

Bromelia antiacantha

One o'clock, two o'clock, three o'clock ... 2014. 50 × 50cm (20 × 20 in).

Margaret Mee, MBE
British botanical artist
(1909–1988)

Artist-cum-lifelong activist Margaret Ursula Mee used her art to draw attention to the flora and environmental plight of the Amazon rainforest in Brazil. Working as botanical artist to São Paulo's Instituto de Botânica in 1958 – a role she gained in her early forties after heading out to Brazil to teach art with second husband Greville Mee – she observed firsthand both the spectacular biodiversity of the region and the devastating effects of large-scale mining and deforestation. These observations were recorded as hundreds of folios of gouache illustrations, often with the natural habitat of plants in the background, and in scores of sketchbooks and diaries, brought together in a major exhibition at Kew in 1988 and the publication *Margaret Mee in Search of Flowers of the Amazon Forest* (Nonesuch Expeditions Ltd, 1988). Sadly, Mee died shortly afterwards in a car crash but she is remembered for her courage and flair.

Rosie Sanders
British botanical artist and printmaker
(1944–)

Award-winning, self-taught botanical artist and printmaker Rosie Sanders paints fruit that looks good enough to eat, as well as luxuriant, often larger-than-life watercolours of flowers and plant details that ache with sensuality. First inspired by the accurary of Kew's herbarium artist in residence Mary Grierson (1912–2012) and the minimalism of John Lindley's great-grandson, the artist Rory McEwen, she has been pushing at the boundaries of botanical representation for more than 40 years – backlighting stems to create a unique luminosity, capturing plants at unusual angles and delighting in realistically observing nature's imperfections. The visceral beauty she so wonderfully captures is exemplified in various publications including *The Apple Book* (Frances Lincoln, 2010) and the large-format *Rosie Sanders' Flowers* (Pavilion Books, 2016) with a text by Swiss botanist Dr Andreas Honegger.

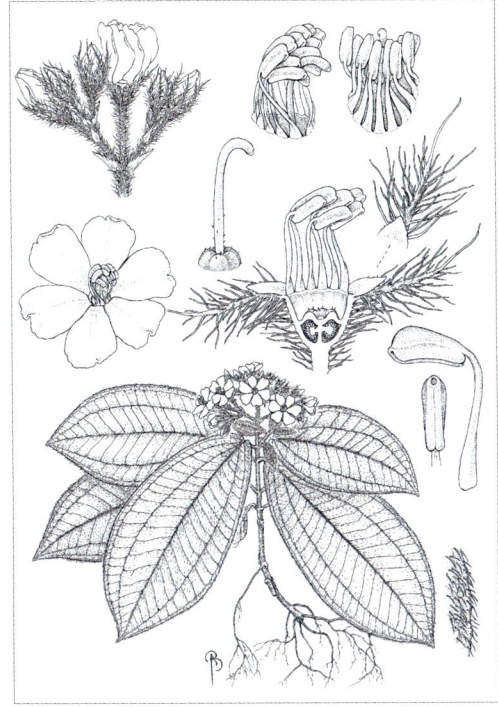

Physeterostemon aonae, 2016. 35.5 × 25.5cm (14 × 10 in).

Rosa rubiginosa

Bobbi Angell
American botanical illustrator and artist
(1955–)

Bobbi Angell deals primarily in monochrome depictions of plant specimens. Graduating with a degree in botany from the University of Vermont in 1977, she has been drawing plants ever since, making her mark with richly detailed pen and ink illustrations and beautifully produced copper etchings. Working with herbarium specimens, many of her illustrations represent species new to science, created through Angell's role as scientific illustrator for the New York Botanical Garden, among other academic institutions. Others graced the weekly gardening Q&A column of *The New York Times* for many years and have appeared on commercial offerings. In order to fulfill this varied illustrative role, Angell cultivates plants in her gardens in southern Vermont and has traveled widely, searching for specimens and sketching 'in the field' in far-flung places such as French Guiana and Ecuador. She is co-author of *A Botanist's Vocabulary* (Timber Press, 2016).

Niki Simpson
British digital botanical artist
(1955–)

Niki Simpson's unique work stems from a 2003 award to investigate digital techniques for botanical illustration, aiming to combine the best of old traditions with the technologies and demands of the Digital Age. Each composite illustration provides a 'comprehensive plant portrait of a plant species, showing the diagnostic and, where space allows, also the characteristic features of that plant'. Simpson uses scanning electron micrographs, digital drawing and scanned hand-drawing, as well as digital photographs, to depict each plant throughout the year, and a voucher herbarium specimen is created for each plant. Traditional features such as scale bars and letters referring to the caption are used, but new features include colour references, time-bar and botanical symbols to indicate the plant's life form and sexual arrangement. When viewed onscreen, zoom tools can be used to explore Simpson's micro-botanical world.

Macoto Murayama

Japanese artist
(1984–)

The Digital Age hurtles forwards at ever greater speeds, yet botanical illustration has remained largely within the confines of drawing and painting. Why? According to experts, while photography is a useful tool it often shows too much where illustration allows the choice to highlight, dim or eliminate certain features or details. It's a philosophy that has certainly helped to keep the fine art of botanical illustration alive, but this doesn't mean that there isn't or shouldn't be room for exploration, especially when the results are as thought-provoking as those produced by Japanese artist Macoto Murayama. Hailing from an architectural design background, he learned how to build architectural diagrams using computer-aided design (CAD) and computer generation (CG) and was inspired to similarly reconstruct plants after a university tutor showed him a number of meticulously detailed natural history illustrations. Ideas about a potential marriage of art, science and technology began to flow and the seeds for Murayama's otherworldly garden of *Inorganic Flora* were sown, with the first crucial step of the process being to head into the field and go back to the live form. As such, Murayama carefully selects and finds his plant specimens, dissects them with a scalpel and then observes them with a magnifying glass. Next come sketches and photographs before plants are reconstructed on the computer to transparent effect using 3DCG software, with final composition produced in Adobe Photoshop. By including plant details and imposing measurements, part names, scale and scientific names, Murayama is referencing both scientific diagrams and architectural blueprints. The final step towards ultimate scrutiny is to print the results on a large-scale printer. Never has the fragrant yet fragile sweet pea looked so regal than via Murayama's circumspective views of *Lathyrus odoratus*, nor a *Phalaenopsis* (moth orchid) looked so explicitly inviting – it might almost be the pollinator's view. As an extension of these works, Murayama also experiments with what he calls Botech Art, a symbiotic portmanteau of 'Botanical Art' and 'Technology'. Some Botech works appear like ghostly jellyfish floating in an inky sea, others as vibrant pulsating mandalas – a digital insight perhaps of a plant's innermost vibrational frequency.

My botanical world . . .

'*The illustration of a flower with overwhelming descriptive information, besides being beautiful and connecting art and science, reveals the elaborate structure, the general principle involved in the formation of the organism. It sheds light on the essence of creation itself.*'

MAKE THE CONNECTION
Find out more about Macoto's botanical world at www.frantic.jp/en; @franticgallery

BOTANICAL ART & ILLUSTRATION

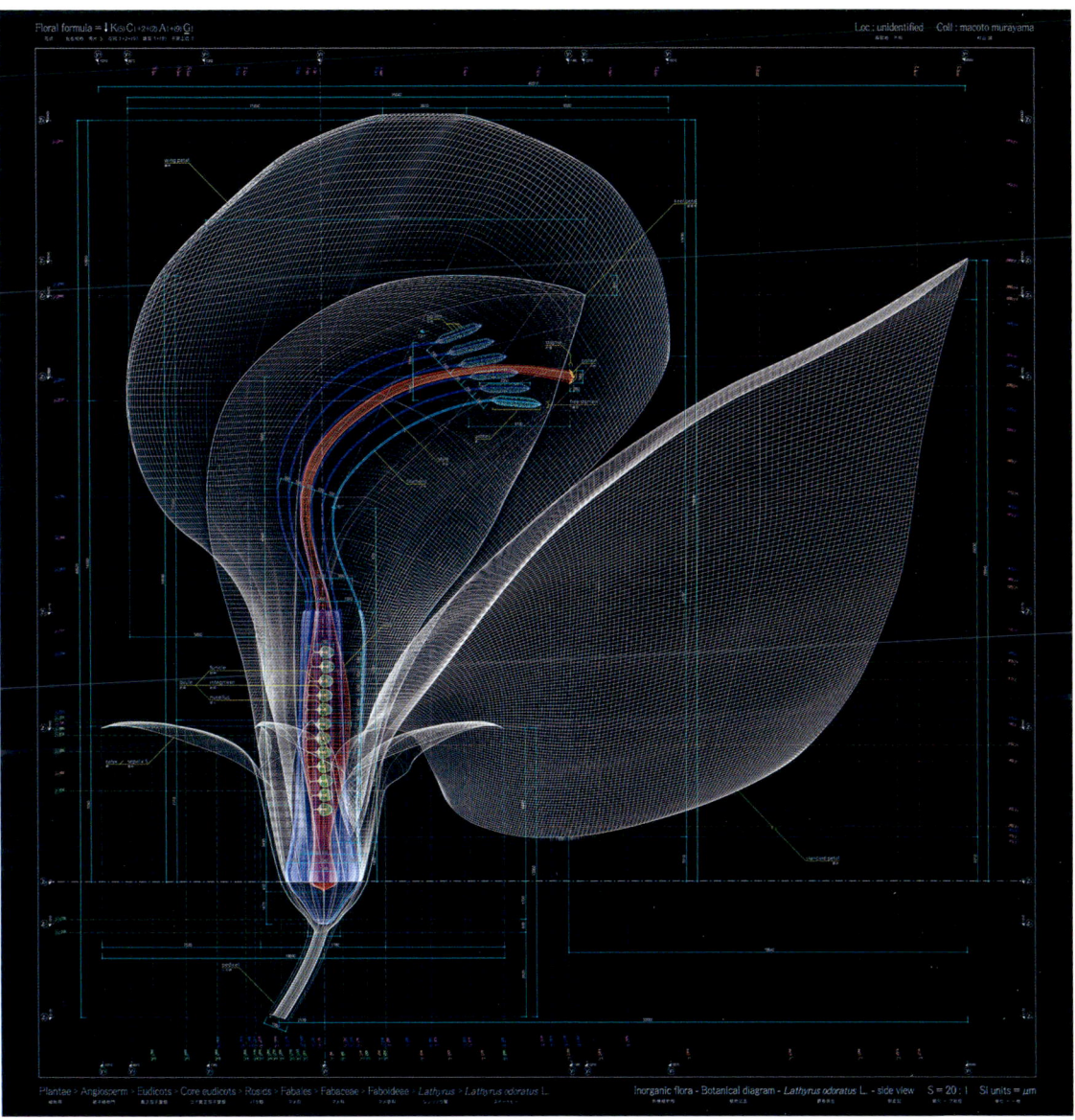

Above Macoto Murayama, *Lathyrus odoratus* L – side view – b, 2012.
Image courtesy of Frantic Gallery. More about the artist: www.frantic.jp.

PHOTOGRAPHY & FILM

The birth of practical photography in 1839 followed by motion picture cameras in the 1890s threw a whole new light on ways of capturing and recording our visions of the world, including the Plant Kingdom. Fast-forward to the twenty-first century and we are well into the throes of the 'Digital Age'. Yet while our senses are continuously bombarded with images, many of them of plants or nature, some will always stand out – for their exquisite composition, imaginative concept, or ability to help us look closer or see the bigger picture.

Karl Blossfeldt
German photographer and artist
(1865–1932)

Today's leading macro photographers such as Minghui Yuan and Joni Niemelä owe much to the pioneering German photographer, artist and teacher Karl Blossfeldt. With the birth of practical photography, generally accepted as 1839, it was still very much in a fledgling state when Karl Blossfeldt first picked up a camera towards the end of the nineteenth century. Unperturbed by a lack of technology, he developed a series of home-made cameras that allowed him to zoom in on plants, flowers and seedheads and observe nature's textures, abstract forms and repetitive patterns in unprecedented magnified detail. Primarily used as teaching aids for architects, sculptors and artists, he finally published a collection of these works in the landmark book *Urformen der Kunst* (*Art Forms in Nature*) in 1928, followed by *Wundergarten der Natur* (*The Wondergarden of Nature*) in 1932, launching a new direction in photography and the observation of plants.

> '*He [Karl Blossfeldt] has done more than his share of that great stock-taking of the inventory of human perception that will alter our image of the world in as yet unforeseen ways . . . These photographs reveal an entire, unsuspected horde of analogies and forms in the existence of plants.*'
>
> Walter Benjamin, German Jewish philosopher (1892–1940), in *News About Flowers* (1928), referencing Karl Blossfeldt's *Art Forms in Nature* (1928)

Anna Atkins
Engish botanist and photographer
(1799–1871)

Thanks to her father's friendship with photography pioneer William Henry Fox Talbot and cyanotype inventor Sir John Herschel, Anna Atkins was well-placed to turn her visions into reality – and take the title of first female photographer to boot. In 1843, just a year after Herschel taught her how to make cyan-blue photograms by placing objects on chemically coated paper, Atkins self-published the first of three volumes of cyanotypes of dried algae in *Photographs of British Algae: Cyanotype Impressions* (1843–1853). In the 1850s she then collaborated with Anne Dixon to produce three more collections of cyanotype photograms including *Cyanotypes of British and Foreign Ferns* (1853) and *Cyanotypes of British and Foreign Flowering Plants* (1854). The process still inspires artists today, including Tim Simpson and Sarah van Gameren of the studio Glithero, and their 'Blueware' range of vases and tiles made using pressed botanicals.

Opposite Karl Blossfeldt, *Nigella Damascena Spinnenkopf*, ca. 1932, printed 1976, Gelatin silver print. The Metropolitan Museum of Art/Warner Communications Inc. Purchase Fund, 1978. Accession number 1978.602.2.
Right Anna Atkins, *Ptilota plumosa*, ca. 1853, cyanotype.

PHOTOGRAPHY & FILM

334 COLLINS BOTANICAL BIBLE

Rob Kesseler
British visual artist and professor
(1951–)

We are living in a time when the boundaries of art and science are becoming more permeable and opportunities for collaboration more common. Artist Rob Kesseler is one such collaborator, working with scientists across a range of specialisms – plant morphology, cell biology and genetics – to explore the Plant Kingdom and the living world at microscopic level, before presenting his findings via artworks, books, lectures and events.

Kesseler first explored the territory of complex patterns and structures that underpin life in 2000, when he realised that microscopy was much underused as a creative source. Encouraged by the possibility of funding from Central Saint Martins, the art college at which he still teaches, he approached plant scientists at the Herbarium and Jodrell Laboratory at the Royal Botanic Gardens, Kew. He was rewarded with a response by palynologist Dr Madeline Harley, who invited him into her lab to view stunning micrographs of pollen made as Head of the Pollen Research Unit. Collaborating with Kesseler provided an ideal opportunity to engage with a wider audience and he was soon up and running on the scanning electron microscope producing images of his own. Further funding from the National Endowment for Science, Technology and the Arts (NESTA) allowed the project to come to fruition, during which time Kesseler and Harley built up enough material for a book. *Pollen: The Hidden Sexuality of Flowers* was published in 2004 with the help of visionary Alexandra Papadakis, followed by *Seeds: Time Capsules of Life* (2006) and *Fruit: Edible, Inedible, Incredible* (2008) in collaboration with Dr Wolfgang Stuppy, a seed morphologist at Kew's Millennium Seed Bank. The journey then continued as Bio-Diversity Fellow of the Gulbenkian Science Institute in 2010. As Kesseler attests, intense looking is fundamental to the creative process, starting with close-up photographs of specimens in their natural environments and quick, water-dispersed sketches using aniline dye and Indian ink before heading to the lab to zoom in further. He then uses these reference points and his own 'chromatic intuition' to colour his magnifications. The end result is a powerful collection of mesmerising iconic images, designed to attract the human eye, much as some flowers attract their pollinators. Indeed, Kesseler's images are hard to resist.

My botanical world . . .

‘*At the age of ten my father gave me a beautiful Victorian brass microscope. It gave me a window into a fantastical world of butterfly wings, pollen grains and plant cells. It ignited a passion for nature that through my recent collaboration with scientists I have been able to share with audiences around the world.*’

Opposite top This asphodel flower (*Asphodelus*) is made up of more than 80 photographs, to capture it in microscopic detail.
Opposite left The pollen of a quince tree (*Cydonia oblonga*).

MAKE THE CONNECTION
Find out more about Rob's botanical world at www.robkesseler.co.uk; @robkesseler

Irving Penn
American photographer
(1917–2009)

Irving Penn is best known for his fashion photography, including the simplistic use of a plain grey or white background against which he carefully posed his models. His oeuvre of flower photographs are similarly arranged: sparsely backgrounded, highly organised and placed face on for maximum impact. This sensuous, unconventionally beautiful oeuvre stemmed from an assignment for the Christmas 1967 edition of American *Vogue*, for which Penn was commissioned by art director Alexander Liberman to photograph flowers. The brief continued for the next seven years, with Penn focusing on one specimen each time, from *Tulips* (1967) to *Begonias* (1973) through *Poppies* (1968), *Peonies* (1969), *Orchids* (1970), *Roses* (1971) and *Lilies* (1972). The photographs were then collectively published as *Flowers* (Harmony, Books 1980), with a dedicated exhibition at Hamiltons Gallery, London in 2015.

Nick Knight, OBE
British fashion photographer
(1958–)

High-concept fashion shoots for leading editorials and designers are only part of the Nick Knight story. In 1992, the polymathic photographer took a year-long break from the front row of fashion to work on a long-term exhibition for the Natural History Museum, London, with the British architect David Chipperfield. Entitled *Plant Power* (1993), the show explored the relationship between humans and plants via visually alluring representations of plant specimens from the museum's on-site herbarium. In 1993, Knight returned to the glamour of the rag trade with a sumptuous ring-flash shot of supermodel Linda Evangelista for British *Vogue*. Behind the scenes, however, trips to the herbarium continued, with 46 of the most striking specimens finally published in *Flora* (Schirmer/Mosel) in 2004. Knight's botanical leanings can also be observed in a series of melting floral photo- manipulations inspired by still life paintings from the Baroque period, and exhibited alongside 15 *Plant Power* works in 2012.

Dain L. Tasker
American radiologist and photographer
(1872–1964)

While floral radiography first occurred around 1913–14, the first images to capture public attention were published by Hazel Engelbrecht and radiologist Dr Dain L. Tasker in the 1930s, the former in the name of botanical research, the latter in prestigious photographic competitions. Tasker's fragile-looking yet innately strong, ghost-like representations of plants and flowers inspired a new wave of artists to continue experimenting with the medium. Perhaps the most well known are those of the American dental scientist Albert G. Richards, who self-published a book of works entitled *The Secret Garden – 100 Floral Radiographs* (1990), and radiologist Steven N. Meyers. Meyers was so taken with the process and the beautiful results, he purchased his own specimen radiography unit, and he produces hundreds of unique still life compositions of flowers each year.

Robert Mapplethorpe
American photographer
(1946–1989)

Male nudes, female bodybuilders, the New York S & M scene of the 1970s, studio portraits of artists and celebrities and delicate flower still lives – Robert Mapplethorpe did them all. 'My whole point is to transcend the subject.... Go beyond the subject somehow, so that the composition, the lighting, all around, reaches a certain point of perfection. That's what I'm doing. Whether it's a cock or a flower, I'm looking at it in the same way. . . .' As such, lilies like leather are afforded the same highly stylised, scrupulously framed treatment, simultaneously adhering to yet challenging classical aesthetic standards, endlessly pushing boundaries. According to Mapplethorpe's lover, the writer Jack Fritscher, Mapplethorpe saw both beauty and the devil in flowers, a subject he explored with extraordinary dedication, from black-and-white Polaroids of roses to vivid dye-transfer prints of orchids. Find the full works in *Mapplethorpe Flora: The Complete Flowers* (Phaidon, 2016).

Above Dr Dain L. Tasker, *Fleur-de-lis*, 1936.
Vintage gelatin silver print, 23.5 × 17.75cm (9¼ × 7⅜ in).
Courtesy of Joseph Bellows Gallery.

PHOTOGRAPHY & FILM

Above Sebastião Salgado, *Macaroni penguins in Zavodovski Island* 2009. ©Sebastiao Salgado/Amazonas/nbpictures. **Opposite** Ansel Adams, photographed in California in 1983.

Sebastião Salgado
Brazilian social documentary photographer (1944–)

Sebastião Salgado was inspired to take up photography in 1970 while working as an economist with the *International Coffee Organisation*, a UN-initiated development project. He began with news assignments before switching to documentary-style projects working for agencies including Sygma, Gamma and Magnum Photos. In 2004, he set off on the third of three self-initiated, long-term photographic projects dealing with global issues. The Genesis project would take Salgado around the world for eight years, with the aim of showing the unblemished faces of nature and humanity. Presented as an internationally touring exhibition of more than 200 works and a book published by Taschen in 2015, his trademark black-and-white images cover wildlife, landscapes, seascapes and indigenous peoples both living amid and working with nature. The spectacular yet haunting works aim to raise public awareness about pressing issues such as climate change, by showing the still-pristine majesty of some regions that nod to how the Earth was and should be.

Ansel Adams
American photographer and environmentalist (1902–1984)

Growing up amid the sand dunes of the Golden Gate, San Francisco, Ansel Adams led a somewhat solitary childhood, spending many hours immersed in nature and the discipline of music, setting the scene for his eventual career as a highly skilled photographer and protector of the wilderness. Adams spent many summers hiking and exploring the Sierra Nevada, acting as official photographer for the Sierra Club's epic outings into the mountains during the late 1920s. By 1934 he was well established as an artist of the Sierra Nevada and defender of Yosemite (see page 181). After co-developing the Zone System of photography as a way to determine proper exposure and adjust the contrast of the final print, Adams became an unparalleled photographer of the sublime, luminous beauty of nature, publishing many works and books, including the landmark *This is the American Earth* (Sierra Club, 1960) with text by writer and curator Nancy Newhall.

> '*The most beautiful experience we can have is the mysterious. It is the fundamental emotion which stands at the cradle of true art and true science. Whoever does not know it and can no longer wonder, no longer marvel, is as good as dead, and his eyes are dimmed.*'
>
> Albert Einstein, German-born theoretical physicist, in *The World As I See It* (The Wisdom Library/Philosophical Library, 1949)

Viviane Sassen
Dutch photographer and artist
(1972–)

Confidently straddling the worlds of fine art and high fashion since the 2000s, the Dutch photographer Viviane Sassen is known for her beautiful yet often unsettling images, where primal poses, mirrors, geometric forms, paint, striking backgrounds, happenstance shadows, organic forms and amateurish Photoshopped superimpositions come together to create quasi-fantastical dreamscapes, deep-rooted in the primal, symbiotic nature of plants and people.

In *Mud and Lotus*, a body of work first exhibited in spring 2017 at the Stevenson Gallery in Johannesburg, Sassen somehow travels to the very core of the botanical world and the human relationship with it. Mud, it transpires, stands for the 'female struggle in an earthly way, a rather organic, growing process, blood, sweat and tears, that isn't perfect – quite the opposite; it's raw and doesn't always smell good'. Lotus, on the other hand, 'stands for the spiritual, beauty, calmness and wisdom. It's about inner strength, and about the inner transformation you go through in life, for instance when you become a mother.'

Indeed, the lotus flower needs mud to survive, its rhizomes fanning out horizontally within the murky regions of shallow ponds, lakes and marshes, acting as anchors for the self-cleaning leaves and magically water-repellent flowers that navigate up above the water. The much-revered lotus can also produce seeds that can exist for thousands of years before germinating.

It's a complexity that's mirrored in Sassen's seemingly abstract images, where female forms are merged and collaged with plant forms, fungi, organic materials such as milk and intense colours, at once pure, sensual, timeless and primordially soiled. For *Pikin Slee* (Prestel, 2014), Sassen travelled to Suriname to document an isolated community founded by former slaves in the heart of the rainforest. The project brought back vivid memories of an early childhood spent in Africa. In *Roxane II* (Oodee, 2017), a favourite female muse cavorts amid technicolour and earthy landscapes, sensually resplendent in the shadows and painterly overlays of nature, the energy of which, in Sassen's work, is never far away.

My botanical world . . .

'*I remember that when I lived in Kenya as a child, I was always looking at Baobabs in wonder, as if they were giants that could come to life at night, roaming the Earth with their huge roots, gently waving their arms in the fresh air. Every child that has ever read 'Max and the Maxi-monsters'* (Where the Wild Things Are, *by Maurice Sendak) knows exactly what I'm talking about when I say I would love to experience my bedroom filling with plants and trees overnight, waking up in a forest of possibilities.*'

MAKE THE CONNECTION
Find out more about Viviane's botanical world at www.vivianesassen.com; @vivianesassenstudio

Caption *Ultramarine*. From *Of Mud and Lotus*, 2017. Archival pigment ink on lustre paper, 40 × 30cm (15¾ × 11¾ in).

PHOTOGRAPHY & FILM

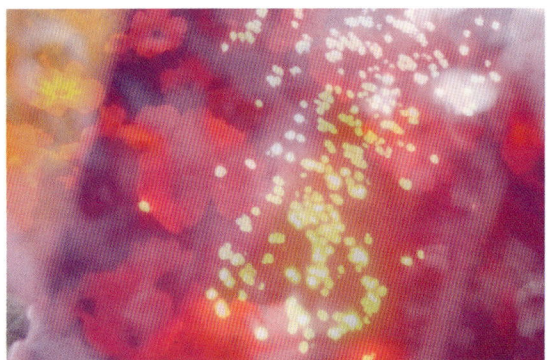

Mystical England, 2014. Digital Film Still.

Samuel Levack and Jennifer Lewandowski
British artists (1978– & 1979–)

Memorandum of Understanding between the Royal Government of Cambodia and the Government of Australia Relating to the Settlement of Refugees in Cambodia. Ministry of Interior, Phnom Penh, Cambodia, 26 September 2014.

Taryn Simon
American multidisciplinary artist (1975–)

Mystical England (2014) is one of a series of films presented by artist duo Samuel Levack and Jennifer Lewandowski, exploring alternative ways of living, poetry and the psychic landscape. Inspired by the early Super 8 works of late British artist, film director and gardener Derek Jarman, such as *Journey to Avebury* (1971), playing on an endless loop with a hypnotic soundtrack provided by their band Das Hund, the contemplative work provides a psychedelic glimpse of a landscape that might well only be accessible via a state of altered consciousness – a sun-drenched, superimposed mirage of dancing flowers and refracted light, alluding to the semi-imagined Utopia of times gone past and the fleeting intangibility of such a rose-tinted vision. Indeed, the pair appear drawn to ancient, mystical landscapes, with Glastonbury, Joshua Tree and Epping Forest featuring in similarly collaged dreamscapes.

Guided by an interest in systems of categorisation and classification, it was only a matter of time before politically enquiring artist Taryn Simon would be drawn to the intricacies of the Plant Kingdom, and our interaction with it. At first glance *Paperwork and the Will of Capital* (2015) presents an arresting display of high-colour photographs of floral bouquets, alongside beautifully mounted herbarium specimens on a series of concrete plinths – but something jars. Further inspection reveals that each photograph is a recreated centrepiece from official signings by countries present at the 1944 United Nations Monetary and Financial Conference, which addressed the globalisation of economies after World War II. Plant specimens from each 'impossible bouquet' were then pressed, dried, mounted and forced against further photographic prints within each concrete press. As a whole, the installation questions the instability of executive decision-making and the precarious nature of survival.

PHOTOGRAPHY & FILM

Haarkon at the Barbican Conservatory, Barbican Centre, London, May 2016.

Haarkon
British photographers
(1987 – & 1987–)

Yorkshire-based photographers and 'accidental bloggers' Magnus Edmondson and India Hobson operate under the moniker Haarkon (inspired by Magnus's brother's middle name), a brand that has taken them to the far corners of the world to document aspects of the botanical world for their much-followed blog and Instagram feed, as well as for individual clients and commissions. For the past few years the couple have been on an international greenhouse tour, from the houseplant decor of their own light-filled home in Sheffield to the Rainforest Biome of the Eden Project in Cornwall to the Hortus Botanicus in Amsterdam. A personal favourite is the Barbican Conservatory in London, where tropical plants cascade down Brutalist balconies, satisfying two Haarkon passions at once: architectural design and greenery. Such greenhouses as this also exemplify the raison d'être of their search – the positive intertwining of man and nature.

Louie Schwartzberg
American cinematographer and director
(1950–)

In 2013 award-winning filmmaker Louie Schwartzberg released *Mysteries of the Unseen World*, using his signature time-lapse cinematography as well as high-speed and nano-photography to transport audiences to invisible worlds: plants creeping towards the sun; a root growing from a seed; the tiny structures on a butterfly's wing. The same year, the US release of Disneynature's *Wings of Life* brought yet more of Schwartzberg's awe-inspiring direction and insightful thoughts on nature to life, focusing on the dramatic yet fragile love story played out between plants and pollinators. The footage is as much about gratitude as it is about nature, two over-riding themes of Schwartzberg's work. 'Beauty and seduction are nature's tools for survival because we protect what we fall in love with,' he muses. 'It opens our hearts and makes us feel we are part of nature and we're not separate from it.' Wonder at his films for their universal rhythms, patterns and beauty; act on them by celebrating life.

PAINTING & DRAWING

Plants and flowers crop up numerous times in the mark-makings of art history – as part of the landscape, as religious symbols, as decorative motifs, in still lives, hyper-realistic figurative portrayals or conceptual abstractions. Some movements stand out for their dedication to flowers or nature, while others include nature-inspired paintings or drawings as part of a wider oeuvre, the case for many late-twentieth- and twenty-first-century artists. In a way these works are the most interesting, revealing as much about the maker as the Plant Kingdom.

Claude Monet
French Impressionist painter
(1840–1926)

It's hard not to think of water lilies when one considers the work of quintessential artist-gardener and Impressionist painter Claude Monet. If not water lilies, then the gently curving arch of the Japanese-style bridge found in his private garden at Giverny. Now visited by hundreds of thousands of people each year, Monet largely limited invitations to his earthly paradise to fellow artists, writers and horticulturalists, who in turn helped to document his time there. This seclusion was largely created in order to focus on his first love – painting – the most iconic of which are surely his *Water Lilies* series, produced between 1902 and 1908. Over time, Monet gradually eliminated distracting features such as river banks to hone in on subtle modulations of light and reflection, illustrating what the great French writer Marcel Proust imagined that he would find if he ever visited Giverny – not so much a garden of flowers as of colours and tones.

Henri Matisse
French artist
(1869–1954)

There exists a wonderful black-and-white image of the artist Henri Matisse's studio that shows a gargantuan *Monstera deliciosa*, its leaves both fanning upwards to the sky and proliferating along the ground. The specimen and the Plant Kingdom at large served as both obvious and subconscious muse to lyrical paintings and colourful paper cut-outs such as *Still Life with Geraniums* (1910), *Garden at Issy* (1917), *The Daisies* (1939), *The Lagoon* (1946), *Violet Leaf on Orange Background* (1947), *Snow Flowers* (1951) and *The Sheaf* (1953). Matisse's cut-outs were largely produced during the last decade of his life, when he was often bedridden and needed help with the evolutionary placement of his colour carvings, referring to *The Parakeet and the Mermaid* (1952) as 'a little garden all around me where I can walk', a treat he also bestowed on the viewers of his artworks.

> '*Happy are those who sing with all their heart, from the bottom of their hearts . . . To find joy in the sky, the trees, the flowers. There are always flowers for those who want to see them.*'
>
> Henri Matisse, from *Jazz* (Tériade, 1947), a limited edition art book of colourful cut paper collages, accompanied by the artist's written thoughts

Opposite *The Waterlily Pond: Green Harmony*, 1899, by Claude Monet (1840-1926). **Right** Henri Matisse at his workshop in Cimiez, Nice, with *The Parakeet and the Mermaid*, 1950. © Walter Carone/Getty images. Artwork: © Succession H. Matisse/DACS 2018.

PAINTING & DRAWING

Top Image in detail: *Cactus No. 51*, 2010. Oil on Canvas, 194 × 200cm (76½ × 78¾ in). **Right** *Untitled* 1212, 2017. Oil on canvas, 180 × 135cm (71 × 53 in). **Below** Image in detail: *Cactus No. 59*, 2011. Oil on canvas, 259.1 × 170cm (102 × 67 in). All images courtesy Johyun Gallery.

Kwang Ho Lee
Korean artist
(1967–)

It's tempting to introduce the South Korean artist Kwang Ho Lee as a master of realism, but on closer inspection, his portraits of people and cacti, and more recent landscapes of the Gotjawal Forest on Jeju Island, are imbued with so much more. For Lee, it's as much about the process as it is the observation, the tactile nature of producing a hyper-realistic end result designed to both lure the viewer in and – certainly in the case of his *Cacti* series – leave them wanting. In order to create his impossibly lifelike renditions of some of the world's spikiest plant forms, Lee first takes a photograph of a specimen before re-observing and drawing it, an artistic intervention that allows him to think about how best to express the inherent qualities and character of the plant. Moving to canvas he then employs the thinning effect of oil medium to capture lighter (more medium) and darker (less medium) areas of his subject, a technique that avoids an undesirable build-up of oil paint, which can happen when tonal mixers are added. The surface may then be scratched with a knife, rubbed or scoured to create levels of physical delicacy or roughness. For the fine hairs and brutal spikes of his cacti, he used a fine rubber-tipped brush or needle to scrape away paint – the complete opposite of how one might imagine a realistic image is created. It's a process that contributes to the strangeness of paintings that at first seem straightforward: cacti, spiky, lifelike. On second glance, or a lingering hypnotic stare, a mode that Lee's paintings are apt to induce, these larger-than-life cacti – some more than 2.5m (8 ft) tall – bristle with look-but-don't-touch sexuality. Reading into his earlier portraits of people in his *Inter-View* series, such intimate, pull-and-push interplay between artist and subject, and then subject and viewer is one that Lee consistently aspires to, even in the face of potential rejection. It's also present in his similarly realistic treatment of the woodland copses of Gotjawal, a forest located on the middle slopes of Halla Mountain, on Jeju Island. Rather than standing outside of his subject, Lee paints from within the landscape of each thorny untitled thicket, turning the idea of an observed subject on its head through the very process of having to paint himself out.

My botanical world . . .
'I'm driven to bring out the tactile qualities of cacti, in a painterly way. With each brushstroke it is as if I am touching the plant.'

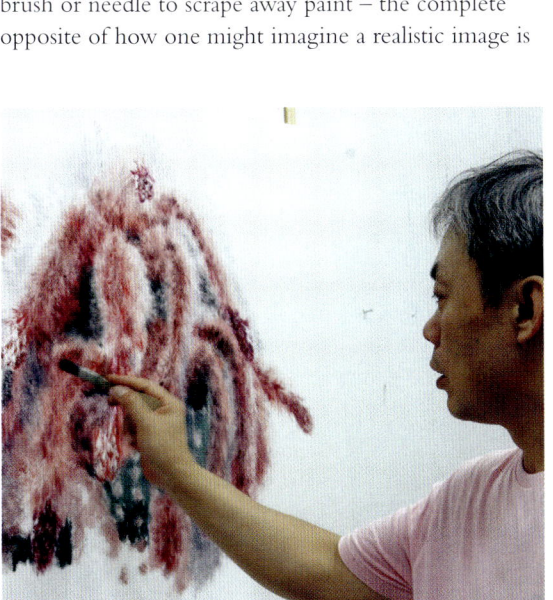

MAKE THE CONNECTION
Find out more about Kwang Ho's botanical world at
www.johyungallery.com; @johyungallery

Frida Kahlo
Mexican painter
(1907–1954)

'I paint flowers so that they will not die,' said the great Mexican artist and style icon Frida Kahlo. The legacy of this mantra appears in numerous Kahlo works from *Self-Portrait with Monkey* (1938) to *Self-Portrait with Thorn Necklace and Hummingbird* (1940). It also inspired a major exhibition devoted to her life and works at the New York Botanical Garden in 2015, a rambling confection of evergreen foliage, potted euphorbias, marigolds, salvias, echeverias, cannas and cacti against recreations of the intense cobalt-blue walls of Kahlo's 'Blue House' and garden, La Casa Azul in the Coyoacán area of Mexico City. Kahlo also worked with photographer Nickolas Muray – a friend, lover and confidant during the 1930s and 1940s – staging herself before Muray's camera with the same virtuosity she brought to her self-portraits, proudly resplendent in traditional dress and signature crown of flowers.

Rachel Ruysch
Dutch still-life painter
(1664–1750)

Daughter of Frederik Ruysch, an eminent professor of anatomy and botany, Rachel Ruysch had much at her disposal from which to compose early still-life drawings, from animal skeletons to botanical specimens. She also inherited her father's ability to depict nature with great accuracy and was apprenticed to the prominent flower painter Willem van Aelst in 1679 at the tender age of 15 years old, receiving tutelage in painting but also the construction of a less formalised bouquet that helped bring a three-dimensional aspect to her work. Despite giving birth to ten children, Ruysch produced hundreds of paintings, catering to an international circle of dedicated patrons. *Flowers in a Vase* (1685) exemplifies Ruysch's often asymmetrical, playful compositions, each deftly observed, jewel-coloured flower unfurling against the intrigue of an inky background. Many works, such as *Still Life with Flowers* (1665) also included the flower of the era – the regal tulip (see page 73).

Hilma af Klint
Swedish artist and mystic
(1862–1944)

The late nineteenth century – an era in which natural science went beyond the visible, with the discovery of electromagnetic waves (1886) and the invention of the x-ray (1895) – also set the scene for a wave of spiritualism. On the back of it rode an esoteric Swedish artist by the name of Hilma af Klint. In the winter of 1905–1906, while conducting séances with a group of four other female artists collectively known as *de Fem* (the Five), she received a 'commission' from a higher power named Amaliel. She was told to paint 'on an astral plane' and represent the 'immortal aspects of man', resulting in 193 extraordinary *Paintings for the Temple* produced between 1906 and 1915. Pre-dating paintings by abstract art pioneer Wassily Kandinsky, her monumental *The Ten Biggest* series (1907) includes geometric shapes, plant-like forms, symbolic colours and motifs that allude to evolution, matter and sex.

Georgia O'Keeffe
American Modernist artist
(1887–1986)

In 1912, Georgia O'Keeffe enrolled on a summer art class at the University of Virginia. There she learned of the innovative ideas of artist and arts educator Arthur Wesley Dow. Influenced by principles of Japanese art, he taught that portrayals of nature should focus on the living force and harmonies of the elements rather than realistic renditions. Georgia O'Keeffe was sufficiently inspired to begin experimenting with abstract compositions based on her personal sensations. Her drawings impressed future husband Alfred Stieglitz enough to exhibit them in his 291 gallery in 1916. He also introduced her to early American Modernist artists such as the photographer Paul Strand. O'Keeffe moved from landscapes to simplified images of leaves, flowers and rocks, before producing the large-scale, magnified, sensual depictions of flowers for which she is most well known, among them *Flower of Life II* (1925), *Black Iris VI* (1936) and *Jimson Weed/White Flower No. 1* (1932).

Above HaK-137 *Tree of Knowledge no 5*, Series W, 1915.
By courtesy of the Hilma af Klint Foundation.
Photo: Albin Dahlstöm/Moderna Museet, Stockholm, Sweden.

Ellsworth Kelly
American painter
(1923–2015)

Many people know Ellsworth Kelly for his rigorous hard-edge abstract painting, but he also produced an extraordinary body of figurative work including beautiful plant drawings. A selection of around 30 of these works first appeared collectively in a 1969 exhibition at the Metropolitan Museum of Art, in New York, situated in the context of his wider paintings and sculpture. In 2012, The Met devoted an entire exhibition to Kelly's depictions of flowers and foliage, drawings that provided a 'bridge to the way of seeing' that brought about his later abstracted paintings. Composing each sketch from life, he would often barely lift the pencil in order to translate the plant's contours onto paper, stating that such executions were 'exact observations of the form of the leaf or flower or fruit seen. Nothing is changed or added.' Via the immediacy of overlapping lines and negative space he created a perfect tranquillity.

‘*Each drawing that I've done, I have found. Meaning, I see a plant I want to draw.*’

Ellsworth Kelly in *Ellsworth Kelly – Plant Drawings*, by German art historian Michael Semff and Metropolitan Museum curator Marla Prather (Schirmer/Mosel, 2011)

Piet Mondrian
Dutch painter and theoretician
(1872–1944)

High priest of geometric abstraction, Piet Mondrian was also compelled to express his softer, more intimate side through an ongoing collaboration with nature. His passionate portraits of flowers amount to some 150 predominantly undated works, posing the question of whether such works were made during an earlier figurative period or in direct contrast to his primary-coloured grids. In 1991, the Sidney Janis Gallery presented an exhibition of 37 of 'Mondrian's Flowers'. The same year poet and essayist David Shapiro published *Mondrian: Flowers* (Harry N. Abrams) helping to lay the foundations for Turner Contemporary's 2014 exploration of *Mondrian and Colour*, and the duality of the naturalistic man and the abstract canvas. Refecting on his output, Mondrian mused: 'I enjoyed painting flowers, not bouquets, but a single flower at a time, in order that I might better express its plastic structure.'

Opposite Ellsworth Kelly, *Tropical Plant, St. Martin*, 1981. Graphite on paper, 60.9 × 45.7cm (24 × 18 inches). Christie's, New York. © Ellsworth Kelly Foundation. **Right** Piet Mondrian, *Chrysanthemum*, 1908–09. Charcoal on paper, 25.4 × 28.7cm (10 × 11¼ inches). Solomon R. Guggenheim Museum, New York; 61.1589. © 2018 Mondrian/Holtzman Trust.

PAINTING & DRAWING

Tom Ellis
British artist
(1973–)

Flowers are an obvious muse for many painters, via the medium of still lives or used more symbolically to inform a story or a vision of the world. 'Flowers with their cellularly thin, skein-like coloured petals are also one of the closest things formally to the painted mark,' says artist Tom Ellis, 'and this similarity of form gives flower painting a strange potency that goes beyond the purely descriptive.' Flowers remain a consistent presence in Ellis's work, a series of large-scale monochromatic lilies appearing in a solo show entitled *The Vacuum* at *Kunsthalle Winterthur*, in 2010, again as *10,000 Forms* in 2017 and most recently through an ongoing exploration in colour. For Ellis, however, it's potentially less about the flower itself than about the genre of flower painting, a time-honoured tradition that specifically hinges on mankind's relationship with, and desire to curate, nature – flowers as a plant's natural embodiment of display, attraction and identity only serving to increase their affinity with the act of painting itself. Hovering somewhere between abstraction and figuration, his 'flower painting versions' appear somewhat restless, the act of making continuously jostling with the form of the plants and flowers themselves, nature and humanity as one. Starting points include formalised bouquets of 'traditional' flowers such as lilies, tulips and roses as immortalised by Dutch flower painters such as Rachel Ruysch (see page 348). More specifically, he has referenced paintings such as *Flowers on a Fountain with a Peacock* (*c.*1700–1710) by the Dutch painter Jan Weenix, a tumbling cascade of flowers, fruits and fowl set within a mythical landscape. It's a work that continued to hold Ellis's interest during the making and installation of his exhibition *The Middle at the Wallace Collection* in 2017, for the 'luscious yet transient nature of its offerings'. Indeed, it's the ephemeral nature of flower forms that attracts – how cycles of beauty and decay such as the drooping of a flowerhead or a wizened leaf can inform a kind of desconstructed brushwork. It's a wabi-sabi approach, a whole-hearted embrace of the imperfections in life that informs the final stage of Ellis's flower version paintings: to reincarnate them as closely approximated iterations, species to species but every evolving flower for itself.

MAKE THE CONNECTION
Find out more about Tom's botanical world at www.tomellisartist.com; @tomellisartist

PAINTING & DRAWING

Top Studio portrait by photographer Rob Murray illustrating works for *The Middle*, 2016 at The Wallace Collection. **Left** *The Flowers (10,000 Forms)*, 2017, Oil and marker pen on canvas. **Opposite** *Arrangement No. 3*, 2018, Oil and marker pen on board.

My botanical world . . .

'*Flower painting is a time-honoured genre that specifically hinges on mankind's relationship with, and desire to curate, nature. As such it is the perfect setting in which to explore our complex push-and-pull relationship with the natural world.*'

PAINTING & DRAWING

Emily Kame Kngwarreye (1910–1996), *Earth's Creation*, 1994. © Emily Kame Kngwarreye /Copyright Agency. Licensed by DACS 2018.

Emily Kame Kngwarreye
Indigenous Australian artist
(1910–1996)

Emily Kame Kngwarreye grew up in a remote desert area some 230 kilometres (143 miles) northeast of Alice Springs known as Utopia – a name afforded to the lands of the Anmatyerre and Alyawarr peoples to whom Kngwarreye belonged, by the first pastoralists who settled there in the 1920s. Like many Aboriginal Australian peoples of the location and time, she worked the land of pastoral properties until such time as Aboriginal land rights were granted in 1976. It was then that she began experimenting with the art of batik, leading to her first canvas paintings in 1988. In the eight years before her death at the age of 86, she produced more than 3,000 intimately expressive paintings – including the monumental *Earth's Creation* (1994) – alluding to the Dreaming, Awelye (women's ceremonies and body design) and the landscape and culture within.

Andy Warhol
American artist, director and producer
(1928–1987)

The great American Pop artist Andy Warhol found the beauty in the everyday but also the darkness in seemingly beautiful things. The 10 screenprints that make up his *Flower* series, created in a two-year era of floral obsession between 1964 and 1965, were initially inspired by a photograph of hibiscus flowers taken by the then executive editor of *Modern Photography* magazine, Patricia Caulfield. Warhol appropriated the image (itself used to depict a new home-processing system by Kodak), cropped and copied it, and enhanced the contrast. Caulfield took legal action (finally settling out of court) allowing Warhol to exhibit his large-scale silkscreen prints at the Leo Castelli Gallery, New York, in 1964. Combining bright colours and American kitsch with an any-way-up square format, Warhol's flowers helped propel him to stardom.

PAINTING & DRAWING

David Hockney (1937–), *The Arrival of Spring in Woldgate, East Yorkshire 2011 (twenty-eleven)*.

David Hockney
British artist
(1937–)

'If you want to replenish your visual thinking, you have to go back to nature, because there's the infinite there, meaning you can't think it up.' So says David Hockney in Bruno Wollheim's documentary, *David Hockney: A Bigger Picture* (2009), a film of Hockney's return from his adopted home in California to paint the East Yorkshire landscape of his childhood. Hockney is filmed working outside for the first time, through the seasons and in all weathers, resulting in a landmark exhibition of the same name at the Royal Academy, London, in 2012. While landscapes have been part of Hockney's oeuvre since the 1960s, paintings such as *The Arrival of Spring in Woldgate, East Yorkshire in 2011 (twenty eleven)* – a mammoth composition of 32 joined-together canvases – took the artform to another level in terms of size and innovation, even more so when juxtaposed with a suite of unique plant-inspired iPad artworks. Together they show the largesse, the eternal, the immediacy and the intimacy of nature.

Katsushika Hokusai
Japanese artist and printmaker
(c.1760–1849)

Best known for his monumental print series *Thirty-six Views of Mount Fuji* (1826–1833) including global hit *The Great Wave off Kanagawa*, the Japanese nineteenth-century artist Katsushika Hokusai also initiated a popular form of *ukiyo-e* ('pictures of the floating world', usually produced as woodblock prints) depicting flowers, animals and birds during this latter stage of his life. Nature had always provided inspiration, but as he reveals in a postscript to his follow-up series *One Hundred Views of Mount Fuji* (c.1834–1835): 'At seventy-three, I began to grasp the structures of birds and beasts, insects and fish, and of the way plants grow. If I go on trying, I will surely understand them still better by the time I am eighty-six, so that by ninety I will have penetrated to their essential nature.' Works such as the joyful *Peony and Canary* (1833–1834) and the beautifully detailed *Chrysanthemum and Bee* (pictured on page 68) and *Poppies* (both c.1832) are testament to his endeavours.

INSPIRED BY NATURE   355

SCULPTURE & INSTALLATION

There are those who fashion sculptures out of wood or natural fibres. More obviously related to the Plant Kingdom are works by artists who interact with botanicals or the landscape such as Azuma Makoto (see page 359) or NILS-UDO (see page 363). At the other end of the spectrum are the sculptural works that seek to portray aspects of nature using other media, from Marc Quinn's bronze orchids (see page 360) to Yayoi Kusama's iconic pumpkins (above), each three-dimensional work providing a window into the multi-dimensional world at large.

Yayoi Kusama
Japanese artist and writer
(1929–)

As a child Yayoi Kusama began to experience vivid hallucinations of flashes of light, auras, dense fields of dots, speaking flowers and engulfing animated patterns that would leap from fabric. Deeply affected by these visions, an unhappy childhood and World War II, she dreamt of personal and creative freedom and enrolled at art school in Kyoto. Frustrated with the Japanese style, she was inspired to stage her own exhibitions and happenings by the European and American avant-garde scene, the all-encompassing polka dot quickly becoming a trademark motif. Moving to New York in 1958, she created her first Mirror/Infinity room in 1965, six of which were displayed in 2017 at the Hirshhorn Museum in Washington DC. Her epic, recreated *All the Eternal Love I have for the Pumpkins* (2016) room pays homage to the radiant energy of Kusama's beloved pumpkins, a quality that seems to affect all who visit it.

Ai Weiwei
Chinese contemporary artist and activist
(1957–)

Natural forms appear in many of the artist and activist Ai Weiwei's visual art work, from the salvaged ancient temple wood used to create his 2006 sculpture *Map of China* to the most widely known of his installations – *Sunflower Seeds*, conceived for the Unilever Series commission at Tate Modern's Turbine Hall, in 2010. To realise his vision, Weiwei produced – with the help of more than 1,600 artisans from the Imperial porcelain-making town of Jingdezhen – 100 million porcelain sunflower seeds, weighing a total 136 tonnes (150 tons), which were distributed over the gargantuan floor to a depth of 10cm (4 in). The layers of seeds, each one a tiny sculpture of its own, hold multiple meanings and nuances, from a common Chinese street snack to memories of hardships and hunger during the Cultural Revolution, to the significance of the individual and the imposing strength of joining together in a united cause.

> 'From a very young age I started to sense that an individual has to set an example in society. Your own acts or behaviour tell the world who you are and at the same time what kind of society you think it should be.'

Ai Weiwei in *Ai Weiwei: Sunflower Seeds*, by Tate Modern curator Juliet Bingham (Tate Publishing, 2010)

Opposite YAYOI KUSAMA with *PUMPKIN*, 2010. Installation view: Aichi Triennale 2010. © YAYOI KUSAMA, 2010. **Right** Ai Weiwei holds sculpted seeds from his Unilever Installation *Sunflower Seeds* at The Tate Modern in London, 2010.

SCULPTURE & INSTALLATION

Top *Exobiotanica 2, Botanical Space Flight.* From *In Bloom*, 2017.
Right *Shiki – Landscape and Beyond*, 2015. **Below** *Shiki 1 x Sandstone 1*, 2017. All photographs by Shunsuke Shiinoki.

Azuma Makoto
Japanese flower artist and botanical sculptor (1976–)

Once you've seen an image of one of Azuma Makoto's stratospheric botanical sculptures, it's hard to imagine that these naturally impossible installations are not still up there, joy-riding through space, liberated from the anchors of their earthly roots. The fourth instalment of Makoto's *In Bloom* series, several large 'bouquets' of carefully arranged flowers weighing 6kg (13 lb) each with a diameter of 1.5m (5 ft), were launched from the Lovelock Desert of Nevada in the United States. In order to get *World Flowers*, *World Plants* and *World Flowers II* into space, each bouquet was attached to a vehicle with a single or double balloon, with a medium-format mirrorless camera to document the journey. Makoto hoped that this device would capture the ever-changing landscape of the flowers from 'ground to heaven' due to changes of light, temperature (down to a predicted -60°C/-76°F), atmospheric pressure and speed of travel. As with the first piece in this series, *Exobiotanica (Botanical Space Flight)*, which propelled a bonsai tree and ikebana floral arrangement of living plants into space, the aim was to breathe new life into a traditional art form, and thereby attach more value to plants and flowers. The images from both projects are not only mind-blowing for the supernatural beauty of the image of plant life floating against the illuminated curve of the Earth and darkened sky, but also for the technical prowess that has quite obviously gone into the making of them. A foray into the Makoto archives turfs up an astonishing array of similarly thought-provoking botanical explorations: decomposing indoor and outdoor flower landscapes alluding to the Buddhist contemplation of life and death; bonsai trees and bouquets plunged 1,000m (3,280 ft) deep into the sea or transported to cultural sites around the world; a tremendous pine tree suspended above a snowfield and a palm tree treated to the same inviability in the desert, their root forms hovering just above the Earth; gorgeous bouquets trapped and decaying in glass cases or blocks of ice; and a performative, ritual ceremony involving the burning of 2,000 flowers in a cave, so releasing their locked beauty. While Makoto may not have realised his former ambition of becoming a rock star, he certainly has elevated floral arrangement to stellar heights of performance, intrigue and drama.

My botanical world . . .

‘My passion lies in the plants themselves. There are countless existing plants in this world, while each of them carries a different expression. My mission is to bring out each plant's hidden potential and beauty, and to make them even more attractive. Every day I meet different new plants. As long as I am interacting with them my passion will never fade away.’

MAKE THE CONNECTION
Find out more about Azuma's botanical world at www.azumamakoto.com; @azumamakoto

Marc Quinn
British artist
(1964–)

Marc Quinn is perhaps best known for using his own blood to create a frozen effigy of his head or for his yogic sculpture of supermodel Kate Moss, but the botanical world has starred just as frequently as controversial human forms. Plants first made it centre stage for an exhibition at the Fondazione Prada, Milan, with a work entitled *Garden* (2000), whereby an impossible garden of the world's plants was plunged into frozen silicone oil, a commentary on the human desire to shape and control the natural world. Quinn further explores this theme through large-scale, full-frontal bronzes of orchids such as *The Engine of Evolution* (2010), the collaged-together hybrids of his *Nurseries of El Dorado* series (2007) and the acid-bright, hyper-realistic yet unsettling assemblages of fruit and flora in flower paintings such as *Out of Body* (2012).

Zadok Ben-David
Israeli artist
(1949–)

London-based artist Zadok Ben-David has a penchant for covering the floors of gallery spaces worldwide with enchanting carpets of thousands of tiny trees, flowers, insects or even people. Each of these forms is handcut from thin sheets of metal, a work of art in its own right. For *Blackfield* (2006–2009), Ben-David etched and then painted 20,000 steel flowers, their forms derived from nineteenth-century Victorian encyclopedias, before placing them in a carpet of sand. By enabling viewers to walk around the installation (made in different shapes and sizes as the exhibition toured around the world), the duality and purpose of Ben-David's work became clear: from one angle was viewed a black field, from the opposite direction, a full-colour version, designed to query our perception of the natural world and interaction with it.

Simon Heijdens
Dutch artist
(1978–)

Imagine standing in a space in which unpredictable natural elements such as the wind, rain or sunshine have been largely planned out. Slowly, almost imperceptibly, a living digital organism of light begins to grow up the wall, tentatively sending out its quivering, illuminated stems and foliage. The specimens appear to grow together but respond in different ways to an apparent 'higher power', each retaining an organic individuality much as plants do in the natural world. Welcome to *Lightweeds* (2004), follow-up to London-based artist Simon Heijdens's pioneering work, *Tree* (2004). Not only do these endearing digital weeds and trees respond to elements caught by environmental sensors outside, they also bend, lose their seeds, drop their leaves or pollinate other walls as people go by. The combination produces a constantly evolving ecosystem that queries both how we use space and what is left of nature.

Robert Smithson
American artist and writer
(1938–1973)

Emerging onto the Pop Art scene of early 1960s New York, the great land artist Robert Smithson first toyed with phantasmagorical drawings and collages before exploring Minimalist and geometric sculpture. Expanding his work out of galleries and into the uncultivated landscape, he eventually produced his pioneering earthwork *Spiral Jetty* (1970) a remarkable 457-metre (1500-foot) long, 4.5-metre (15-foot) wide, counter-clockwise coil of black basalt rocks and earth displaced in Great Salt Lake near Rozel Point, Utah. The jetty was submerged by rising waters in 1972 but revealed by continued drought in 2000, the surrounding waters often changing colour due to the presence of beta carotene–rich algae. Writing in *Artforum* in 1973, shortly before his untimely death, Smithson compared land art to the British picturesque garden movement, that spoke not just of the beautiful and the sublime but the manipulation of real land, nature and politics.

Above Simon Heijdens, *Lightweeds*, Van Abbe Museum, Eindhoven, The Netherlands, 2017.

SCULPTURE & INSTALLATION

'What I wanted was to live, act and work in symbiosis with nature in the closest possible way. The living nature itself, all the phenomena that are characteristic of it, were all of a sudden potential issues. The sphere of nature simultaneously became the sphere of art, in which I inscribed myself.'

NILS-UDO in 'Nature Visions', from *Art Nature Dialogues: Interviews with Environmental Artists*, by writer and art critic John K. Grande (State University of New York Press, 2004)

SCULPTURE & INSTALLATION

NILS-UDO
Bavarian artist
(1937–)

Like Robert Smithson, NILS-UDO has been making environmental art since the 1960s, his site-specific installations displaying a great sensitivity towards the botanical world while recognising the paradox of his 'potential utopias' – by intervening with nature, he harms what he touches, even while trying to work in parallel with it. Playing with organic and inorganic materials available on site – *Clemson Clay Nest* (2005) – or plantings of specimens – *Birch Tree Planting* (1975) – which are then subjected to the will of nature, NILS-UDO's work confronts the very notion of art as distinct from nature. Thus a sweet chestnut leaf provides a raft for a collection of spirea blossoms, a hedge provides the backdrop for a line of red flowers taking a walk, or berries spill out of tree crevices. As nature does its thing, blowing petals away or subjecting works to seasonal decay, photographic evidence remains. NILS-UDO's legacy is also visible in the work of fellow environmental artists such as Andy Goldsworthy.

Opposite NILS-UDO, CRACK IN LAVA, FLOWER PETALS CALLED TONGUES OF FIRE, 'Ile de La Réunion', 1990. Pigment print, 150 × 150cm (59 × 59 in), Ed 8+3 ap.

herman de vries
Dutch artist
(1931–)

The fact that herman de vries stipulates that his name be spelled entirely in lowercase to avoid hierarchies, says much about his values and artistic leanings. Part of the ZERO movement in the late 1950s and early 1960s, he walks an elegant line between the pursuit of simplicity and economy and an inventor's approach to materials and language. In the 1970s he turned primarily to nature, presenting its materials, processes and phenomena as both vital and tantamount to human existence. In 2015, de vries was commissioned to represent the Netherlands at the Venice Biennale for which he produced *to be all ways to be*, which explored relationships between nature and culture. Works included *from earth: everywhere* consisting of rubbings from his 'earth museum' – a collection of over 8,000 earth samples from around the world since 1978 – plus *108 pound rosa damascena*, using the olfactory allure of medicinal roses to enhance visitors' experience of the sights and sounds of the installation.

Below herman da vries 'from the laguna of venice - a journal'. 123 × (35 × 25 cm/13¾ × 10 in). Private collection, Germany.

INSPIRED BY NATURE 363

SCULPTURE & INSTALLATION

Rebecca Louise Law
British installation artist
(1980–)

Walking into a Rebecca Louise Law installation feels akin to gazing at a beautiful constellation of flowers, the artworks for which she is most known made up of cascades of natural materials – namely flora – suspended from the ceilings of museums, galleries, foyers and private spaces around the world. The effect is both ethereal and touching, and wonderfully hypnotic, compelling the viewer to look up (not so readily done in the days of digital downplay) and engage their gaze with what is tantamount to an upside-down meadow. Indeed, it was a meadow that first gave Law the idea for the work that she has been creating since 2003, more specifically a Fenland field of ox-eye daisies that her gardener father insisted the family view one sunny day during her teenage years. At first nonplussed about the spectacle, Law was unwittingly struck by the vitality of the moment, the field at its perfect best – a phenomenon that would probably last for only a day or two. Determined to recreate and share the physicality and sensuality of the experience, she created her first floral installation while at art college, an idea that stemmed from a foray into flower painting and an obsessive need to explore and play with colour.

Looking for a palette that wasn't paint, she was struck by another group of flowers – the dahlias of her father's nursery garden. The ever-changing palette of the natural world had been there all along and, as the breadth of Law's artwork shows, appears to keep on giving: the vibrancy of fresh flowers when they are first hung, the fading colours of petals as they dry in suspension or where pressed specimens are prepared in advance, each one strung by hand on a thin strand of copper wire, and preserved in a state of decay by their own oils. By observing visitors to installations such as *The Flower Garden Display'd* (2014) at The Garden Museum in London, Law also realised that her artworks went deeper than just experiments in colour, they played with the relationship between humanity and nature, and concepts of beauty and preservation.

A book of her works, *Rebecca Louise Law: Life in Death* (Kew Publishing, 2017) pays homage to her vision, published to coincide with a major installation of 375,000 garlanded preserved flowers shown alongside preserved Ancient Egyptian funeral garlands from Kew's own collection, dating back to 1300 BCE – a nod to our timeless tryst with nature.

My botanical world . . .

'*My relationship with the botanical world was founded in my childhood. Every natural material I discovered was a gift from this earth, a mystery and a wonder. The simplicity of the love that I had with nature as a child is what I search for now as an adult through my artwork.*'

MAKE THE CONNECTION
Find out more about Rebecca's botanical world at
www.rebeccalouiselaw.com; @rebeccalouiselaw

SCULPTURE & INSTALLATION

Top *Still life*, Katherine Mager and Broadway Gallery, 2016. **Below** *Life in Death*, Charles Emerson, 2017. **Left** *The Beauty of Decay 1*, Rachel Warne, 2016. **Opposite** Studio portrait, photo by Fabio Affuso.

SCULPTURE & INSTALLATION

Flower construction No. 82. 170 × 130 × 10cm (67 × 51 × 4 in)

The Cactus House at Hignell Gallery, Mayfair, 2017.

Anne ten Donkelaar
Dutch artist
(1979–)

The timeless suspension of nature is a recurring theme of Anne ten Donkelaar's work, from the reconstructed wings of *Broken Butterflies* (2011) and otherworldly *Flower Constructions* (2011) to the ethereal *Underwater Ballet* photographic series (2017). 'Imagine a big bang, a firework of flower seeds thrown into space. What would happen? New fragile flowers arise, new flower planets start evolving, planets where no one has ever been.' So goes ten Donkelaar's own description of the delicate flower installations for which she is most known, each pressed flower and cut-out floral element meticulously pinned onto boards to create odd yet beautiful 3D collages. Thus cacti nestle with orchids and fritillaria, strange green fungi emerge with tulips, and foxgloves defy gravity to grow downwards – gentle reminders of the strangeness of beauty and nature.

Ben Russell
British artist
(1986–)

Ben Russell's hothoused 'plant gang' of alabaster, onyx and Portland limestone sculptures, exhibited in The Cactus House at the Hignell Gallery, London, in 2017, were inspired by the unique way in which light passes through the flesh of a cactus. By working with natural mediums such as alabaster, which emits a similar translucency to cacti when drenched with light, juxtaposed with Portland stone and onyx interpretations, he was able to produce a range of complementary and contrasting shapes and textures, each one sanded and polished to a luminous sheen. Drawing on the notion of a sculpture's environment or 'House' as being as much an artwork as the piece itself, Ben then invited botanical artists from Conservatory Archive in Hackney to create the perfect environment for his works: a living, breathing 'Cactus House', as it were.

Will Cruickshank. Studio works 2016–2017. Silo Studio, Essex.

Will Cruickshank
British artist
(1974–)

A self-imposed residency in three disused grain silos on a farm in Essex means that nature is never far from Will Cruickshank's work. Using the space to develop machines and processes via which he can manipulate materials – a makeshift lathe using a cement mixer to turn large logs, or a weaving machine with which to mathematically envelop totems of concrete, plaster or wood in thread – creates a constantly fluctuating push and pull between material, machine and maker. Cruickshank's move towards semi-mechanical production was embraced as a novice, with 'incorrect equipment', materials found and explored – the exposure of growth rings in a hunk of wood, for example – but also produced such as sawdust from carving a tree trunk or leftover threads from weavings. It's an evolutionary approach, one drawn perhaps from the very push and pull of nature itself.

Camille Henrot
French artist
(1978–)

Camille Henrot's work can appear dizzyingly varied – as in the exhausting yet hypnotic film *Grosse Fatigue* (2013) – but the ultimate aim appears to be the visual bringing together of information so that it can be better understood. For her 2012 show *Is it possible to be a revolutionary and like flowers?*, Henrot translated a personal library of books into the language of ikebana (a Japanese form of symbolic flower arrangement), twisting botanical names and historical plant use to her advantage. The title of the show is taken from a line in Marcel Liebman's book *Leninism Under Lenin* (1975). On pondering his own question about flowers and action, one of Lenin's lieutenants states: 'You start by loving flowers and soon you want to live like a landowner, who, stretched out lazily in a hammock, in the midst of his magnificent garden, reads French novels and is waited on by obsequious footmen.'

CRAFT, DESIGN & STYLE

People have been plundering the Plant Kingdom for motifs and materials since the Dawn of Civilisation, yet somehow 'botanical style' always seems like such a fresh new thing. Today, social media has helped create a huge 'trend' for all things botanical, from must-visit green spaces and awe-inspiring floral installations to blooming textiles and cult curations such as Plants on Pink (see page 378). If you want to engage with the botanical world, it's all about finding your botanical niche, as the artists, designers and makers to follow reveal.

CRAFT, DESIGN & STYLE

William Morris
English designer, artist and social activist (1834–1896)

Lucy Augé
British artist, Bath, UK 1988–

The Arts and Crafts Movement's main influencer William Morris is the quintessential purveyor of botanical style, his plant-inspired textiles and wallpaper furnishing the homes of pattern aficionados since the mid-nineteenth century. Inspired by friendships with Pre-Raphaelite artists such as Dante Gabriel Rossetti and Edward Burne-Jones, he abandoned a fledgling career in architecture to become an artist, co-founding the company that would eventually become Morris & Co. The following year he designed the first of 50 wallpapers – *Trellis* (1862) followed by *Daisy* (1864) and classics such as *Chrysanthemum* (1877). Although influenced by medieval art and herbals, the main sources for Morris's 600-strong collection of designs was nature itself, observed in his gardens or on country walks: '. . . any decoration is futile', he claimed, '. . . when it does not remind you of something beyond itself.'

Waking from a nap one day, Bath-based artist Lucy Augé was transfixed by a pattern of leaf shadows cast against a pile of picture frames in her countryside studio. The shadows led her not only to a deeper investigation of the trees in her immediate vicinity but also to a similarly mesmerising body of work exhibited at plant-based concept store Botany Shop in East London. The underlying beauty of *Tree Shadows* (2017), and indeed all of Augé's work – from the wonderful diversity and spontaneity of *500 Flowers* (2015) to *Moving into Abstraction* (2018) at The Garden Museum, London – is a willingness to let nature choreograph the timing and shape of the ultimate form. The parallel mastery of it perhaps is in the choice of materials – the simplest combination of Chinese fat brush, calligraphy ink and paper – and a deftly fluid style of draughtsmanship with which to capture the ephemerality of the moment.

'*Nature is painting for us, day after day, pictures of infinite beauty if only we have eyes to see them.*'

John Ruskin, British art critic, draughtsman and social thinker (1819–1900)

Opposite Wallpaper sample with a chrysanthemum (genus *Chrysanthemum*) and acanthus leaf (genus *Acanthus*) design by William Morris, woodblock print, 1877. **Right** Lucy Augé, *Marigold* (genus *Tagetes*), from *500 Flowers: Observation into nature and the artist's process* (2015).

INSPIRED BY NATURE 369

CRAFT, DESIGN & STYLE

Jenny Kiker
American artist
(1982–)

Botanical art comes in many guises. Jenny Kiker's 'living patterns' represent a new brand of plant documentary. One that seeks to portray the accuracy of the plant – in Jenny's case through the medium of watercolour – but imbue the work with an overriding sense of playfulness and flow. There's joy and vibrancy in each brush-stroke and an unashamed delight in celebrating the aesthetic appeal of the Plant Kingdom in all its verdant, stylistic stride.

Living Pattern was launched in 2013 as a way for Jenny to connect with nature, encourage others to do the same and, as a cultural response, serve as a reminder of what's real. Trained in illustration and painting with a background in fashion and visual display with brands such as Free People and Urban Outfitters, she learned the value of compelling creativity in a brand – but also that the fit wasn't quite right. One look at some of Jenny's most iconic artworks, her *Monstera Deliciosa* or *Maidenhair Fern*, confirms that painting botanicals most certainly is.

Although Jenny hails from a very nature-centric family – she always grew plants with her Nana and her parents kept gardens – she didn't start hoarding her own plants until college, where they served as the constant subjects of numerous still lives. Drawn at times to the peripatetic lifestyle, she continued to paint along the way, inspired by the plants and flowers of California, the arid climates of San Diego and the greenhouse-like environment of Florida where she now lives and works. Her work is also influenced by nineteenth-century botanical art, American painter Georgia O'Keeffe (1887–1986), Australian pattern designer Florence Broadhurst (1899–1977) and the natural-world mosaics of the American artist-designer Christopher Marley (1969–present day).

With botanicals now the focus of Jenny's artwork, her home has become something of a working conservatory, where specimens can be observed directly but also imagined in their natural habitat. Her approach celebrates the beauty of plants but also brings longevity to an all too fragile world, lightly raising awareness about the need to conserve our natural world – each exquisitely brush-stroked plant placed in the collective consciousness to live another day.

My botanical world . . .

'I've always been drawn to plants. Their slow motion instills a sense of patience in me that I can't find anywhere else. I love to keep them happy and because of the care they get, they always give back so much.'

MAKE THE CONNECTION
Find out more about Jenny's botanical world at www.livingpattern.net; @livingpattern

CRAFT, DESIGN & STYLE

Top Jenny in her conservatory-cum-studio with a super-size Swiss cheese plant (*Monstera deliciosa*). **Below** Finding desert tones and colour-matching Saguaro cacti and succulents. **Left** The living pattern of a sword fern (*Polystichum munitum*) is brought to life in numerous shades of green. **Opposite** A flowering *Notocactus* (now *Parodia* cactus) slowly reveals its spines.

INSPIRED BY NATURE 371

Seana Gavin, *Sirius Life*, 2011, from *Alternate Dimensions* solo show at B store, London.

Seana Gavin
British collage artist

Growing up in the supreme shadow of the Catskill mountains, near New York, now-London-based artist Seana Gavin consistently seeks out nature to inform her fantasy collages. Kew Gardens, the Kyoto Garden in Holland Park and the JG Ballard-esque conservatory of the Barbican Centre are firm favourites, as is the Jardin de Cactus on the volcanic island of Lanzarote – a surreal garden of 4,500 different species of cacti curated by the artist, architect and nature champion César Manrique in 1970 in a disused amphitheatre-style quarry. Gavin's response to all this greenery and the overload of imagery and visual noise of the Modern Age, is to create psychedelic, Hieronymus Bosch-style dreamscapes such as *Fairyville*; *Homage to Richard Dadd* (2011) – a nod to the Victorian painter's visions of the supernatural.

Mary Delany
English artist
(1700–1788)

Subjected to an unhappy marriage to an elderly Cornish MP at the tender age of 17 years, Mary Delany née Granville was left a widow in her early twenties. Some two decades later, against the wishes of her family, she married Dr Patrick Delany, who first encouraged her to follow her artistic pursuits. Upon his death in 1768, she met the botanical artist Georg Dionysius Ehret (see page 313) and the globe-trotting naturalists Sir Joseph Banks and Daniel Solander. The fortuitous introduction resulted in Delany's unique masterpiece, *Flora Delanica* (1771–1783), a florilegium of exquisite 'paper mosaicks' such as *Papaver somniferum*, *The Opium Poppy* (1776), whereby handcut pieces of coloured paper (200 pieces per flower in some cases) were glued onto a black ink background to create a truly unique scientific catalogue of plant specimens. See *Passiflora laurifolia* on page 44.

Embrace, 2012. Paper Collage, 15.5 × 11.5cm (6½ × 4½ in).

Nature printed oak leaf (*Quercus*).

Lynn Hatzius
German-British artist and illustrator
(1979–)

From surreal collages and prints of plants and people, to commissioned covers or column illustrations for nature-themed books or periodicals – including the wonderful designs for the cover and chapter openers of this book – it appears that artist and illustrator Lynn Hatzius always has one foot in the natural world. This may owe something to a multi-cultural background with varied landscapes, changing biomes and ways of seeing and crafting all part of the inspiring mix. Or it may just be intrinsic – a stronge urge to relocate to the countryside in adult life means that she now lives and works by a lake in rural Sweden. Either way, what transpires creatively are beautiful multi-media amalgamations of the fabric of life in all its wonderful duality – flora and fauna, birth and decay, plants and people, nature and humanity, and past and present providing rich 'threads' with which to 'weave'.

Pia Östlund
Swedish artist and printmaker
(1975–)

In 2016, printmaker Pia Östlund and author Simon Prett published a book that told the tale of Östlund's quest to revive and master the lost art of nature printing – in part to raise funds for the digitisation of specimens in the Chelsea Physic Garden. In *The Nature-Printer* (The Timpress, 2016), Prett describes Östlund's journey through fields of ferns and 'meetings' with nature-printing experts past and present to create not just a fascinating tale of discovery and invention but also a body of beautiful artwork. Taking a process that was developed during the Middle Ages to help illustrate herbals that was further advanced during the nineteenth century by Austrian printer Alois Auer, the results are captivating echoes of plant forms so true to life that they might almost be the specimens themselves.

Leopold and Rudolf Blaschka
Bohemian glass artists
(1822–1895 & 1857–1939)

The Harvard Museum of Natural History (HMNH) was established in 1998 to consolidate the collections of the Harvard Botanical Museum (founded in 1890 and now administered by the Harvard University Herbaria), the Mineralogical & Geological Museum (founded 1901) and the Museum of Comparative Zoology (founded 1859). Arguably its most prized exhibit is the *Glass Flowers: The Ware Collection of Blaschka Glass Models of Plants*. Created by Dresden-based father and son glass artists Leopold and Rudolf Blaschka between 1886 and 1936, the collection has been on display at Harvard University since the late 1880s. Born into a family of glass experts, a heritage that Leopold credited for their artistry and skill, the pair created more than 4,000 highly accurate, exquisitely crafted models of plant species from blooming flowers to rotten apples. The Blaschkas had previously won acclaim within the scientific community for their stunning, anatomically correct models of invertebrate animals which were highly sought after by museums and educational institutions.

Tiffanie Turner
American botanical sculptor
(1970–)

Architect training helped inform the work of botanical sculptor Tiffanie Turner, whose expertly constructed paper flowers bring the largesse and often overlooked detail of nature to the fore. The conceptual artist Tom Friedman, abstract sculptor Lee Bontecou and Dutch master painters are all cited as inspirations alongside the rhythms and patterns found in nature as well as the 'missteps and irregularities caused by decay, rot, wilt, dormancy, death and genetic and viral mutations like phyllody, petalody and fasciation'. Although created to some degree of botanical accuracy, the works are often preternaturally large (taking between 200 and 400 hours to complete) and open to aspects of fading beauty and human intervention with wilting, decaying and artistically-interpreted petals sitting side by side. Turner's book *The Fine Art of Paper Flowers* (Ten Speed Press, 2017) is a wonderful guide to her art and craft.

> 'Many people think that we have some secret apparatus by which we can squeeze glass suddenly into these forms, but it is not so. We have tact. My son Rudolf has more than I have, because he is my son, and tact increases in every generation.'
>
> Leopold Blaschka

Opposite *Malus pumila* with apple scab (Model 812), Rudolf Blaschka, 1932. The Ware Collection of Blaschka Glass Models of Plants, Harvard University Herbaria. **Right** Tiffanie Turner, *Cremon Mum*, 2016. Photographed by Scott Chernis, 2016. Paper mâché and Italian crepe paper, 132-cm (52-in) diameter, 36-cm (14-in) deep.

Flora Starkey
British florist

There are hundreds of inspiring floral designers past and present, but London-based Flora Starkey stands out for her unconventional blooms and the cross-pollination of her practice into the fashion, art, film and editorial worlds. Taking inspiration from Dutch Masters and Renaissance artists, Victorian flower photographs and the English country garden, her bespoke displays are romantic and sumptuous, often made using homegrown flowers and foliage to add an extra layer of authentic seasonality. Starkey's mother was a florist so it was a natural progression to the trade that she has very much made her own, albeit via a period designing and running her own fashion label. The two fields are by no means mutually exclusive, however, sharing a similar design process and principles. Yet Starkey found that flowers afforded a more spontaneous and instant approach, one she likens more to painting than making clothes. It's a medium that allowed her to be more home-focused while still offering the potential to go wild. An incredible colourist, her designs were used at the opening of the *Savage Beauty* (2015) exhibition, a retrospective of Alexander McQueen's creations at the Victoria and Albert Museum, London. Alexander McQueen is just one stellar client on an impressive roster that also includes Givenchy, Gucci, Louis Vuitton, Preen and Frieze London art fair. Starkey's ethereal designs were also chosen to adorn the London Gate at the RHS Chelsea Flower Show 2016 in honour of the 90th birthday of Her Majesty The Queen, comprising six crowning moments at the top of each of the six gate pillars. Blooms included roses, clematis and foxgloves with flowers and foliage spilling over gate pillars and climbing up iron railings. Her editorial line-up is similarly impressive, with installations gracing the pages of British *Vogue* and *AnOther* magazine. Blessed with an aptronymic first name – Flora after the Roman goddess of flowers – a background in fashion design, and another top client in the form of photographer Nick Knight's SHOWstudio, Starkey was also the perfect choice to curate his *Fashion Flora* (2017), exploring florals in fashion throughout the decades. Needless to say, Starkey also designed the flowers that accompanied specially commissioned illustrations and Knight's own floral forays (see page 336).

MAKE THE CONNECTION
Find out more about Flora's botanical world at www.florastarkey.com; @florastarkey

CRAFT, DESIGN & STYLE

My botanical world . . .

'*Flowers and plants have the ability to lift your mood instantly – they bring pleasure in such a simple and reassuring way. It doesn't have to be a decadent arrangement of expensive blooms, more often than not it's a few stems picked from the garden or the hedgerow that make me happiest. Nature is an ever changing and constant inspiration that I'm always grateful for, flowers are really nothing short of everyday miracles.*'

Top A still from *Spring Rites* (2018), a digital installation film depicting the beauty and magic of flowers from seed to bloom. **Left** A floral arrangement from the opening gala for *Alexander McQueen: Savage Beauty* (2015) at the V&A museum in London. **Opposite** Studio portrait by photographer Yolanda Chiaramello.

INSPIRED BY NATURE

Agave Americana, Plants on Pink.

Loose Leaf, *Monstera Hover Wreath,* Tokyo, Japan, 2017.

Lotte van Baalen
Dutch artist and curator

Once you're party to the Plants on Pink phenomenon, conceptualised by Dutch artist and curator Lotte van Baalen, it's hard not to seek out a pink surface on which to place a houseplant, or become obsessed by finding such an occurrence in the 'wild'. The next step is to post your image of pink backgrounded botanical onto Instagram with the hashtag #plantsonpink for the chance of featuring in Baalen's online gallery @plantsonpink – and a viewing by her 100,000-strong worldwide community. Inspired by her artist parents, van Baalen studied visual arts and has always been inspired by colours and combinations. A 'very pink and green phase' led to the formation of her now wildly popular brand, in turn helping to elevate the colour pink, houseplants and van Baalen to unexpected stardom. It's a combination that brings people together and makes them smile – no mean feat.

Wona Bae and Charlie Lawler
Korean and Tasmanian designers
(1976– & 1980–)

Husband and wife team Wona Bae and Charlie Lawler seem pre-destined to set up their internationally-renowned botanical design studio Loose Leaf. Bae, whose name translates as 'the best seedling', grew up on a South Korean flower farm, learning the family trade from a young age, while Lawler's family operated a plant nursery in Tasmania. The German-trained master florist and Master of Kokozi (Korean-style ikebana) and designer (Lawler) now collaborate to create awe-inspiring permanent and temporary natural sculptures such as their signature Monstera chandeliers and huge wreaths. By bringing the outside in through natural design and the inventive use of materials such as plants and flowers – as illustrated in their book *Loose Leaf: Plants Flowers Projects Inspiration* (Hardie Grant Books, 2016) – the team hope to highlight the nobility of nature.

Guelder, Hebe, Love in the Mist taken from *Flowers* series by Ali Mobasser and Miria Harris, 2013.

Miria Harris
British landscape, garden and floral designer
(1975–)

Landscape design and floristry are often viewed as separate entities, one working with the wider environment, the other honing in on flowers for a largely interior setting. London-based designer Miria Harris highlights their oneness, primarily creating high-end garden designs for private and public clients (see page 152), while also undertaking choice floral design commissions. Such works include a series of 'plant planets', called *Flowers*, with photography artist Ali Mobasser, an installation for Aspinal of London and an evolving display for The Marksman Public House and Dining Room near Columbia Flower Market in London's East End. Combining an art curator's eye for composition (a previous career) with a landscape designer's knowledge of how plants go together in nature, Harris's floral works remind us that macro detail is elemental to landscape, too.

Justina Blakeney
American designer and artist
(1979–)

'Jungalow' style is a way of life for Justina Blakeney, comprising a wild but cosy meeting of colour, pattern and plants, shared via her art, designs, interiors and much-followed blog www.jungalow.com. Explored through her book *The New Bohemians* (Stewart, Tabori & Chang, 2015) and *The New Bohemians Handbook* (Abrams Books, 2017), Blakeney's covetable aesthetic emerged from a multicultural upbringing in California, cemented by a degree in World Arts and Culture and extensive global explorations. Although infinitely more well-travelled than Morris, the Los Angeles-based artist and designer has a similar ethos towards amalgamating nature into her creations, from surface pattern designs of the palms, cacti and agave plants of her native landscape to the tropical plant 'gangs' that helped relaunch the humble houseplant as a must-have style icon.

CRAFT, DESIGN & STYLE

Above Vita Sackville-West, 1958 by Cecil Beaton (LB1316).
©The Cecil Beaton Studio Archive at Sotheby's. **Opposite** Aerial view of Central Park, New York, which was co-designed by Frederick Law Olmsted.

Vita Sackville-West

English poet, novelist and garden designer
(1892–1962)

The ranks of famous female gardeners would not be complete without homage to writer Vita Sackville-West. Like Flora, the Roman goddess of plants before her, planting peers such as Gertrude Jekyll (1843–1932), Ellen Biddle Shipman (1869–1950) and Beatrix Farrand (1872–1959) and more contemporary horticultural heroines such as Beth Chatto, Carol Klein, Alys Fowler and Nancy Goslee Power, Sackville-West is forever entwined with the botanical world, most notably through the garden she passionately brought to life with her husband Harold Nicolson at Sissinghurst Castle, Kent. Within this landscape they created a series of 'rooms', each with a different character or theme. Indeed, two of her most famous poems, *The Land* (1926) and *The Garden* (1946) are dedicated to the natural world, while *Some Flowers* (1937) uses beautifully expressive language to pay homage to a selection of her favourite blooms.

Frederick Law Olmsted

American landscape architect
(1822–1903)

The 'Father of American landscape architecture', Frederick Law Olmsted, will forever be associated with New York City's Central Park. Interwoven into his winning design was the firm belief that the common green space must be accessible to all. His design ethic was similarly magnanimous, working with the subordination of details such as pathways, trees, pastures or bridges so that the whole effect of the space took precedence. Primarily inspired by pastoral and picturesque styles he thus combined vast expanses of green, lakes and trees with rocky, broken terrain and naturalistic creepers and shrubs, through which visitors could walk or proceed by carriage. A keen conservationist, Olmsted also helped to establish America's first state park, the Niagara Reservation, preserving the epic waterfalls and surrounding landscape and biodiversity for generations to come.

Ornamental grasses beside pond, Southhampton, NY, USA.

The Garden of Cosmic Speculation, Dumfries, Scotland.

Wolfgang Oehme and James van Sweden
German and American landscape architects (1930–2011 & 1935–2013)

'At eleven o'clock a man in a raincoat, dragging a lawn-mower, tapped at my front door and said that Mr Gatsby had sent him over to cut my grass.' So relates narrator Nick Carraway in F. Scott Fitzgerald's landmark novel *The Great Gatsby* (1925), an episode that illustrates both how much Jay Gatsby wanted to impress Daisy Buchanan and the rise of the lawn (and mower) as a symbol of status and wealth. This carefully manicured vision, a key feature of the suburbs of latter-day post-war America, did not sit well with landscape architect partners Wolfgang Oehme and James van Sweden who, in the early 1960s, created the New American Garden style of landscape architecture, a celebration of the seasonal splendour and ecology of the American meadows via 'tapestry-like plantings'. Thus prairie-style planting began.

Charles Alexander Jencks
American theorist and landscape designer (1939–)

Famous in the 1980s as a theorist of Postmodernism, American-born Charles Alexander Jencks switched to landscape design – or landform art – when he and his second wife, the late garden designer and Chinese garden expert Maggie Keswick Jencks, decided to transform their family home at Portrack House, near Dumfries in Scotland. What they came up with was the monumental Garden of Cosmic Speculation, begun in the late 1980s and now open for one day a year. Inspired by a swimming hole that his wife had dug for their children, Jencks decided to use the space to create a new grammar of landscape design, one that explored ideas about nature, science, mathematics and discovery. The result is a truly unique conglomeration of features such as The Universe Cascade, the twisting waves of Snake Mound and The Black Hole Terrace.

Artscience Museum, Singapore.

Bosco Verticale (Vertical Forest), Lombardy, Milan, Italy.

Moshe Safdie

Israeli-Canadian architect and theorist
(1938–)

Bold shapes, large-scale urban projects and integrated green spaces are the benchmarks of internationally-renowned architect Moshe Safdie. Born in Israel and raised in Canada, he first burst onto the architectural scene in 1967 with his modular Habitat 67 housing complex – a prefab design where each residence got its own roof garden – created for the World Expo in Montreal. Perhaps the most obvious botanical reference in his work, however, comes in the form of The ArtScience Museum at Marina Bay Sands, Singapore. Nodding to the elegant unfolding of the lotus flower, the spectacular building comprises a base that is surrounded by the Bay's water and a giant lily pond, plus a flower-like structure of ten skylighted petals, which rise towards the sky at varying heights. The dish-like roof also harvests rainwater, which then falls through the centre of the museum in a giant cascade.

Stefano Boeri

Italian architect and urban planner
(1956–)

Imagine a skyscraper that was also a garden. That's exactly what Stefano Boeri did when he designed his *Bosco Verticale* (Vertical Forest). Although many people had proposed such a vision, Boeri took it to fruition with two towers clad in 730 specially-cultivated trees, 11,000 groundcover plants and 5,000 shrubs. The vision was made possible by a team of botanists and engineers who had to deal with factors such as wind resistance, weight, the plants' nutritional demands and the potential of growth versus residents' views. The effect is optimistically beautiful and environmentally beneficial, using plants to naturally absorb dust, produce oxygen and help humidify indoor air and insulate the building. Boeri now hopes to build entire forest cities to help combat pollution in China and other countries, and bring greenery into the lives of more people.

CRAFT, DESIGN & STYLE

'*Snatching the eternal out of the desperately fleeting is the great magic trick of human existence.*'

Tennessee Williams writing in 'The Timeless World of a Play', a preface to his play *The Rose Tattoo*, first produced in 1951

Above Rachel Dein, *Rose, Daisy, Plantain and Achillea*, spring 2017. Plaster and clay, 25 × 25cm (10 × 10 in), 1.5-cm (⅝-in) deep.
Opposite Phoebe Cummings, *Triumph of the Immaterial*, (clay, water). Winner of the Woman's Hour Craft Prize, Victoria & Albert Museum, 2017. Photo by Sylvain Deleu.

Rachel Dein
British artist
(1969–)

In 2017, artist Rachel Dein was commissioned by Hidcote Manor Gardens, Gloucestershire, to document the changing seasons of its Lawrence Johnson-designed borders and 'outdoor rooms'. Renowned for her evocatively beautiful plaster and concrete casts of pressed botanicals, Dein set about permanently encapsulating ephemeral moments in the life cycle of various plants and flowers from the arts and crafts garden, employing a bespoke process artfully developed from a glass-casting technique. As with previous commissions, including Chelsea Physic Garden and Le Manoir aux Quat' Saisons for the Jardin Blanc Restaurant at Chelsea Flower Show 2017, the power of Dein's work is in the skilfully captured, intricate detail of each naturalistically-arranged specimen. The results are not only exquisitely beautiful, the poetic transience, ephemerality and yet tenacity of the Plant Kingdom is captured for all time.

Phoebe Cummings
British artist
(1981–)

Themes of earthiness and ephemerality are also explored via the ceramic installations of British inter-disciplinary artist Phoebe Cummings. Working with raw, unfired clay to create temporary site-specific pieces, Cummings's work responds to the natural world and lasts only for the duration of an exhibition. Thus, *Triumph of the Immaterial* (2017), a fountain-piece made over five weeks from layers of exquisitely sculpted, air-dried clay flowers and foliage that was subjected to a flow of water that gradually eroded it, running for a few minutes each day for the six month span of the exhibition. The absurdity of investing time and energy into something that will break back down into the raw material is not lost on Cummings, indeed, it's all part of the performative mix – a response to a preoccupation with owning things and the emotions drawn from fleeting entities such as nature.

Sonya Patel Ellis
British writer, editor and artist
(1973–)

The first flowers I ever pressed were those of my childhood, inserted between the pages of my other love – books. My knowledge of nature is certainly tied to this early pursuit, coupled with the feeling that flowers and plants hold the key to understanding just about everything. The world of herbaria and botanical art has always entranced me; a place where science and art naturally combine and discovery is intrinsic. The concept of disciplined methodology also appeals, the twin forces of constraint and poetry in motion. Such were my thoughts in 2013 when I first sowed the seed for The Herbarium Project, now part of A Botanical World (www.abotanicalworld.com). On a personal level, I was aware that writing and editing were off the cards for the duration of my second maternity leave. I did, however, want to create a means by which I could bring my writing and artwork together for the future. Nature, a lifelong obsession, was the obvious answer. I duly set about making my own traditional, large-scale slatted herbarium press and pressing and documenting plant specimens from my garden for a year. Taking a naturalistic approach, I was astounded by the features of plants and flowers that careful, herbarium-quality pressing reveals: the tendrils of a sweet pea, the phallic seedhead of a California poppy, the star-like umbels of fennel (see pages 10, 174 and 147 respectively). I was also dedicated to the labelling of specimens using common and botanical names and recording environmental factors, a task that revealed even more about the plants and flowers around me. Presenting my discoveries was the next step, with much research into conservational mounting techniques and backgrounds. Some mounts stayed true to the herbarium format, others ventured off to explore aspects of kitsch and Victoriana – a journey that resulted in my now-signature black background. Exhibitions, events, workshops and private commissions followed including *The Earth Laughs in Flowers* (2014) at which I displayed my *Herbarium Wall* (2013); *Hortus Uptonensis* (2015), a homage to botanist Dr John Fothergill in West Ham Park's bandstand; *Seeing Floral – The Art of Florilegia* at The Garden Museum (2015); *Herb Garden at The Marksman* (2016) with landscape designer Miria Harris – and through continued writings, *Collins Botanical Bible* (2018), the ultimate 'herbarium' of plants and people.

My botanical world . . .

'*For me nature, and thus the world, is all about connection – how plants relate to each other, how people relate to plants, those platforms where art and science intermingle and converge. Pressing botanicals reveals the minutiae of nature but also highlights the bigger picture.*'

MAKE THE CONNECTION
Find out more about Sonya's botanical world at www.abotanicalworld.com; @abotanicalworld

CRAFT, DESIGN & STYLE

Top A preface for *Herb Garden at The Marksman* (2016) illustrates the diversity of the culinary herb garden. **Below** The coiled tendrils of the old-fashioned sweet pea (*Lathyrus odoratus*) 'Prima Donna' from *The Earth Laughs in Flowers* (2014) compel the viewer to look closer. **Left** The papery petals of corn poppy (*Papaver rhoeas*) created for *Hortus Uptonensis* in West Ham Park (2015) are perfect for pressing. **Opposite** Studio portrait by photographer Carmel King (www.carmelking.com).

A page from Emily Dickinson's herbarium, (Houghton Library, Harvard University, Dickinson MS AM 1118.11).

Little Meadow lino print, 25.4 × 20.3cm (10 × 8 in).

Emily Dickinson
American poet
(1830–1886)

The celebrated American poet is well known for lyrical musings such as *May-Flower* (1896), but her love of nature is most evident through the herbarium she made when she was just 14 years old. Corresponding with former Amherst Academy classmate Abiah Root, she wrote: 'Have you made an herbarium yet? I hope you will if you have not, it would be such a treasure to you; most all the girls are making one. If you do, perhaps I can make some additions to it from flowers growing around here.' A common pursuit of mid-nineteenth-century women, Dickinson took pressed botanicals to another level with 424 elegantly juxtaposed specimens arranged on 66 pages with handwritten labels. Emily Dickinson's Herbarium can be viewed at Harvard Houghton Library, online or via a beautiful cloth-bound facsimile produced in 2006.

Mike Pollard and Rika Yamasaki
London-based artist-maker duo

There's something very personal about pressing flowers that artist-maker duo Mike Pollard and Rika Yamasaki portray so well through their MR Studio London portfolio of artworks, cards and prints. It's obvious that each tiny sprig of foliage or little bloom has been hand-picked from a place that they have formed a relationship with. This includes specimens from their garden and walks through the forest nearby, plus weeds or plants that otherwise might be overlooked. Their much-followed brand, MR Studio London, came about via a shared love of nature and the search for a material that they could both really explore, Pollard for its miniaturised sculptural form, Yamasaki for its fairytale, feminine quality. The pair also produce digital, lino or riso prints of their pressings, adding permanence to the tender sprigs of nature to which they are drawn.

CRAFT, DESIGN & STYLE

Colocasia, Richmond, 2011.

Stitched Mallow: moved together we make one. From *Leaf Poems* (2017).

Anne Blackwell Thompson
American artist and plantswoman
(1963–)

Blackwell Botanicals, set up by artist and plantswoman Anne Blackwell Thompson in 2010, began life as a celebration of the immense diversity of America's southern landscape. Following an apprenticeship with pressed botanicals maestro Stuart Thornton in Turin, Italy (think huge pressed gunnera leaves on outsized canvases), Thompson directed her own artistic training towards harvesting and pressing specimens from the swamps of South Carolina, the gardens of Virginia's countryside, the waters of the Atlantic Ocean and the mountains of the Shenandoah Valley near her home. Her work now adorns the walls of private collectors and public spaces throughout Virginia and beyond (as far afield as Hawaii), with beautifully mounted large-scale leaves as just one of her desirable signature pieces. As a growing number of interior designers are finding, it's the perfect way to bring the outside in.

Nicola Cunningham
British artist, aromatherapist and arts educator
(1977–)

Growing up in the wilds of the Isle of Man, just off the west coast of England, artist and aromatherapist Nicola Cunningham quite literally lives and breathes nature. As a child she would pick and press foliage and flora from the hedgerows and fields of her beloved island, the magic and mystique of its fairy-like glens and mist-shrouded coastlines providing much fodder for songs and poems to go with. Today, Cunningham divides her time between a busy aromatherapy practice (find remedies on pages 288 and 295), teaching children how to freely express themselves through art and poetry, and producing artworks that play with human relationships and the natural world. One such offering is her *Leaf Poem* series (2015–2017), made using fresh plant matter that is often manipulated, folded or stitched to bring out the multi-layered, biosemiotic potential of plants.

ACKNOWLEDGEMENTS

A tremendous amount of time, effort and expertise goes into creating a book of this size and collaborative ilk, and thus there are numerous people to thank . . .

Enormous gratitude is foremost extended to the wonderful team at Harper Collins: Hazel Eriksson for her unfailing enthusiasm, professionalism, collaborative attitude and patient editing of words, pictures and not a small amount of Latin – it is a complete joy to work with this lady; Helena Caldon for similarly good-natured attention to consistency, style, tone, botanical language and conversions, an invaluable service; Eleanor Ridsdale for the lovely, clean design and letting the images and quotes sing throughout; Jo Carlill for her contribution to the sourcing of more than 300 botanically-inspired images, one of the most crucial aspects of the book; Elizabeth Woabank for help with editorial planning and content; Anne Rieley, and Rachel Malig for fastidious proofreading and editorial work; and Gareth Butterworth for design assistance through the latter part of the process.

A huge amount of thanks also goes to botanist, botanical author and taxonomic expert Professor Clive A. Stace, D.Sc., author of *New Flora of the British Isles* (Cambridge University Press, 2010) and co-author of *Alien Plants* (William Collins, 2015) among other volumes, our consultant on the book. Where my skills lie in combining words and pictures to create a compelling narrative, driven by my passion for plants, Clive's esteemed expertise in all things botanical has been crucial in terms of the accuracy of such a presentation. In short, he has made this book possible.

Also at the top of my thanks list is Lynn Hatzius (www.lynnhatzius.com), dear friend, illustrated publishing companion, fellow mother, incredible listener and a woman of many talents including illustration, printmaking and collage. Lynn conjured up the white-on-black silhouetted diagrams found at key twists and turns within the book, along with each and every plant detail on the cover, transforming inspired briefs and collaborative brainstorms into unique illustrations with effortless talent and grace.

The collaborative and generous nature of this book is also apparent from numerous expert contributions of words and images from fellow plant lovers around the world. As follows (in order of appearance):

Chapter 3 – Growing & Gathering

Thanks to landscape designer Miria Harris (www.miriaharris.com) for planting suggestions and striking images of some of the plants in her wonderful gardens, as well as Marksman images in chapter 4 and pre-planning brainstorms; Josefina Oddsberg from Bee Urban, Sweden (www.beeurban.se) for advice on planting for pollinators; House of Hackney (houseofhackney.com) for *Palmeral* interior loveliness; and illustrator Lindsey Carr (www.littlerobot.org.uk) for the beautiful *Garden* illustration used on the 'Planting for Biodiversity' pages.

Chapter 4 – Botanical Recipes

Thanks to chef Tom Harris of The Marksman Public House and Restaurant (www.marksmanpublichouse.com) for many years of culinary inspiration and laughter plus introductory quotes and recipes for autumn and winter. Recipes, quotes about their botanical worlds, and several images were also kindly donated by the following: Robin Harford, (www.eatweeds.co.uk); Bernadett Vanek (www.bornunderthesun.com); John Rensten (www.foragelondon.co.uk), also courtesy of Emma Bishop at Macmillan Publishing; Mark Diacono (www.otterfarm.co.uk), also courtesy of Dan Herron at PFD and Claire Weatherhead at Bloomsbury Publishing; Claire Ptak (www.violetcakes.com), also courtesy of Adriann at Violet Bakery Café plus Rowan Yapp at Penguin Random House; Dina Falconi (www.botanicalartspress.com); Nina Olsson (www.nourishatelier.com); Sarah Witt (www.sarahwitt.net/hdtk); Matt Hoyle at Nobu 57, New York (www.noburestaurants.com); Kitty Travers (www.lagrotteices.tumblr.com); Skye Gyngell (www.springrestaurant.co.uk), also courtesy of Evie Horsell at Gerber Communications and Quadrille

Publishing Ltd.; Gill Meller (www.gillmeller.com), also courtesy of Emma Marijewycz at Hardie Grant; Adele Nozedar (www.breconbeaconsforaging.com); Paola Gavin; Matt and Lentil Purbeck (www.grownandgathered.com) also courtesy of Ashley Carr at Macmillan Australia; Pascal Baudar (www.urbanoutdoorskills.com); Alan McQuillan (www.bloomsburydistilleryspiritco.co.uk); Amy Zavatto (www.amyzavatto.com); Charisse Baker (www.eastlondonjuice.com), also for a remedy in chapter 5; and Rawia Baitamal Edwards (www.secretsage.com).

Chapter 5 – Botanical Remedies

Thanks to Nicola Cunningham (www.stemapothecarystore.com) for remedies plus the therapeutic massaging of my writing wrists, shoulders and arms, and artwork in chapter 6. Plus the wonderful Rosemary Gladstar (www.sagemountain.com) – an inspiration to so many around the world – also courtesy of Storey Publishing; Dr J J Pursell (www.fettlebotanic.com); Julia Lawless (www.aqua-oleum.co.uk); Deborah Hanecamp (www.mamamedicine.nyc), also courtesy of Claire Austin at Mama Medicine; Kim Walker and Vicky Chown (www.handmadeapothecary.co.uk), also courtesy of Sophie Allen at Kyle Books; and Marlene Adelmann (www.herbalacademy.com), also courtesy of Amber Meyers at The Herbal Academy.

Chapter 6 – Inspired by Nature

Thanks to Gavin Turk (www.gavinturk.com), also courtesy of Mira Bajagic at Gavin Turk Studio; Katie Scott (www.katie-scott.com), also courtesy of Lisa Edwards at Big Picture Press; Jess Shepherd (www.inkyleaves.com); Rachel Pedder Smith (www.rachelpeddersmith.com); Rosie Sanders (www.rosiesanders.com); Bobbi Angell (www.bobbiangell.com); Niki Simpson (www.nikisimpson.co.uk); Macoto Murayama (www.frantic.jp/en), also courtesy of Rodion Trofimchenko at Frantic Gallery; the brilliantly helpful and charming Rob Kesseler (www.robkesseler.co.uk), also courtesy of Papadakis Publishing; Dain L. Tasker; Sebastião Salgado; Magnus Edmonson and India Hobson (www.haarkon.co.uk); Viviane Sassen (www.vivianesassen.com), also courtesy of Peter-Frank Heuseveldt at Viviane Sassen studio; Taryn Simon (www.tarynsimon.com), also courtesy of Kerin Sulock at Taryn Simon studio; Nick Knight (www.nickknight.com); Sam Levack and Jennifer Lewandowski (www.levacklewandowski.com); Kwang-Ho Lee (www.johyungallery.com), also courtesy of Boram Yun at Johyun Gallery; the Hilma Af Klint Foundation; Stockholm's Moderna Museet, Tom Ellis (www.tomellisartist.com); Azuma Makoto (www.azumamakoto.com), also courtesy of Eri Narita at Azuma Makoto Studio; Simon Heijdens (www.simonheijdens.com); Lilian & Co Seegers for studio herman de vries; NILS-UDO; Rebecca Louise Law (www.rebeccalouiselaw.com), also courtesy of Olivia Mary Deane at Rebecca Louise Law studio; Anne Ten Donkelaar (www.anneten.nl); Ben Russell (www.benrussell.com); Will Cruickshank (www.willcruickshank.com); Lucy Augé (www.lucyauge.co.uk); Jenny Kiker (www.livingpattern.net); Seana Gavin (www.seanagavin.com); Pia Ostlund; Tiffanie Turner (www.papelsf.com); Flora Starkey (www.florastarkey.com), also courtesy of Charlie at Flora Starkey studio; Wona Bae and Charlie Lawler (www.looseleafs.com.au); Lotte van Baalen (www.plantsonpink.com); Rachel Dein (www.racheldein.com); Phoebe Cummings (www.phoebecummings.com); Mike Pollard and Rika Ramasaki (www.mrstudiolondon.com); and Anne Blackwell Thompson (www.blackwell-botanicals.com).

General Images & Research

Thanks to all who have generously contributed images – they really do make the book – most especially the Biodiversity Heritage Library (www.biodiversitylibrary.org); The New York Public Library Digital Collections (www.digitalcollections.nypl.org); Plant Curator (www.plantcurator.com); Internet Archive (www.archive.org); as well as Getty Images, Alamy, and Bridgeman Images. Also to Kew Botanic Gardens, the RHS and The British Library for a wealth of ongoing research and inspiration: written, visual, collected and generously shared.

Last but absolutely not least, a few personal thanks to my nearest and dearest . . .

To the various contingents of Patel and Ellis, notably my fabulous husband, the artist Tom Ellis (see page 352) and my gorgeous wild boys Sylvester and Iggy for supporting, helping and mischievously distracting me in equal measures, and for giving me a garden to sow and grow my ideas in; my mum and dad, Maureen and Dinker Patel for instilling in me, my siblings Anjali and Robbie and our respective children, a life-long, action-packed love of nature, art, culture, science, travel and of course books, and for providing various beautiful gardens for us to create in; and my lovely mum-in-law and builder of houses, Liz Ellis, maker of another gorgeous garden in which we continue to play, plus direct access to the wonders of our shared love, Epping Forest.

To my incredibly supportive friends here there and everywhere, who have listened to me banging on about all things botanical for months, most especially Louise and Rob Murray, for a multitude of corner discussions, cuppas, cheering on and office-sharing respectively; the fellow mums, dads and wider community of Forest Gate who keep me going and help keep my kids occupied, you know who you are; Nancy Waters (www.itemsofnote.co.uk) for half-time editorial encouragement, especially on the image front; Mic and Mary Clarke for my first exhibition at CoffeE7 and a much-missed home-from-home in which to work; Mark Searle for ongoing encouragement, Matthew Biggs for taking the time to look closer at my early pressed botanicals and thus pitching my *Gardens Illustrated* profile; Carmel King for my lovely portrait on page 386; Jen Lewandowski for lyrical fun and wordplay always; and Lucy Hellberg for coming home, making the parallel push worth it – keep shooting for the stars.

And finally thanks to Radio 6 for providing much of the soundtrack; Wanstead Flats for giving me the space to run off the day's writing, editing and picture research; and David Attenborough, my all time plant-inspired hero. My botanical world is significantly amplified, energised and enhanced by all three.

FURTHER READING

Selected books, magazines, journals and websites to help you explore the Plant Kingdom and look more deeply into your botanical world.

The Plant Kingdom

Botanicum (Welcome to the Museum), Kathy Willis and Katie Scott (Big Picture Press, 2016)

Kingdom of Plants: A Journey Through Their Evolution, Will Benson (Collins, 2012)

Plants: A Very Short Introduction, Timothy Walker (OUP Oxford, 2012)

Plants of the World: An Illustrated Encyclopedia of Vascular Plant Families, Maarten J.M. Christenhusz, Michael F. Fay and Mark W. Chase (Kew Publishing, 2017)

The Evolution of Plants, Kathy Willis and Jennifer McElwain (OUP Oxford, 2013)

The Private Life of Plants, David Attenborough (BBC Books, 1995)

The State of the World's Plants report (1st and 2nd editions, Royal Botanic Gardens, Kew, 2016 and 2017)

Plants and People

50 Plants That Changed the Course of History, Bill Laws (David & Charles, 2010)

Botanical Shakespeare: An Illustrated Compendium of All the Flowers, Fruits, Herbs, Trees, Seeds, and Grasses Cited by the World's Greatest Playwright, Gerit Quealy and Sumie Hasegawa Collins (Harper Design, 2017)

Flora: The Art of Plant Exploration, Sandra Knapp (Natural History Museum, 2014)

Life on Air, David Attenborough (BBC Books, 2010)

On the Origin of Species, Charles Darwin (first published 1859)

Plant Lore and Legend: The Wisdom and Wonder of Plants and Flowers Revealed, Ruth Binney and Freya Dangerfield (Rydon Publishing, 2016)

Plant Love: The Scandalous Truth About the Sex Life of Plants, Michael Allaby (Filbert Press, 2016)

Plants for People, Anna Lewington (Eden Project Books, 2003)

Plants: From Roots to Riches, Kathy Willis and Carolyn Fry (John Murray, 2014)

Second Nature: A Gardener's Education, Michael Pollan (Bloomsbury, 1997)

Seeds, Sex and Civilization: How the Hidden Life of Plants has Shaped Our World, Peter Thompson (Thames and Hudson Ltd, 2010)

Sex, Botany & Empire: The Story of Carl Linnaeus and Joseph Banks, Patricia Fara (Icon Books, 2017)

The Botany of Desire: A Plant's-eye View of the World, Michael Pollan (Bloomsbury, 2002)

The Cabaret of Plants: Botany and the Imagination, Richard Mabey (Profile Books, 2015)

The Invention of Nature: The Adventures of Alexander von Humboldt, the Lost Hero of Science, Andrea Wulf (John Murray, 2015)

The Origin of the Species: By Means of Natural Selection, Charles Darwin (Senate, 2010)

The Perfumier and the Stinkhorn, Richard Mabey (Profile Books, 2011)

The Plant Hunters, Carolyn Fry (Andre Deutsch, 2017)

The Reason for Flowers: Their History, Culture, Biology, and How They Change Our Lives, Stephen Buchmann (Scribner, 2015)

The Selfish Gene, Richard Dawkins (first published 1976; OUP Oxford, 2016)

The Secret Life of Plants: a Fascinating Account of the Physical, Emotional, and Spiritual Relations Between Plants and Man, Peter Tompkins and Christopher Bird (HarperPerennial, 1989)

The Silk Roads: A New History of the World, Peter Frankopan (Bloomsbury, 2016)

Wicked Plants, Amy Stewart (Timber Press, 2010)

Ecology and the Environment

'Hamatreya', *Poems*, Ralph Waldo Emerson (first published 1847)

Climate Change: A Very Short Introduction, Mark Maslin (OUP Oxford, 2014)

Deserts: A Very Short Introduction, Nick Middleton (OUP Oxford, 2009)

Eden: Updated 15th Anniversary Edition, Tim Smit (Eden Project Books, 2016)

Environmental Economics: A Very Short Introduction, Stephen Smith (OUP Oxford, 2016)

Environmental Politics: A Very Short Introduction, Andrew Dobson (OUP Oxford, 2011)

Feral: Rewilding the Land, Sea and Human Life, George Monbiot (Penguin, 2014)

Forests: A Very Short Introduction, Jaboury Ghazoul (OUP Oxford, 2015)

Four Changes, Gary Snyder (first published 1969)

Gaia: A New Look at Life on Earth, James Lovelock (OUP Oxford, 1979)

Inheritors of the Earth: How Nature is Thriving in an Age of Extinction, Chris D. Thomas (Allen Lane, 2017)

Plant Conservation: Why it Matters and How it Works, Timothy Walker (Timber Press, 2013)

Practice of the Wild, Gary Snyder (Counterpoint, 2010)

The Biosphere and the Bioregion: Essential Writings of Peter Berg, Cheryll Glotfelty and Eve Quesnel (Routledge, 2014)

The Ecology of Plants, Jessica Gurevitch, Samuel M. Scheiner and Gordon A. Fox (OUP USA, 2006)

The Gary Snyder Reader: Prose, Poetry and Translations, 1552–1988, Gary Snyder (Counterpoint, 1999)

The Hidden Life of Trees: What they feel, how they communicate: Discoveries from a secret world, Peter Wohlleben (William Collins, 2017)

The Man Who Planted Trees, Jean Giono (first published 1953)

The Plant Messiah: Adventures in Search of the World's Rarest Species, Carlos Magdalena (Viking, 2017)

The World As I See It, Albert Einstein (first published 1949)

Why People Need Plants, Carlton Wood and Nicolette Habgood (Royal Botanic Gardens, Kew, 2010)

Landscape

Landmarks, Robert Macfarlane (Penguin, 2016)

Landscape and Memory, Simon Schama (Harper Perennial, 2004)

My first summer in the Sierra: And Selected Essays, John Muir (First published, Houghton Mifflin, 1911)

Silent Spring, Rachel Carson (Penguin Classics, 2000)

The Land, Vita Sackville-West (first published William Heinemann, 1926)

The Old Ways: A Journey on Foot, Robert Macfarlane (Penguin, 2013)

The Wild Places, Robert Macfarlane (Granta Books, 2017)

The World of Laura Ingalls Wilder, Marta McDowell (Timber Press, 2017)

This Changes Everything, Naomi Klein (Penguin, 2015)

Walden: Or, Life in the Woods, Henry David Thoreau (First published, Ticknor and Fields, Boston, 1854)

Botany and Classification

A Botanist's Vocabulary: 1300 Terms Explained and Illustrated (Science for Gardeners), Susan K. Pell and Bobbi Angell (Timber Press, 2016)

Anatomy of Plants by Nehemiah Grew (first published 1682)

Anatomy of Seed Plants, Katharine Esau (first published John Wiley and Sons, 1960)

Botany Coloring Book, Paul Young (Collins Reference, 1999)

Botany for the Artist: An Inspirational Guide to Drawing Plants, Sarah Simblet (DK, 2010)

Botany in a Day, Thomas J Elpel (HOPS Press, 2013)

De Causis Plantarum (Causes of Plants), Theophrastus (written c. 350–287 BCE)

De Historia Stirpium Commentarii Insignes (Notable Commentaries on the History of Plants), Leonhart Fuchs (first published, 1542)

De Materia Medica, Pedanius Disocorides (written 50–70 AD)

De plantis libri XVI, Andrea Cesalpino (first published 1583)

Eléments de botanique, Joseph Pitton de Tournefort (first published 1694)

English Botany, James Edward Smith and James Sowerby (first published 1791–1814)

Historia Plantarum (Enquiry into Plants), Theophrastus (written c. 350–287 BCE)

Historia Plantarum (The History of Plants), John Ray (first published, 1686)

How Plants Work: The Science Behind the Amazing Things Plants Do, Linda K. Chalker-Scott (Science for Gardeners, 2015)

Micrographia, Robert Hooke (first published, The Royal Society, 1665)

Naturalis Historia, Pliny the Elder (written 77–79 CE)

New Illustration of the Sexual System of Carolus von Linnaeus including (part 3) the Temple of Flora, or Garden of Nature, Robert John Thornton (first published 1807)

Philosophia Botanica, Carl Linnaeus (first published 1751)

Pinax theatri botanici (Illustrated Exposition of Plants), Gaspard Bauhin (1623)

Plant Anatomy, Katharine Esau (first published, Wiley, 1953)

Plant Families: A Guide for Gardeners and Botanists, Ross Bayton (University of Chicago Press, 2017)

Plant Names Simplified: Their Pronunciation Derivation and Meaning, A.T. Johnson and H.A. Smith (Old Pond Publishing, 2008)

RHS Botany for Gardeners: The Art and Science of Gardening Explained & Explored (Mitchell Beazley, 2013)

RHS Geneaology for Gardeners: Plant Families Explored & Explained (Mitchell Beazley, 2017)

RHS Latin for Gardeners: Over 3,000 Plant Names Explained and Explored (Mitchell Beazley, 2012)

RHS Practical Latin for Gardeners: More than 1,500 Essential Plant Names and the Secrets They Contain (Mitchell Beazley, 2016)

Species Plantarum (The Species of Plants), Carl Linnaeus (first published 1753)

Systema Naturae (The System of Nature), Carl Linnaeus (first published 1735)

The Book of Seeds: A lifesize guide to six hundred species from around the world, Dr Paul Smith (Ivy Press, 2018)

The Naming of Names: The Search for Order in the World of Plants, Anna Pavord (Bloomsbury, 2005)

The Secret Life of Trees, Colin Tudge (Penguin, 2006)

The Triumph of Seeds: How Grains, Nuts, Kernels, Pulses, and Pips Conquered the Plant Kingdom and Shaped Human History, Thor Hansen (Basic Books, 2016)

What a Plant Knows: A Field Guide to the Senses, Daniel Chamovitz (Scientific American, 2013)

Pollinators

Collin's Beekeeper's Bible: Bees, honey, recipes and other home uses, Philip Et Al Mccabe (Collins, 2010)

Pollination Power, Heather Angel (Canadian Museum of Nature, 2015)

Garden Interest and History

Around the World in 80 Gardens, Monty Don (W&N, 2008)

Beatrix Potter's Gardening Life: The Plants and Places That Inspired the Classic Children's Tales, Marta McDowell (Timber Press, 2013)

Colour in the Flower Garden, Gertrude Jekyll (first published, Country Life, 1908)

Derek Jarman's Garden, Derek Jarman (Thames & Hudson, 1995)

Dreamscapes: Inspiration and Beauty in Gardens Near and Far, Claire Takacs (Hardie Grant Books, 2018)

Frida Kahlo's Garden, Adriana Zavala and New York Botanical Gardens (Prestel, 2015)

Fundacion Cesar Manrique, Lanzarote, Simon Marchan Fiz and Pedro Martinez de Albornoz (Edition Axel Menges, 1997)

Gardenista, Michell Slatalla (Artisan, 2016)

Biodynamic Gardening (DK, 2015)

Gardens of the High Line: Elevating the Nature of Modern Landscapes, Piet Oudolf and Rick Darke (Fontaine Uitgevers B.V., 2017)

Hummelo: A Journey Through a Plantsman's Life, Piet Oudolf and Noel Kingsbury (Monacelli Press, 2015)

Life in the Garden, Penelope Lively (Fig Tree, 2017)

Matisse's Garden, Samantha Friedman (Harry N Abrams, 2014)

Outstanding American Gardens: A Celebration: 25 Years of the Garden Conservancy, Page Dickey and Marion Brenner (Stewart, Tabori & Chang, 2015)

Paradise Gardens: the world's most beautiful Islamic gardens, Monty Don and Derry Moore (Two Roads, 2018)

Private Paradise: Contemporary American Gardens, Charlotte M. Freize (Moncelli Press, 2011)

The Curious Gardener, Anna Pavord (Bloomsbury, 2011)

The Garden, Vita Sackville-West (first published, Michael Joseph, 1946)

The Gardener's Dictionary, Philip Miller (first published 1731)

The Gardens of the British Working Class, Margaret Willes (Yale University Press, 2015)

The Japanese Garden, Sophie Walker (Phaidon Press, 2017)

The Wild Garden, William Robinson (first published 1870)

Vita Sackville-West's Sissinghurst: The Creation of a Garden, Vita Sackville-West and Sarah Raven (Virago, 2014)

Wood and Garden, Gertrude Jekyll (Longmans, Green and Co., 1899)

Gardening and Growing

Creating a Forest Garden: Working with nature to grow edible crops, Martin Crawford (Green Books, 2010)

Designing with Plants, Piet Oudolf and Noel Kingsbury (Timber Press, 1999)

Down to Earth: Gardening Wisdom, Monty Don (DK, 2017)

Garden Design Bible: 40 great off-the-peg designs, Tim Newbury (Hamlyn, 2016)

Garden Design: A Book of Ideas, Heidi Howcroft and Marianne Majerus (Mitchell Beazley, 2015)

Garden Renovation: Transform Your Yard Into the Garden of Your Dreams, Bobbie Schwartz (Timber Press, 2017)

Garden Revolution: How Our Landscapes Can Be a Source of Environmental Change, Larry Weaner (Timber Press, 2016)

Grow Your Own Garden: How to propagate all your own plants, Carol Klein (BBC Books, 2010)

Grown and Gathered: Traditional Living Made Modern, Matt and Lentil Purbrick (Pan Macmillan Australia, 2016)

Home Ground: Sanctuary in the City, Dan Pearson (Conran, 2011)

Jekka's Complete Herb Book, Jekka McVicar (Kyle Books, 2009)

Making a Garden: Successful gardening by nature's rules, Carol Klein (Mitchell Beazley, 2015)

Miniature Moss Gardens: Create your Own Japanese Container Gardens (Tuttle Publishing, 2017)

Natural Selection: a year in the garden, Dan Pearson (Guardian Faber Publishing, 2017)

New Wild Garden: Natural-style planting and practicalities, Ian Hodgson (Frances Lincoln, 2016)

One Hundred Flowers, Harold Feinstein (Little, Brown, 2000)

Organic Gardening: The Natural No-dig Way, Charles Dowding (Green Books, 2013)

Plant Love, Alys Fowler (Kyle Books, 2017)

Planting Design: Gardens in Time and Space, Piet Oudolf and Noel Kingsbury (Timber Press, 2005)

Planting in a Post-Wild World: Designing Plant Communities for Resilient Landscapes, Thomas Rainer and Claudia West (Timber Press, 2015)

Planting: A New Perspective, Piet Oudolf and Noel Kingsbury (Timber Press, 2013)

RHS Complete Gardener's manual: How to Dig, Sow, Plant and Grow (DK, 2011)

RHS Encyclopedia of Garden Design (DK, 2012)

RHS Encyclopedia of Gardening, Christopher Brickell (DK, 2012)

RHS Encyclopedia of Plants and Flowers, Christopher Brickell (DK, 2010)

RHS How to Plant a Garden: Design tricks, ideas and planting schemes for year-round interest, Matt James and The Royal Horticultural Society (Mitchell Beazley, 2016)

RHS Lessons from Great Gardeners: Forty Gardening Icons and What They Teach Us, Matthew Biggs (Mitchell Beazley, 2015)

RHS What Plant Where Encyclopedia (DK, 2013)

The Bold Dry Garden: Lessons from the Ruth Bancroft Garden, Johanna Silver (Timber Press, 2016)

The Complete Gardener, Monty Don (DK, 2009)

The Earth Care Manual: A Permaculture Handbook for Britain and Other Temperate Climates (Permanent Publications, 2016)

The Garden: A Year at Home Farm, Dan Pearson (Ebury Press, 2001)

The Magical World of Moss Gardening, Annie Martin (Timber Press, 2015)

The New Kitchen Garden: How to Grow Some of What you Eat No Matter Where You Live, Mark Diacono (Headline Home, 2015)

The Thoughtful Gardener: An Intelligent Approach to Garden Design, Jinny Blom (Jacqui Small LLP, 2017)

The Well-Tended Perennial Garden: The Essential Guide to Planting and Pruning Techniques, Tracy DiSabato-Aust (Timber Press, 2017)

Indoor Gardening and Plant Craft

Designing with Succulents, Debra Lee Baldwin (Timber Press, 2017)

House of Plants: Living with Succulents, Air Plants and Cacti, Rose Ray and Caro Langton (Frances Lincoln, 2016)

How Not to Kill Your Plants, Nik Southern (Hodder & Stoughton, 2017)

Living with Plants: A Guide to Indoor Gardening, Sophie Lee (Hardie Grant Books, 2017)

Prick: Cacti and Succulents: Choosing, Styling, Caring, Gynelle Leon (Mitchell Beazley, 2017)

Succulents Simplified: Growing, Designing, and Crafting with 100 Easy-Care Varieties, Debra Lee Baldwin (Timber Press, 2013)

Terrarium Craft: Create 50 Magical, Miniature Worlds, Amy Bryant Aiello and Kate Bryant (Timber Press, 2017)

Terrariums – Gardens Under Glass: Designing, Creating, and Planting Modern Indoor Gardens, Maria Colletti (Cool Springs Press, 2015)

Terrariums & Kokedama, Alyson Mowat (Kyle Books, 2017)

Urban Botanics: An Indoor Plant Guide for Modern Gardeners (Aurum Press, 2017)

Urban Jungle: Living and Styling with Plants, Igor Josifovic and Judith de Graaff (Callwey, 2016)

Flowers and Floral design

A Victorian Flower Dictionary: The Language of Flowers Companion, Mandy Kirkby (Ballantine Books, 2011)

At Home in the Garden, Carolyn Roehm (Potter Style, 2015)

Branches & Blooms, Alethea Harampolis, Jill Rizzo (Artisan Division of Workman Publishing, 2017)

Bringing Nature Home: Floral Arrangements Throughout the Seasons, Ngoc Minh Ngo (Rizzoli International Publications, 2012)

Discovering the Meaning of Flowers: Love Found, Love Lost, Love Restored, Shane Connolly (Clearview, 2017)

Exploring Ikebana, Ilse Beunen (Stichting Kunstboek, 2015)

Flora Magnifica: The Art of Flowers in Four Seasons, Makoto Azuma and Shunsuke Shiinoki (Thames and Hudson, 2018)

Flower Decoration, Constance Spry (first published 1934)

Flower Decoration in the House, Gertrude Jekyll (first published, Country Life, 1907)

Flowers Every Day: Inspired Florals for Home, Gifts and Gatherings, Florence Kennedy (Pavilion Books, 2017)

Flowers, Carolyne Roehm (Clarkson Potter, 2012)

Flowerevolution: Blooming into Your Full Potential with the Magic of Flowers, Louie Schwartzberg, Katie Hess (Hay House UK, 2016)

Foraged Flora: A Year of Gathering and Arranging Wild Plants and Flowers, Louesa Roebuck and Sarah Lonsdale (Ten Speed Press, 2016)

Ikebana: The Art of Arranging Flowers, Shozo Sato (Tuttle Shokai Inc, 2013)

In Bloom: Creating and Living with Flowers, Ngoc Minh Ngo (Rizzoli International Publications, 2016)

Martha's Flowers: A Practical Guide to Growing, Gathering, and Enjoying, Martha Stewart (Clarkson Potter, 2018)

Rachel Ashwell: My Floral Affair: Whimsical Spaces and Beautiful Florals, Rachel Ashwell (CICO Books, 2018)

Selina Lake: Garden Style: Inspirational Styling for your Outside Space, Selina Lake, 2018 (Ryland, Peters & Small, 2018)

Some Flowers, Vita Sackville-West (first published, Cobden-Sanderson, 1937)

The Book of Orchids: A life-size guide to six hundred species from around the world, Maarten Christenhusz (Ivy Press, 2017)

The Cutting Garden: Growing and Arranging Garden Flowers, Sarah Raven and Penelope Hobhouse (Frances Lincoln, 2013)

The Floret Farm's Cut Flower Garden: Grow, Harvest, Arrange, Erin Benzakein (Chronicle Books, 2017)

The Flower Appreciation Society: An A to Z of All Things Floral, Anna Day and Ellie Jauncey (Sphere, 2015)

The Flower Chef: A Modern Guide to Do-It-Yourself Floral Arrangements, Carly Cylinder (Grand Central Publishing, 2016)

The Flower Farmer's Year: How to Grow Cut Flowers for Pleasure and Profit, Georgie Newbery (Green Books, 2014)

The Flower Recipe Book, Alethea Harampolis (Artisan, 2013)

The Flower Workshop: Lessons in Arranging Blooms, Branches, Fruits, and Foraged Materials, Ariella Chezar and Julie Michaels (Ten Speed Press, 2016)

The Flowers Art and Bouquets (Classics), Sixtine Dubly (Assouline Publishing, 2016)

The Surprising Life of Constance Spry, Sue Shephard (Pan, 2011)

The Tulip, Anna Pavord (Bloomsbury, 2000)

The Wreath Recipe Book, Alethea Harampolis (Artisan, 2014)

Wildflowers of Britain: Over a Thousand Species by Photographic Identification, Roger Phillips and Sheila Grant (Pan Books, 1983)

Foraging and Feasting

Edible Wild Plants: Eastern / Central North America (Peterson Field Guides), Lee Allen Peterson and Roger Tory Peterson (Houghton Mifflin Harcourt, 1999)

Food for Free, Richard Mabey (Collins, 2012)

Forager's Cocktails: Botanical Mixology with Fresh Ingredients, Amy Zavatto (HarperCollins, 2015)

Foraging & Feasting: A Field Guide and Wild Food Cookbook, Dina Falconi and Wendy Hollender (Botanical Arts Press LLC, 2013)

The Eatweeds Cookbook, Robin Harford (Eatweeds, 2015)

The Edible City: A Year of Wild Food, John Rensten (Boxtree, 2016)

The Forager Handbook, Miles Irving (Ebury Press, 2009)

The Forager's Harvest: A Guide to Identifying, Harvesting, and Preparing Edible Wild Plants, Samuel Thayer (Foragers Harvest Press, 2006)

The Foragers Handbook, Miles Irving (Ebury Press, 2009)

The Garden Forager: Edible Delights in your Own Back Yard, Adele Nozedar (Square Peg, 2015)

The Hedgerow Handbook: Recipes, Remedies and Rituals by Adele Nozedar (Square Peg, 2012)

The New Wildcrafted Cuisine: Exploring the Exotic Gastronomy of Local Terroir, Pascal Baudar (Chelsea Green Publishing Co, 2016)

The Wildcrafted Cocktail, Ellen Zachos (Storey Publishing, 2017)

The Wildcrafting Brewer: Creating Unique Drinks and Boozy Concoctions from Nature's Ingredients, Pascal Baudar (Chelsea Green Publishing Co, 2018)

Wild Cocktails from the Midnight Apothecary, Lottie Muir (CICO Books, 2015)

Wild Food, Roger Phillips (Pan Books 1983; Macmillan, 2014)

Wild Fruits, Henry David Thoreau, Bradley P. Dean (first published, W. W. Norton, 2000)

Food and Drink

26 Grains, Alex Hely-Hutchinson (Square Peg, 2016)

A Year at Otter Farm, Mark Diacono (Bloomsbury, 2014)

A Year in My Kitchen, Skye Gyngell (Quadrille, 2006)

Bowls of Goodness: Vibrant Vegetarian Recipes Full of Nourishment, Nina Olsson (Kyle Books, 2017)

Cooking with Flowers: Sweet and Savory Recipes with Rose Petals, Lilacs, Lavender and Other Edible Flowers, Miche Bacher (Quirk Books, 2013)

Edible Flower Garden, Rosalind Creasy (Periplus Editions, 1999)

*Fern Verrow: Recipes from the Farm Kitche*n, Harry Astley, Jane Scotter (Quadrille, 2015)

Gather, Gill Meller (Quadrille, 2016)

Green Kitchen at Home, David Frankiel and Luise Vindahl (Hardie Grant Books, 2017)

Hazana: Jewish Vegetarian Cooking, Paola Gavin (Quadrille, 2017)

Jekka's Herb Cookbook, Jekka McVicar (Firefly Books, 2012)

Kew on a Plate with Raymond Blanc, Royal Botanic Gardens, Kew, and Raymond Blanc (Headline Home, 2015)

Kew's Teas, Tonics and Tipples: Inspiring Botanical Drinks to Excite Your Tastebuds, Royal Botanic Gardens, Kew (Kew Publishing, 2015)

On Vegetables, Jeremy Fox, Noah Galuten and David Chang (Phaidon Press, 2017)

Petal, Leaf, Seed, Lia Leendertz (Kyle Books, 2016)

Plants Taste Better: Delicious plant-based recipes, from root to fruit, Richard Buckley (Jacqui Small LLP, 2018)

Plenty, Yotam Ottolenghi (Ebury Press, 2010)

River Cottage Much More Veg: 175 delicious plant-based vegan recipes, Hugh Fearnley-Whittingstall (Bloomsbury, 2017)

River Cottage Every Day!, Hugh Fearnley-Whittingstall (Bloomsbury, 2011)

Spring, Skye Gyngell (Quadrille, 2015)

Tender: Volume 1, A cook and his vegetable patch, Nigel Slater (Fourth Estate, 2009)

The Drunken Botanist, Amy Stewart (Timber Press, 2013)

The Edible Flower Garden: From Garden to Kitchen: Choosing, Growing and Cooking Edible Flowers, Kathy Brown (Hermes House, 2014)

The First Mess Cookbook: Vibrant Plant-Based Recipes to Eat Well Through the Seasons, Professor of Chemistry Laura Wright (Avery Publishing Group, 2017)

The Flavour Thesaurus, Niki Segnit (Bloomsbury, 2010)

The Herb and Flower Cookbook: Plant, Grow and Eat, Pip McCormac (Quadrille, 2014)

The Modern Cook's Year, Anna Jones (Fourth Estate, 2017)

The New Vegetarian Cooking for Everyone, Deborah Madison (Ten Speed Press, 2014)

The Violet Bakery Cookbook, Claire Ptak (Square Peg, 2015)

Vegan in 7: Delicious plant-based recipes in 7 ingredients or fewer, Rita Serano (Kyle Books, 2017)

Vegan: The Cookbook, Jean-Christian Jury (Phaidon Press, 2017)

Vegetable Literacy, Deborah Madison (Ten Speed Press, 2013)

Herbals, Health and Wellbeing

600 Aromatherapy Recipes for Beauty, Health & Home, Beth Jones (CreateSpace Independent Publishing Platform, 2014)

A Curious Herbal, Elizabeth Blackwell (first published 1737–1739)

A Modern Herbal, Mrs M. Grieve (first published 1931 / Dorset Press, 1992)

A Pukka's Life, Sebastian Pole (Quadrille, 2011)

All Natural Beauty: Organic and Homemade Beauty Products, Nici Hofer, Karin Berndl (Hardie Grant Books, 2016)

Ayurveda Beginner's Guide: Essential Ayurvedic Principles and Practices to Balance and Heal Naturally, Susan Weis-Bohlen (Althea Press, 2018)

Botanical Beauty: 80 Essential Recipes for Natural Spa Products, Aubre Andrus (Curious Fox, 2017)

Complete Herbal, Nicholas Culpeper (first published 1653)

Grow Your Own Drugs, James Wong (BBC Books, 2009)

Hedgerow Medicine: Harvest and Make Your Own Herbal Remedies, Julie Bruton-Seal and Matthew Seal (Merlin Unwin Books, 2008)

Herball, or Generall Historie of Plantes, John Gerard (first published, John Norton, 1597)

Herbarium, Caz Hildebrand (Thames and Hudson Ltd, 2016)

Infuse: Herbal teas to cleanse, nourish and heal, Paula Grainger and Karen Sullivan (Hamlyn, 2016)

Natural Beauty Alchemy, Fifi M. Maacaron (Countryman Press, 2015)

Natural Beauty: 35 step-by-step projects for homemade beauty, Karen Gilbert (CICO Books, 2013)

Neal's Yard Remedies Beauty Book, Susan Curtis, Fran Johnson, Pat Thomas (DK, 2015)

Plant-powered Beauty: The essential guide to using natural ingredients for health, wellness, and personal skincare, Amy Calper and Christina Daigneault (Benbella Books, 2018)

Pseudo-Apuleius Herbarius, Pseudo-Apuleius (compiled in the 4th century; first printed 1481)

Rosemary Gladstar's Herbal Recipes for Vibrant Health: 175 Teas, Tonics, Oils, Salves, Tinctures and Other Natural Remedies for the Entire Family, Rosemary Gladstar (Storey Publishing LLC, 2008)

Rosemary Gladstar's Medicinal Herbs: A Beginner's Guide, Rosemary Gladstar (Storey Publishing LLC, 2012)

The Art of Aromatherapy, Robert Tisserand (CW Daniel, 1977)

The Complete Book of Ayurvedic Home Remedies: A comprehensive guide to the ancient healing of India, Vasant Lad (Piatkus, 2006)

The Complete Aromatherapy and Essential Oils Sourcebook, Julia Lawless (HarperCollins, 2018)

The Encyclopedia of Essential Oils: The Complete guide to the use of aromatic oils in aromatherapy, herbalism, health and well-being, Julia Lawless (Harper Thorsons, 2014)

The Gardener's Companion to Medicinal Plants: An A-Z of Healing Plants and Home Remedies, Royal Botanic Gardens and Jason Irving (Frances Lincoln, 2017)

The Green Witch: Your Complete Guide to the Natural Magic of Herbs, Flowers, Essential Oils and More (Adams Media, 2017)

The Handmade Apothecary, Kim Walker and Vicky Chown (Kyle Books, 2017)

The Herbal Apothecary, Dr J J Pursell (Timber, 2016)

The Herball's Guide to Botanical Drinks Michael Isted (Jacqui Small LLP, 2018)

The Illustrated Encyclopedia of Healing Remedies, C. Norman Shealy (HarperCollins, 2009)

The Illustrated Herbal Handbook, Juliette de Bairacli Levy (Faber & Faber, 1982)

The Inner Beauty Bible, Laurey Simmons, Louis Weinstock (Harper Thorsons, 2017)

The Nature Fix: Why Nature Makes Us Happier, Healthier, and More Creative, Florence Williams (W.W Norton & Company, 2017)

The Earth Has a Soul: C. G. Jung's Writing on Nature, Technology and Modern Life, C. G. Jung and Meredith Sabini (North Atlantic Books, 2002)

The Woman's Herbal Apothecary, Dr J J Pursell (Fair Winds Press, 2018)

Botanical Art and Illustration

A Garden for Eternity: The Codex Liechtenstein, H.W. Lack (Benteli Verlag, 1999)

A New Flowering: 1000 Years of Botanical Art, Shirley Sherwood (Ashmolean Museum, 2014)

A Specimen of the Botany of New Holland, James Edward Smith and James Sowerby (first published 1793–1795)

Art Forms in Nature: Prints of Ernst Haeckel, Olaf Breidbach, Irenaeus Eibl-Eibesfeldt and Richard Hartmann (Prestel, 1998)

Bartram's Travels, William Bartram (first published, James & Johnson, 1791)

Basilius Besler's Florilegium: The Book of Plants, Klaus Walter Littger and Werner Dressendörfer (Taschen GmbH, 2015)

Botanical Art from the Golden Age of Scientific Discovery, Anna Laurent (University of Chicago Press, 2016)

Botanical Drawing in Color: A Basic Guide to Mastering Realistic Form and Naturalistic Color, Wendy Hollender (Watson-Guptill, 2010)

Botanical Drawing: A Step-By-Step Guide to Drawing Flowers, Vegetables, Fruit and Other Plant Life, Penny Brown (Search Press, 2018)

Botanical Line Drawing: 200 Step-by-Step Flowers, Leaves, Cacti, Succulents, and Other Items Found in Nature, Peggy Dean (Pigeon Letters, 2017)

Botanical Painting with Coloured Pencils, Ann Swan (Collins, 2009)

Botanical Sketchbook, Mary Ann Scott and Margaret Stevens (American Artist, 2010)

Codex Liechtenstein, Norbert Boccius (first published, 1776)

Dodel-Port Atlas, Dr Arnold Dodel-Port and Carolina Dodel-Port (published 1878–1883)

Emily Dickinson's Herbarium (Harvard University Press, 2006)

Flora Delanica (Mrs Delany's Flora), Mary Delany (1771–1783)

Flora Graeca, John Sibthorp and Ferdinand Bauer (1806–1840)

Flora Londonensis, William Curtis and James Sowerby (first published 1777–1798)

Herbarium Handbook 3rd Edition, L. Forman and D Bridson (Royal Botanic Gardens, Kew, 2000)

Hortus Esytettensis, Basilius Besler (first published 1613)

Illustrationes Florae Novae Hollandiae, Ferdinand Bauer (1813)

Joseph Banks's Florilegium: Botanical Treasures from Cook's First Voyage, Mel Gooding, Joe Studholme and David Mabberley (Thames and Hudson Ltd, 2017)

Kew: Golden Age of Botanical Art, Martyn Rix (Andre Deutsch, 2018)

Kunstformen der Natur (Art Forms in Nature), Ernst Haeckel (first published 1899–1904)

Les Liliacées (The Lilies), Joseph Redouté (published 1802–1816)

Les Roses (The Roses), Joseph Redouté (published 1817–1824)

Mrs Delany and Her Circle, Mark Laird, Alicia Weisberg-Roberts (Yale University Press, 2009)

Natural History of Carolina, Florida and the Bahama Islands, Marc Catesby (first published, 1729–1747)

Pierre-Joseph Redouté – the Raphael of Flowers, Pieter Baas, Terry Van Druten, Pascale Heurtel, Alain Pougetoux and Pierre-Joseph Redouté (NAI Publishers, 2013)

Plant: Exploring the Botanical World, Phaidon Editors (Phaidon Press, 2016)

Rory McEwen: The Colors of Reality, Martyn Rix (Kew Publishing, 2015)

Rosarum monographia; or a botanical history of Roses by John Lindley (first published, J Ridgeway, 1820)

Rosie Sanders' Flowers: A Celebration of Botanical Art, Rosie Sanders (Batsford Ltd, 2016)

The Apple Book, Rosie Sanders and Harry Baker (Frances Lincoln, 2010)

The Art of Botanical & Bird Illustration: An artist's guide to drawing and illustrating realistic flora, fauna, and botanical scenes from nature, Mindy Lighthipe (Walter Foster Publishing, 2017)

The Art of Botanical Illustration, Wilfred Blunt and William T. Stearn (Antique Collectors' Club, 2015)

The Art of Botanical Painting, Margaret Stevens, The Society of Botanical Artists (Harper Thorsons, 2015)

The Botanical Treasury, Christopher Mills (Andre Deutsch, 2016)

The Botanical Wall Chart: Art from the golden age of scientific discovery, Anna Laurent (Ilex Press, 2016)

The Gardener's Chronicle, John Lindley (founded 1841–1986)

The Kew Book of Botanical Illustration, Christabel King (Search Press, 2015)

The Nature-Printer, Pia Ostlund and Simon Prett (The Timpress, 2016)

The Paper Garden: Mrs Delaney Begins Her Life's Work at 72, Molly Peacock (Bloomsbury, 2012)

Art and Design

Ai Weiwei, Sunflower Seeds, Ai Weiwei, Juliet Bingham (Tate Publishing, 2010)

Art Nature Dialogues: Interviews with Environmental Artists, John K. Grande and Edward Lucie-Smith (State University of New York Press, 2004)

Balance: Art and Nature, John K. Grande (Black Rose Books, 2006)

David Hockney: A Bigger Picture, Tim Barringer (Royal Academy of Arts, 2012)

Ellsworth Kelly: Plant Drawings, Marla Prather and Michael Semff (Schirmer / Mosel Verlag GmbH, 2016)

Flower Power: The Meaning of Flowers in Asian Art, Dany Chan and Jay Xu (Asian Art Museum of San Francisco, 2017)

Flower: Paintings by 40 Great Artists, Celia Fisher (Frances Lincoln, 2012)

Georgia O'Keeffe: One Hundred Flowers, Georgia O'Keeffe and Nicholas Callaway (Random House Value Pub, 1995)

Henri Matisse: The Cut Outs, Karl Buchberg (Tate Publishing, 2014)

herman de vries: chance and change, Mel Gooding (Thames and Hudson Ltd, 2006)

Hilma af Klint: Painting the Unseen, Daniel Birnbaum and Hans-Ulrich Obrist (Verlad de Buchhandlung Walther Konig, 2016)

In Paul Klee's Enchanted Garden, Michael Baungartner (Hatje Catnz, 2008)

Islamic Geometric Design, Eric Broug (Thames and Hudson Ltd, 2013)

John Ruskin: Arts and Observer, Christopher Newall (Paul Holberton Publishing, 2014)

Land Art, Ben Tufnell (Tate Publishing, 2006)

Land Art, The Earth as Canvas, Michael Lailach (Taschen GmbH, 2007)

Liber Abaci (The Book of Calculation), Leonardo of Pisa or Fibonacci (first published 1202)

Mark Dion: Misadventures of a 21st-Century Naturalist, Ruth Erickson (Yale University Press, 2017)

Matisse in the Studio, Elle McBreem (MFA Publications – The Museum of Fine Arts, Boston, 2017)

Mondrian: Flowers, David Shapiro (Abrams Books, 1991)

Nature and Its Symbols (Guide to Imagery), Impelluso (Getty Publications, 2006)

Nature Morte: Contemporary artists reinvigorate the Still-Life tradition, Michael Petry (Thames and Hudson Ltd, 2016)

Of Green Leaf Bird and Flower: Artist's Books and the Natural World, Elizabeth Fairman (Yale University Press, 2014)

Painting the Modern Garden: From Monet to Matisse (Royal Academy of Arts, 2015)

Rebecca Louise Law: Life in Death, Rebecca Louise Law (Royal Botanic Gardens, Kew, 2017)

The Beauty of Numbers in Nature, Ian Stewart (The MIT Press, 2017)

V&A Pattern: India Florals, Sue Stronge and R. Crill (V&A Publishing, 2009)

V&A Pattern: Kimono, Anna Jackson (V&A Publishing, 2010)

V&A Pattern: Liberty's, Anna Brumer (V&A Publishing, 2012)

V&A Pattern: The Fifties, Sue Prichard (V&A Publishing, 2009)

V&A Pattern: William Morris, Linda Parry (V&A Publishing, 2009)

Van Gogh's Flowers, Judith Bumpus (Phaidon Press, 1998)

Wabi-Sabi: For Artists, Designers, Poets & Philosophers, Leonard Koren (Imperfect Publishing, 2008)

Ways of Seeing, John Berger (Penguin, 2008)

William Morris: Artist, Craftsman, Pioneer, Rosalind Ormiston and Nicholas M. Wells (Flame Tree Publishing, 2010)

You Say You Want A Revolution?: Records and Rebels 1966–1970, Victoria Broackes and Geoffrey Marsh (V&A, 2016)

Photography and Film

Cyanotypes of British and Foreign Ferns, Anna Atkins (published 1853)

Cyanotypes of British and Foreign Flowering Plants, Anna Atkins (published 1854)

Flowers, Irving Penn (first published, Harmony, 1980; revised 1988)

Fruit: Edible, Inedible, Incredible, Rob Kesseler and Wolfgang Stuppy (first published 2008; Papadakis, 2013)

Karl Blossfeldt (The Complete Published Work), Hans Christian Adam (Taschen Gmbh, 2014)

Mapplethorpe Flora: The Complete Flowers, Mark Holborn and Dimitri Levas (Phaidon Press, 2016)

Nick Knight: Flora, Nick Knight and Sandra Knapp (first published 1997; Schirmer/Mosel Verlag GmbH, 2014)

Paperwork and the Will of Capital, Taryn Simon (Hatje Cantz, 2015)

Photographs of British Algae: Cyanotype Impressions, Anna Atkins (published 1843–1853)

Pollen: The Hidden Sexuality of Flowers, Rob Kesseler and Dr Madeline Harley (first published 2004; Papadakis, 2014)

Sebastião Salgado: GENESIS, Sebastião Salgado, Lélia Wanick Salgado (Taschen Gmbh, 2013)

Seeds: Time Capsules of Life, Rob Kesseler and Wolfgang Stuppy (first published 2006; Papadakis, 2014)

This is the American Earth, Ansel Adams and Nancy Newhall (first published, Sierra Club, 1960)

Tim Walker: The Storyteller, Robin Muir and Tim Walker (Thames and Hudson Ltd, 2012)

Urformen der Kunst (Art Forms in Nature), Karl Blossfeldt (first published, Ernst Wasmuth, 1928)

Viviane Sassen: In & Out of Fashion, Charlotte Cotton and Nanda Van Den Berg (Prestel, 2012)

Wundergarten der Natur (The Wondergarden of Nature), Karl Blossfeldt (published Verlag für Kunstwissenschaft, 1932)

Botanical Style and Nature Craft

The New Bohemian's Handbook: Come Home to Good Vibes, Justina Blakeney (Abrams, 2017)

At Home with Plants, Ian Drummond and Kara O'Reilly (Mitchell Beazley, 2017)

Botanical Colour at your Fingertips, Rebecca Desnos (Botanical Colour at Your Fingertips, 2016)

Botanical Inks: Plant-To-Print Dyes, Techniques and Projects, Babs Behan (Quadrille, 2018)

Botanical Style: Inspirational decorating with nature, plants and florals, Selina Lake (Ryland, Peters & Small, 2016)

Embroidered Garden Flowers: Botanical Motifs for Needle and Thread, Kazuko Aoki (Shambhala Publications Inc, 2017)

Loose Leaf: Plants Flowers Projects, Wona Bae and Charlie Lawler (Hardie Grant Books, 2016)

Paper to Petal, Rebecca Thuss and Patrick Farrell (Potter Craft, 2013)

Plant Craft: 30 Projects that Add Natural Style to Your Home, Caitlin Atkinson (Timber Press, 2016)

The Fine Art of Paper Flowers: A Guide to Making Beautiful and Lifelike Botanicals, Tiffanie Turner (Watson Guptill, 2017)

The Modern Natural Dyer: A Comprehensive Guide to Dyeing Silk, Wool, Linen, and Cotton at Home, Kristine Vejar (STC Craft / A Melanie Falick Book, 2015)

The New Bohemians: Cool and Collected Homes, Justina Blakeney (Stewart, Tabori & Chang, 2015)

The Wild Dyer: A guide to natural dyes & the art of patchwork & stitch, Abigail Booth (Kyle Books, 2017)

Wild Color: The Complete Guide to Making and Using Natural Dyes, Jenny Dean (Potter Craft, 2010)

Kids

A First Book of Nature, Nicola Davis and Mark Hearld (Walker Books, 2014)

A Forest, Marc Martin (Templar Publishing, 2015)

A River, Marc Martin (Templar Publishing, 2016)

Bee: Nature's Tiny Miracle, Patricia Hegarty (Little Tiger Kids, 2017)

Botanicum Activity Book, Professor Katherine Willis, Katie Scott and Wendy Bartlett (Big Picture Press, 2017)

Garden Crafts for Children, Dawn Isaac (Cico, 2012)

Grow It, Eat It (DK Children, 2008)

How to Raise a Wild Child, Professor Scott D. Samson (Mariner Books, 2016)

I Am an Artist, Marta Altes (Macmillan Children's Books, 2014)

Little House on the Prairie, Laura Ingalls Wilder (first published Harper & Brothers 1932–1943; HarperCollins, 2008)

Magical Jungle, Johanna Bansford (Virgin Books, 2016)

Oh Say Can You Seed? All About Flowering Plants (Cat in the Hat's Learning Library), Bonnie Worth and Aristides Ruiz (Random House Books for Young Readers, 2001)

Plant-Powered Families,: Over 100 Kid-Tested, Whole-Foods Vegan Recipes, Dreena Burton (BenBella Books, 2015)

RHS Ready, Steady, Grow!, Royal Horticultural Society (DK Children, 2010)

Secret Garden, Johanna Bansford (Laurence King, 2013)

The Complete Book of the Flower Fairies, Cicely Mary Barker (Warne, 2002)

The Kew Gardens Children's Cookbook, Caroline Craig and Joe Archer (Wayland, 2016)

The Little Gardener, Emily Hughes (Flying Eye Books, 2015)

The Lost Words, Robert Macfarlane and Jackie Morris (Hamish Hamilton, 2017)

The Stick Book: Loads of Things you can make or do with a stick, Fiona Danks, Jo Schofield (Frances Lincoln, 2012)

The Tale of Peter Rabbit, Beatrix Potter (Frederick Warne & Co., 1902; Warne, 2002)

The Tiny Seed, Eric Carle (Puffin, 1997)

The Very Hungry Caterpillar, Eric Carle (first published, World Publishing Company, 1969; Puffin, 1994)

Tree: Seasons Come, Seasons Go, Patricia Hegarty and Britta Teckentrup (Little Tiger Kids, 2015)

Where the Wild Things Are, Maurice Sendak (first published 1963; Red Fox, 2000)

Wild, Emily Hughes (Flying Eye Books, 2015)

Magazines, Journals and Websites

Botanic Garden Conservation International: BGCI (www.bgci.org)

Botanical Art & Artists (www.botanicalartandartists.com

Curtis's Botanical Magazine also known as *The Botanical Magazine; or Flower Garden Displayed* (founded 1787; www.kew.org))

Garden Collage Magazine (www.gardencollage.com)

Garden Design magazine (www.gardendesign.com)

Gardenista (www.gardenista.com)

Gardens Illustrated magazine (www.gardensillustrated.com)

International Union for Conservation of Nature: ICUN (www.icun.org)

National Geographic magazine (www.nationalgeographic.com)

New Scientist magazine (www.newscientist.com)

Pleasure Garden magazine (www.pleasuregardenmagazine.com)

Rake's Progress Magazine (www.rakesprogressmagazine.com)

The American Gardener magazine (American Horticultural Society; www.ahsgardening.org)

The Garden magazine (Royal Horticultural Society; www.rhs.org.uk)

The Garden Edit (www.thegardenedit.com)

The Plant List (www.theplantlist.org)

The Plant Magazine (www.theplantmagazine.com)

The Planthunter (www.theplanthunter.com.au)

World Land Trust (www.worldlandtrust.org)

INDEX

Use this index to navigate *Collins Botanical Bible* and find the plants and subjects that interest you most. Page numbers in italics indicate images.

10,000 Forms (exhibition, 2017) 352
Abies (fir) 52
Acacia dealbata (mimosa) 97
Acacia senegal (Senegal gum acacia) 55
Academy of Athens 66
acai palm (*Euterpe precatoria*) 51
Acer campestre (field maple) 97
Acer (maple) 45, 51, 101, 102
Acer nigrum (black maple) 130
Acer palmatum (Japanese maple) 141, 175
Acer pseudoplatanus (sycamore) 179
Acer saccharinum (silver maple) 56
Acer saccharum (sugar maple) 168
Achillea millefolium (yarrow) 138, 171, 273
acorn (*Quercus*) 164, *165*
activated charcoal 264
Adams, Ansel 80, 181, *339*
Adansonia (baobab) 55
adaptation 94, 116
Adiantum (maidenhair fern) 35
Aesculus hippocastanum (horse chestnut) 97, 179
African arrowroot (*Canna indica*) 311
Afzelia genus 327
agave Americana (*Agave Americana*) 155, *378*
agave, blue (*Agave tequilana*) 56
Ai Weiwei *357*
  Map of China 357
  Sunflower Seeds 357
air plant (*Tillandsia*) 42
Albucasis 171
Alcea rosea (hollyhock) *166*, 167
Alchemilla vulgaris (lady's mantle) 171
alcohol 264, 270
alder (*Alnus glutinosa*) 98
alder tree, Italian (*Alnus cordata*) 108
Alexander the Great 58, 66
Alexanders (*Smyrnium olusatrum*) 168
algae 24, 25, 27
Alhambra, Granada 70
Alissa and Isaac Sutton Collection, Pittsburg 326
Allium ampeloprasum cultivar (leek) 90, 229
Allium canadense (wild onion) 130
Allium cepa (onion) 42, 65, 94, 209, 234, 273, 277
Allium cepa var. *aggregatum* (shallot) 42, *131*, 137

Allium cepa var. *cepa* (scallions) 189
Allium genus (onion, shallot, leek, chives) 130
Allium hollandicum (ornamental shallot) *130*
Allium sativum (garlic) 42, 94, 204, 206, 209, 230, 234, 273, 277
Allium sativum var. *ophioscorodon* (hardneck garlic) *199*
Allium sativum var. *sativum* (softneck garlic) 198
Allium schoenoprasum (chives) 42
Allium ursinum (wild garlic/ramsons) 51, 167, *192*, 193
allspice 243
almond (*Prunus dulcis* syn. *Prunus amygdalus*) 194, 206, 219, 237, 249, 251, 253, 292, 293, 302
Alnus cordata (Italian alder tree) 108
Alnus glutinosa (alder) 98
aloe (*Aloe*) 42, *43*
aloe, spiral (*Aloe polyphylla*) 311
Aloe vera (Barbados aloe) 141, 264
Aloysia citrodora (lemon verbena) 241
Alpine aster (*Aster alpinus*) 138
Alpine cyclamen (*Cyclamen alpinum*) 141
Alpinia zerumbet (shell ginger) *4*
Althaea officinalis (marshmallow) 167, 291
Amazon water lily (*Victoria amazonica*) 51, 70
Amborella trichopoda (understory shrub or small tree) 23, 41, 49
Amborellales 49
American beech (*Fagus grandifolia*) 97
American hazelnut (*Corylus americana*) 164
American oak (*Quercus alba*) 51
Ammi majus (bishop's flower) 94
Amorphophallus titanum (titan arum) *19*, 75
Ananas comosus (pineapple) 42, 250
Anatomy of Plants (Grew) 317
ancestral green algae (*Streptophyta*) 24, 25
Ancient Egypt 58, 60, 62, 65, 66, 76, 80, 155
Ancient Greece 65, 66, 86, 91, 155
Andropogon gerardi (big bluestem grass) 55
anemone, Japanese (*Anemone x hybrida* 'September Charm') 155

anemone, wood (*Anemone nemorosa*) 51
Anemone x hybrida 'September Charm' (Japanese anemone) 155
Angell, Bobbi *329*
Angiosperm Phylogeny Group (APG) 126, 326
angiosperms (flowering plants) 24, 25, 40–1, 49, 101, 126
Angkor Wat, Cambodia 83
Anigozanthos manglesii (kangaroo paw) 55
Animal Kingdom 22, 24, 27, 126
animal-pollination 98
Animalium (2014) 314
Annona muricata (soursop) 46
Annona reticulata (custard apple) 46
annua (annual) 133
annuals 137
The Ansel Adams Wilderness, Sierra Nevada *180*
Anthoceros agrestis (field hornwort) 30
Anthocerotophyta (hornworts) 24, 25, 30, 33
Anthriscus sylvestris (cow parsley) 124, 172
anthropomorphism 118
Anthurium (flamingo flower) *117*
Antiba rosaeodora (rosewood) 295
Antirrhinum (snapdragons) 179, 311
Apiaceae (carrot/celery/parsley family) 129
Apium 129
Apium crispum Mill. (parsley) 129
Apium graveolens var. *rapaceum* (celeriac) 229
Apium petroselinum L. (parsley) 129
The Apple Book (Sanders) 328
apple, Braeburn (*Malus domestica* 'Braeburn') 134
apple, crab *Malus sulvestris* 163
apple (*Malus pumila*) 65, 102, *135*, 205, 217, 250, 251
apricot (*Prunus armeniaca*) 249, 251
aquatic algae 52
aquatic liverwort (*Riccia fluitans*) 30
Aquilegia (columbine) *135*, 138, 175
Arabidopsis thaliana (Thale-cress) 137
Araucaria araucana (monkey puzzle tree) 38
Arber, Agnes 76

Arbutus unedo (strawberry tree) 157
archaea (single celled organisms) 24, 27, 126
Archaefructus (fossil plant) 41
Archaeopteris 36
Arctic lupine (*Lupinus arcticus*) 52
Arctic moss (*Calliergon giganteum*) 52
Arctium (wild burdock) 168
Areca catechu (betel palm) *136*
Argentinian vervain (*Verbena bonariensis*) 157
aril seed 101
Aristotle 22, 91, 123
Armoracia rusticana (horseradish) 138, 277
arrowroot, African (*Canna indica*) 311
The Art of Botanical Illustration (Stearn & Blunt) 312
Art Forms in Nature (Blossfeldt) 333
Art Forms in Nature (Haeckel) 311, 319
Art Nature Dialogues: Interviews with Environmental Artists (Grande) 362
Arts and Crafts movement 80, 155, 174, 369
The ArtScience Museum, Marina Bay Sands (Singapore) 383
arugula/rocket (*Eruca sativa*) 206, 207
arum lily/wild arum (*Arum maculatum*) 324
Arum maculatum (lords-and-ladies) 98
arvensis (of the field) 133
Asarum caudatum (western wild ginger) 46
Asclepias (milkweed) 55
ash (*Fraxinus excelsior*) 97
ashwagandha 278
Asian rice (*Oryza sativa*) 42, 56, 62, 102, 137
Asian yams 65
Asima triloba (common paw paw) 46
asparagus fern (*Asparagus setaceus*) 35
asparagus, garden (*Asparagus officinalis*) 138, 189
Asphodelus 334
Asplenium nidus (bird's nest fern) 35
Aster alpinus (Alpine aster) 138
aster (*Aster*) 55
aster or daisy family (Asteraceae) 54, 129

Astragalus canadensis (Canada milkvetch) 55
Astrophytum ornatum (monk's hood cactus) 141
Atacama desert (South America) 56
Athyrium angustum f. *rubellum* (lady fern with red stems, 'Lady in Red') 134
Athyrium filix-femina (lady fern) 35, 51, 175
Atkins, Anna 80, *333*
Atlantic ivy (*Hedera hibernica*) 141
Atriplex prostrata (spear-leaved orache) 97
Atropa belladonna (deadly nightshade) 102
Attenborough, David 80
aubergines/eggplants (*Solanum melongena*) 65, 104, 130
Aubriet, Claude 127
Augé, Lucy *369*
Austrobaileya scandens (woody liana) 49
Austrobaileyales 49
autogamy 98
autumn plants/recipes 216–27
Avena sativa (oats) 249
avens (*Geum* 'Totally Tangerine') *154*, 155
avocado (*Persea americana*) 46, 65, 118, *248*, 249, 250, 284
Ayurvedic medicine 65, 261
Ayutthaya Historical Park, Thailand 83

Baalen, Lotte van *378*
bacteria (single celled organisms) 22, 27, 126
Bae, Wona *378*
Baileya (desert marigold) 56
Baines, Thomas 49
Baker, Charisse 187, 253, 299
bald cypress tree (*Taxodium distichum*) 57
bamboo (Bambusoideae) 93, 97
bamboo, Indian timber (*Bambusa tulda*) 51
bamboo palm (*Dypsis lutescens*) 155
bamboo shoots 65
Bambusa tulda (Indian timber bamboo) 51
Bambusa vulgaris (common bamboo) 42
banana, cultivated (*Musa acuminata*) 42, 65, 102, 118, 138, 249, 250, 254
banana family (Musaceae) 129
bananas, wild (*Musa*) 51
Banks's Florilegium 104
Banks, Sir Joseph 75, 76, 323
banyan, Indian (*Ficus benghalensis*) 93
banyan trees (*Ficus*) 51
baobab (*Adansonia*) 55

Baptisia australis (blue wild indigo) 55
Baragwanathia 34
Barbados aloe (*Aloe vera*) 141
barberry 45
Barbican Conservatory, London 343, 372
barley 65
barrel cactus (*Echinocactus/Ferocactus*) 56
Barron, Jenny 47
Bartram, John 79, 316
Bartram, William *316*
Bartram's Garden, Philadelphia 75
basil, sweet (*Ocimum basilicum*) 141, 204
Bateson, William 143
Baudar, Pascal 187
Bauer, Ferdinand *316*
Bauer, Franz 79, *316*
Bauhin, Gaspard 123
bay 65
bay laurel (*Laurus nobilis*) 46, 97
bay leaf 190, 191
bean family (Fabaceae) 129
beans 65
beansprouts 212
bearded iris (*Iris germanica*) 152
beauty 290
  Body Butter 293
  Clay Mask 292
  Floral/Herbal Hydrosol 293
  ingredients 291
  Melissa Lip Balm 296
  NettleFacial Toner and Balancing Facial Oil 295
  Ocean Potion Mask 299
  Skin Scrub 292
Bee Urban Sweden 157
beech, American (*Fagus grandifolia*) 97
beech (*Fagus sylvatica*) 51, 94, 101, 138
bees 157
beeswax 288, 296, 302
beet (*Beta vulgaris* subsp. *vulgaris*) 93, 217, 234, *235*, 250
beet/chard (*Beta vulgaris* subsp. *vulgaris*) 93, 217
beetroot (*Beta vulgaris* cultivars) 217
begonia (*Begonia*) 98, 101
Beihai Park, Beijing 69
bellflower family (Campanulaceae) 129
Bellis perennis (English daisy) 72, 167, 311
Ben-David, Zadok 360
Blackfield 360
Benjamin, Walter 333
bentonite 292
bergamot (*Citrus bergamia*) 206, 281
Bermuda grass (*Cynodon dactylon*) 55
berries 102
A Berry Feast (Snyder) 181

Bertholletia (Brazil nut) 51
Besler, Basilius 65, *74*, 75, 76, *318*, 319
Beta vulgaris cultivars (beetroot) 217
Beta vulgaris subsp. *vulgaris* (Swiss chard/beet) 93, 217, 234, *235*, 250
Betula (birch) 97
Betula papyrifera (paper birch) 51
Betula pendula (silver birch) 151
Bibb lettuce (butterhead) 212
bicarbonate of soda 264
biennial plants 137
big bluestem grass (*Andropogon gerardi*) 55
bigleaf magnolia (*Magnolia macrophylla*) 133
bilberry/whortleberry (*Vaccinium myrtillus*) 163
bindweed (*Convolvulus*) 311
Bingham, Juliet 357
binomial nomenclature *120*, 121, 123, 124, 313
biodiversity planting 158
biofuels 143
biome 19, 50–57
birch (*Betula*) 97
birch, paper (*Betula papyrifera*) 51
bird's nest fern (*Asplenium nidus*) 35
biriba (*Rollinia deliciosa*) 46
Birkenhead Park, Merseyside 58
bishop's flower (*Ammi majus*) 94
bitter orange (*Citrus aurantium*) 64
bitter vetch 65
black maple (*Acer nigrum*) 130
black nightshade (*Solanum nigrum*) 130
black pepper (*Piper nigrum*) 46, 143, 190, 194, 198, 203, 204, 209, 210, 225, 233, 234, 241, 285
black-eyed susan (*Rudbeckia fulgida* var. *sullivantii* 'Goldsturm') 155
blackberry (*Rubus fruticosus*) 162, 163, 217, 218, *224*, 225
blackcurrant (*Ribes nigrum*) 203
Blackwell, Alexander 317
Blackwell, Elizabeth 76, *285*, *317*
bladderwrack 27
Blake, Sir Peter 308
Tulips after Van Brussel, Flowers in a Vase 308
Blakeney, Justina 80, *379*
The New Bohemians 379
The New Bohemians Handbook 379
Blaschka, Leopold and Rudolf *374*, 375
Blauer, Ferdinand 79
blazing star (*Liatris*) 55
Bloedel Reserve, Washington State 31

blood orange (*citrus x sinensis*) 220, 229, 231
Blossfeldt, Karl 80, *332*, 333
blue agave (*Agave tequilana*) 56
blue Egyptian water lily (*Nymphaea caerulea*) 49
blue grama grass (*Bouteloua gracilis*) 55
blue wild indigo (*Baptisia australis*) 55
bluebell, British (*Hyacinthoides non-scripta*) 51
blueberry (*Vaccinium corymbosum*) 151, 238, 249, 250, 251
bluegrass (*Poa*) 93
Blunt, Wilfrid 312
Bocclus, Dr Norbert 316
Boeri, Stefano 308, *383*
Bonnard, Pierre 70
borage (*Borago officinalis*) 45, 93, 157, 205
Boraginaceae (forget-me-not family) 129
borealis (from the north) 133
Bosschaert, Ambrosius 72
Boston, Bernie 83
Flower Power 73
Boston fern (*Nephrolepis exaltata*) 35
Boswellia carteri (frankincense) 288, 291
Botanic Garden Conservation International (BGCI) 75, 83
botanic gardens 75
botanical art/illustration 79, 308, 312–31
Botanical Register 77
botanical style 80
The Botanical Wall Chart (Laurent) 324
botanical world 11, 146, 149, 386
Botanicum (Scott, 2016) 83, *127*, *314*, *315*
botanists 91
A Botanist's Vocabulary (Angell & Pell) 329
botany 86, 88–9, 90, 92
Botany for the Artist (Sumblet) 121
Bouffier, Elzéard 149
bougainvillea (*Bougainvillea*) 51
Bouteloua curtipendula (sideoats grama) 55
Bouteloua dactyloides (buffalo grass) 55
Bouteloua gracilis (blue grama grass) 55
Brachythecium rutabulum (rough-stalked feather moss) 31
bracken (*Pteridium*) 35, 115
bract 117
Braeburn apple (*Malus domestica* 'Braeburn') 134
Brassica genus (broccoli, cauliflower, cabbage, turnip, rapeseed) 130

INDEX 401

Brassica oleracea (cabbage) 94, 137, 189, *228*, 229
Brassica oleracea cultivar (broccoli) 65, 229, 311
Brassica oleracea cultivar (Brussels sprouts) 102, 229
Brassica oleracea cultivar (kale) 72, 229, 230, 250
Brassica rapa subsp. *rapa* (turnip) 229, *236*, 237
Brassicaceae (cabbage family) 129
Brazil nut (*Bertholletia*) 51
bread tree (*Encephalartos altensteinii*) 36, *37*
Breughel the Elder, Jan 72
bristlecone pine (*Pinus longaeva*) 138
British bluebell (*Hyacinthoides non-scripta*) 51
brittlebush (*Encelia farinosa*) 56
broad bean (*Vicia faba*) 203
Broadhurst, Florence 370
broccoli (*Brassica oleracea* cultivar) 65, 229, 311
Bromelia antiacantha 328
bromeliads (Bromeliaceae) 51
Brompton stock (*Matthiola incana*) 137
brown algae (*Phaeophyta*) *24*, *27*
Brown Brothers Continental Nursery Catalog (1909) 135
Brown, Lancelot 'Capability' 72
Brussels sprouts (*Brassica oleracea* cultivar) 102, 229
Bryophyta (mosses) *24*, *25*, *31*, *33*, 52
bryophytes (non-vascular land plants) *24*, *29*
buckwheat 212
buffalo grass (*Bouteloua dactyloides*) 55
bulbs 42, 65
bull bay or southern magnolia (*Magnolia grandiflora*) *47*, 51, 133
bulrush/cattail (*Typha minima*) 158
burdock, pickled root 212
burdock, wild (*Arctium*) 168
Bursera graveolens (Palo santo) 281
Bursera microphylla (elephant tree) 56
butcher's broom (*Ruscus aculeatus*) 94
buttercup (Ranunculaceae) 45, 126, 129, 311
butterflies 157
butterhead (Bibb lettuce) 212
butternut squash (*Cucurbita moschata*) 102, *216*, 217
butters 263, 293

C. americana (American hazelnut) 164
The Cabaret of Plants (Mabey) 70
cabbage (*Brassica oleracea*) 94, 137, 212, 222, *228*, 229, 230
cabbage family (Brassicaceae) 129
cabbage, Isle of Man (*Coincya monensis* subsp. *monensis*) 134
Caboma aquatica (fanwort) 49
cacao *252*, 253, 287
Cactaceae family *314*
cacti 56, *346*
cactus, barrel (*Echinocactus*/*Ferocactus*) 56
Cactus House, Hignell Gallery (London) *366*
cactus, organ pipe (*Stenocereus thurberi*) 56
cactus, prickly pear (*Opuntia*) *45*, 56, *95*
Cahuilla Indians 211
Calendula officinalis (pot marigold) 171, 282, *286*, 287, 291, 292
Californian poppy (*Eschscholzia california*) 11, *174*, 175
calla lily (*Zantedeschia*) *43*, 93
Calliergon giganteum (Arctic moss) 52
calliopsis/golden tickseed (*Coreopsis tinctoria*) *176*, *177*
Calluna vulgaris (heather) 97, *140*
Cambridge University Botanic Garden 91
Camellia sinensis (tea) 45, 62, *63*, 130
Camellia sinensis var. *sinensis* (Chinese tea) 134
Campanula medium (Canterbury bells) 137
Campanulaceae (bellflower family) 129
camphor 46
camphor laurel (*Cinnamomium camphora*) 36
Canada milkvetch (*Astragalus canadensis*) 55
Canada wild rye (*Elymus canadensis*) 55
Cananga odorata (ylang ylang) 46, 301
candelabra tree (*Euphorbia ingens*) 55
Canella winterana (wild cinnamon) 46
Canellales 46
Canna indica (African arrowroot) 311
Cannabis sativa (hemp) 97
cannellini beans 234
Canterbury bells (*Campanula medium*) 137
capitulum 98
Carboniferous (coal-bearing) period 28, 34, 36
cardamom (*Elettaria cardamomum*) 226, 243, 246, 284, 301
carnation (*Dianthus caryophyllus*) 45, 72

carnation/pink family (Caryophyllaceae) 129
carpet moss (*Mnium hornum*) 51
Carpinus (European hornbeam) 101
Carr, Lindsey
Garden 159
carrier oils 263
carrot (*Daucus carota* subsp. *sativus*) 65, 93, 102, 134, 137, 212, 249, 251
carrot, wild (*Daucus carota*) 98, 124
Carya glabra (pignut) 164
Carya (hickory nuts) 51, 164
Carya illinoiensis (pecan) 51, 164
Carya lacinios (shelbbark) 164
Carya ovata (shagbark) 164
Caryophyllaceae (carnation/pink family) 129
caryopsis/grain 102
Casa Azul, Mexico City 70
casava (*Manihot esculenta*) 93
cashews 102
Castanea sativa (sweet chestnut) 164, 217
castile soap 264
castor-oil plant (*Ricinus communis*) 101, *321*
Catesby, Marc 76, 79
Catharanthus roseus (rosy periwinkle) 51
catkins 108
Catteleya labiata (ruby-lipped cattleya orchid) *40*
Caulfield, Patricia 354
cauliflower 65, 102
Causes of Plants (Theophrastus) 66
cayenne 277
cedar (*Cedrus libani*) 94
cedarwood (*Cedrus atlantica*) 281, 295
Cedrus atlantica (cedarwood) 281
Cedrus libani (cedar) 94
celeriac (*Apium graveolens* var. *rapaceum*) 229
celery 102
cells *104*, 115
Centaurea cyanus (cornflower) 137, *172*
Central Park, New York *381*
cereals 65
Cesalpino, Andrea 123
Cézanne, Paul 306
Chamaedorea elegans (parlour palm) 141
chamomile (*Chamaemelum nobile*) 273, 292, 293
chamomile, German (*Matricaria chamomilla*) *272*, 273
chamomile, Roman (*Chamaemelum nobile*) 273
Chamovitz, Daniel 118
Charales (stoneworts or rockweeds) 28
charcoal, activated 264

chard, Swiss (*Beta vulgaris* subsp. *vulgaris*) 93, 217, 234, *235*, 250
Charophyta *see* green algae
Chase, Mary Agnes 76, *77*
Chatto, Beth 381
Chelsea Flower Show 72, 151, 376
Chelsea Physic Garden 69, 75, 317
cherry (*Prunus* cultivars) 203, 205
cherry, wild (*Prunus avium*) 163
chervil 207
chestnut, sweet (*Castanea sativa*) 164
chia (*Salvia hispanica*) 238, 249, 251
chickpeas 65, 102
chickweed (*Stellaria media*) 137, 168, 206
chicory, common (*Cichorium intybus*) 138, 225, 233
chicory, red (radicchio) *232*, 233
children, naturecraft for 178
Chilopsis (desert willow) 56
Chincona officinalis (quinine) 51, 62, 261
Chinese herbalism 261
Chinese meadow rue (*Thalictrum delavayi*) 152, *153*
Chinese tea (*Camellia sinensis* var. *sinensis*) 134
Chinese wisteria (*Wisteria sinensis*) 138, 152
chives (*Allium schoenoprasum*) 42, 198
chlorophyll a and b 27
Chlorophyta see green algae
Chlorophytum comosum see spider plant
chloroplasts 27, 104
chocolate *see* cocoa bean tree
Chown, Vicky 296
Christmas tree *see* Norway spruce (*Picea abies*)
chromosomes 115
chrysanthemum (*Chrysanthemum*) *68*, 72, 80, *128*
Churchill, Morss 278
Cichorium intybus (common chicory) 138
cinchona tree (*Cinchona*) 51
Cinnamomium camphora (camphor laurel) 36
cinnamon, true (*Cinnamomum verum*/*cinnamomum zeylanicum*) 46, 219, 226, 254, 301, 302
Citron (*Citrus medica*) 64
Citrullus lanatus var. *lanatus* (watermelon) 137
Citrus 64, 65
Citrus aurantium (bitter orange) 64
Citrus aurantium subsp. *amara* (neroli) 291
Citrus bergamia (bergamot) 206, 281

Citrus (lime) *240*, 241, 246
Citrus limon (lemon) 97, 102, 190, 194, 198, 204, 212, 218, 226, 230, 231, 233, 237, 242, 246, 264, 301, 303, 304, *305*
Citrus reticulata (mandarin) 134, 245, 301, 302
Citrus sinensis (sweet orange) *64*, 101, 237, 242, *245*, 246, 304
Citrus x limon (Meyer lemon) 134
Citrus x paradisi (grapefruit) 102, 301, 304
citrus x sinensis (blood orange) *220*, 229
clades 126
cladode 94
Cladonia rangiferina (reindeer lichen) 52
clary sage (*Salvia sclarea*) 138, 291
classification 122–3, 126
clays 263
cleavers/goosegrass (*Galium aparine*) 101, 171
Clifford, George 79
cloud rainforest 51
cloudberry (*Rubus chamaemorus*) 52
clove (*Syzygium aromaticum*) 226, 243, 301
clover (*Trifolium repens*) 97, 179
club mosses (*Lycopodiophyta*) 24, 25, 33, 34, 36
coco de mer (*Lodoicea maldivica*) 42, *100*, 101
cocoa bean (*Theobroma cacao*) 51, 130, *252*, 253
cocoa butter 293
coconut (*Cocos nucifera*) 42, 93, 94, 101, 249, 250, 251, 253, 268, *269*, 271, 292, 293
Code of Nomenclature for Cultivated Plants (ICNCP) 134
Codex Liechtenstein 76
coffee (*Coffea*) 51, 62, 63
Cohen, Amanda 184
Coincya monensis subsp. *monensis* (Isle of Man cabbage) 134
Coleridge, Samuel Taylor 316
Collishaw, Matt, *Tulip Mania* 308
Colour in the Flower Garden (Jekyll) 72
columbine (*Aquilegia*) *135*, 138, 175
Columbus, Christopher 66
comfrey (*Symphytum officinale*) 291
Commiphora myrrha (myrrh) 65, 281
common bamboo (*Bambusa vulgaris*) 42
common chicory (*Cichorium intybus*) 138
common dogwood (*Cornus sanguinea*) 98

common grapevine (*Vitis vinifera*) 62, 152
common hazel (*Corylus avellana*) 164
common jasmine (*Jasminum officinale*) 45, 138, 152
common laburnum (*Laburnum anagyroides*) 97
common lime (or linden) (*Tilia x europaea*) 51, *106*, 164, *165*, 281
common madder (*Rubia tinctorum*) 176
common myrtle (*Myrtus communis*) 151
common (or corn or field) poppy (*Papaver rhoeas*) 11, 45, 101, *137*, 157, 172
common reed (*Phragmites australis*) 179
common sunflower (*Flos Solis Maior*) 74
common tumbleweed (*Kali tragus*) 56
common zinnia (*Zinnia elegans*) 137
Complete Herbal (Culpeper) 76
coneflower (*Echinacea*) *54*, 55, 138, *139*, 157, 172, 273
coneflower (*Rudbeckia*) 54
conifers (*Pinophyta*) 24, 25, 28, 38
Conium maculatum (hemlock) 124
conkers 179
Conservatory Archives 366
Cook, James 66, 323
Coreopsis tinctoria (golden tickseed/calliopsis) 176, *177*
coriander (*Coriandrum sativum*) 97, *125*, 246
corn 94
corn (or common or field) poppy (*Papaver rhoeas*) 11, 45, 101, *137*, 157, 172
corn (*Zea mays* subsp. *mays*) 93
cornflower (*Centaurea cyanus*) 137, 172, 205
cornmeal 210
Cornus alba (Siberian dogwood) 141
Cornus sanguinea (common dogwood) 98
corpse lily (*Rafflesia arnoldii*) 51
Correns, Karl Franz Joseph 143
Cortaderia selloana (pampas grass) 55
Corylus avellana (hazelnut) 102, 164
Corylus maxima 'Purpurea' (filbert) 158, 164
corymb 98
cosmos (*Cosmos bipinnatus*) 11, 98, 175
Cotinus coggyria (European smoke bush) 94, 155
cotton (*Gossypium*) 45, 62, *63*, 101, 102

courgette (*Cucurbita pepo*) 45, 184, 203, 204
cow parsley (*Anthriscus sylvestris*) 124, 172
cow parsnip (*Heracleum maximum*) *125*
cowslip (*Primula veris*) 94
crab apple (*Malus sylvestris*) 163
craft, design and style 368–89
Craig-Martin, Michael, *Tulips (after Mapplethorpe)* 308
cranberry (*Vaccinium oxycoccus*) 52, 163, 231
Crataegus monogyna (hawthorn) 281
crescent-cup liverwort (*Lunularia cruciata*) 30
Cretaceous period 35
crocosmia (*Crocosmia x crocosmiiflora*) 115, 175
crocus (*Crocus*) 94
crocus, early (*Crocus tommasinianus*) 157
Crocus sativus (saffron) 281
crocus, smooth (*Crocus laevigatus*) 141
crocus, spring (*Crocus vernus*) 51
Cruickshank, Will 367
Crusades 58, 65
Cryptogamae (non-floral plants) 126
Crystal Palace, London 70
cucumber (*Cucumis sativus*) 204, 212, *248*, 291
Cucurbita maxima (pumpkin) 65, 94, 206, 217
Cucurbita moschata (butternut squash) 102, 217
Cucurbita pepo (courgette) 45, 184, 203
Cucurbita pepo var. *pepo* (squash) 65, 94
Cucurbitaceae (gourd family) 129, *216*
Culpeper, Nicholas 76
cultivar 134
cultivated squash/pumpkin (*Cucurbita maxima*) 65, 94, 217
Cummings, Phoebe 385
Triumph of the Immaterial 385
Cunningham, Nicola 288, 295, *389*
Leaf Poem series 389
Cupresses sempervirens (cypress) 288
Curcuma longa (turmeric) 51, 254, 273
A Curious Herbal (Blackwell) 76, 317
currants 238
Curtis, William 76
Curtis's Botanical Magazine (aka *The Botanical Magazine; or Flower Garden Displayed*) 76, 77, 79, 323
custard apple (*Annona reticulata*) 46
Cyadophyta see cycads

cyanobacteria 24, 25, *26*, 27
Cyanotypes of British and Foreign Ferns (Atkins) 333
Cyanotypes of British and Foreign Flowering Plants (Atkins) 333
cycads (*Cyadophyta*) 24, 25, 28, 38
Cycas revoluta (Japanese sago palm) *38*
cyclamen, Alpine (*Cyclamen alpinum*) 141
cycle of life 136–8
Cydonia oblonga (quince) 217, *317*, *334*
Cymbopogon citratus (lemongrass) 288, 301
cyme 98
Cynara cardunculus var. *scolymus* (globe artichoke) *129*
Cynodon dactylon (Bermuda grass) 55
cypress (*Cupresses sempervirens*) 288

daffodil (*Narcissus*) 42, *92*, 94, 115, 138, 172
dahlia (*Dahlia*) 72, 93, 138, 141, 176
daisy, English (*Bellis perennis*) 72, 167, 311
daisy, oxeye (*Leucanthemum vulgare*) 55, 172
daisy, Tahoka (*Machaeranthera tanacetifolia*) 55
Dalea purpurea (prairie-clover) 55
Damask rose (*Rosa x damascena*) 121, 124
damiana 283
damson (*Prunus domestica* subsp. *insititia*) 217
dandelion (*Taraxacum officinale*) 93, *100*, 101, 115, 168, *169*, 311, *328*
Darwin, Charles 41, 58, 70, 75, 76, 86, 91, 116, 118, 123, 143, 326
Darwin, Erasmus 91
Darwin, Robert 91
date (*Phoenix dactylifera*) 229, 238, *239*
Daucus carota subsp. *sativus* (carrot) 65, 93, 102, 134, 137, 212, 249, 251
Daucus carota (wild carrot) 98, 175
David Hockney: A Bigger Picture (film) 355
Dawsonia (moss) 31
day lily 206
dayflower leaf 206
De Causis Plantarum (*Causes of Plants*) (Theophrastus) 123
De Historia Plantarum (*On Plants*) (Theophrastus) 123
De Historia Stirpium (Fuchs) 79
De Materia Medica (Dioscorides) 65, *259*

De plantis libri XVI (Cesalpino) 123
Dead Sea salts 282
deadly nightshade (*Atropa belladonna*) 102
deciduous forest 51
decomposition 116
Dein, Rachel 385
  Rose, Daisy, Plantain and Achille 384
Delany, Mary 372
  Flora Delanica 372
  Papaver somniferum, The Opium Poppy 372
  Passiflora laurifolia 44
Delany, Dr Patrick 372
Delessert, Madame 76
delphinium (*Delphinium*) 94
dentate leaf 97
Deschampsia cespitosa (tufted hair grass) 158
desert 56
desert cacti 80
desert ironwood (*Olneya tesota*) 56
desert lily (*Hesperocallis*) 56
desert marigolds (*Baileya*) 56
desert sage (*Salvia eremostachya*) 56
desert willow (*Chilopsis*) 56
Desfontaines, René 313
devil's claw (*Harpagophytum procumbens*) 101
Devonian period 28, 34, 35, 36
Diacono, Mark 187, 198
diamond-leaf willow (*Salix pulchra*) 52
Dianthus caryophyllus (carnation) 45
Dickinson, Emily *388*
  'May-Flower' *388*
Dicksonia/Cyathea (larger tree fern species) 35
Digitalis (foxglove) 45
Digitalis purpurea (foxglove) 45, 98, 137
digitata (leaves like a hand) 133
dill 198, 204, 208, 209
'dinosaur food' (*Lepidozamia peroffskyana*) 38
Dioscorides, Pedanius 65, 79, *259*
Diospyros kaki (persimmon) 97
discoid seed 101
dock leaf (*Rumes obtusifolius*) 179
Dodel-Port Atlas 324, *325*
Dodel-Port, Carolina 324
Dodel-Port, Dr Arnold 324
dog rose (*Rosa canina*) 97, 102, 311
dogwood, common (*Cornus sanguinea*) 98
dogwood, Siberian (*Cornus alba*) 141
Donkelaar, Anne ten *366*
  Broken Butterflies 366
  Flower Constructions 366
Dow, Arthur Wesley 348

drawing *see* painting and drawing
Drimys winteri (winter's bark) 46
drinks and bowls
  berry almond smoothie 250
  boost juice 251
  chia breakfast bowl 251
  cocktails 241
  DIY gin (martini) 246
  elderflower cordial 242
  green machine 250
  grenadine 242
  juices, smoothies ingredients 249
  mandarin shrub 244–5
  orange bitters 243
  sloe gin 243
drupes 102
Dryopteris filix-mas (male fern) 35, 138, 155
duckweed 115
dwarf shrubs 52
dwarf umbrella tree (*Schefflera arboricola*) 97
dwarf water lily (*Nymphaea candida*) 114
Dyonaea muscipula (Venus flytrap) 118
Dypsis lutescens (bamboo palm) 155

The Earth Laughs in Flowers (exhibition, 2014) 386
Ebers papyrus 65
Echinacea purpurea (purple coneflower) *54*, 55, 138, *139*, 157, 172, 273
Echinocactus/Ferocactus (barrel cactus) 56
edible plants 42, 45, 51, 65, 162–9
Edwards, Rawia Baitalmal 187, 254
Edwards, Sydenham 77
eggplants/aubergines (*Solanum melongena*) 65, 104, 130
Egyptian lotus, white (*Nymphaea lotus*) 49
Egyptian water lily, blue (*Nymphaea caerulea*) 49
Ehret, George Dionysius 79, *313*
Eichler, August W. 126
einkorn wheat 65
Einstein, Albert 339
Elaeis (oil palm) 62
elaiosome seed 101
elderberries 226, *227*, 238, 278
elderflower (*Sambucus nigra*) 167, 197, 214, 241, 242, 273, *279*
elegans (elegant) 133
elephant grass (*Pennisetum purpureum*) 55
elephant tree (*Bursera microphylla*) 56
Elettaria cardamomum (cardamom) 226, 243, 246, 284, 301
Ellis, Peter 146

Ellis, Sonya Patel 6–11, *386*, *387*
  California Poppy (*Eschscholzia californica*) *174*
  Fennel (*Foeniculum vulgare*) *146*
  Herb Garden at The Marksman *386*, *387*
  Herbarium Wall 386
  Hortus Uptonensis 386, *387*
  Seeing Floral - The Art of Florilegia 386
  Sweet Pea (*Lathyrus odoratus*) 'Prima Donna' *10*, *387*
Ellis, Tom *352*, *353*
Ellsworth Kelly – Plant Drawings (Semff) 351
Ely, Helena Rutherford 72
Elymus canadensis (Canada wild rye) 55
Embryophytes *see* land plants
Emerson, Ralph Waldo 83, 181
emmer wheat 65
Encelia farinosa (brittlebush) 56
Encephalartos altensteinii see bread tree
Encyclopaedia of Essential Oils (2014) 263
Encyclopedia of Natural History (Wilhelm) 177
English Botany 79
English daisy (*Bellis perennis*) 167
English lavender (*Lavandula angustifolia*) 138, 151
English oak (*Quercus robur*) 51
English walnut (*Juglans regia*) 51, 97
Enquiry into Plants (Theophrastus) 62
Ephedra ('jointfirs') 39
Epping Forest, Essex 342
Equisetophyta see horsetails
Equisetum arvense (horsetail) 34
Equisetum giganteum (southern giant horsetail) 34
Ericaceae (heather family) 129
Eruca sativa (rocket/arugula) 207
Eschscholzia california (California poppy) 11, 175
essential oils 263, 302, 303
eucalyptus (*Eucalyptus globulus*) 45, 97, 273
eucalyptus (*Eucalyptus smithii*) 288, 296, 303
Eucalyptus genus 130
Eucalyptus pruinosa (silver-leaved box) *316*
eudicots 25, 45
Eukarya 126
eukaryotes *24*, 25, 27
Euphorbia ingens (Peking spurge, candelabra tree, poinsetta) 55, 130
European grapevine (*Vitis vinifera*) 62
European holly (*Ilex aquifolium*) 51, 98, 138
European hornbean (*Carpinus*) 101
European mistletoe (*Viscum album*) 51

European smoke bush (*Cotinus coggyria*) 94
European yew (*Taxus baccata*) 101
Euterpe precatoria (acai palm) 51
evening primrose (*Oenothera biennis*) 171
Expedition 46 (2016) 83
exploration 66

Fabaceae (bean family) 129, 158
Fagus 101
Fagus grandifolia (American beech) 97
Fagus sylvatica (beech) 51, 94, 101, 138
Falconi, Dina 187, 206
false bird-of-paradise (*Heliconia*) 114
family (*familia*) 129
fanwort (*Caboma aquatica*) 49
Farrand, Beatrix 381
Fashion Flora (exhibition, 2017) 376
Fedtschenko, Olga 124
fennel, Florence (*Foeniculum vulgare Azoricum* Group) 223
fennel (*Foeniculum vulgarae*) 45, 138, *147*, 151, 164, 194, 203, 214, 219, 222, 273
fenugreek (*Trigonella foenum-graecum*) 101
fern, fiddlehead form *32*, 35
fern, male (*Dryopteris filix-mas*) 35, 138, 155
ferns (*Pteridophyta*) *24*, 25, 28, *32*, 33, 35, 36, *88*, *100*
Ferrer, Estanislao Gonzáles 70
Ferula galbaniflua (galbanum) 281
feverfew (*Tanacetum parthenium*) 171
Fibonacci 311
Ficus benghalensis (Indian banyan) 93
Ficus carica (fig) 51, *200*, 201, 203
Ficus (fig trees incl. stranglers/ banyan trees) 51, *82*
Ficus lyrata (fiddle-leaf fig) 45
fiddle-leaf fig (*Ficus lyrata*) 45
field hornwort (*Anthoceros agrestis*) 30
field horsetail (*Equisetum arvense*) 34
field maple (*Acer campestre*) 97
field (or common or corn) poppy (*Papaver rhoeas*) 11, 45, 101, 137, 157, 172
fig (*Ficus carica*) 51, *200*, 203, 214
filbert (*Corylus maxima* 'Purpurea') 158, 164
film *see* photography and film
fir (*Abies*) 52
Fitch, Walter Hood *305*, *323*
Fittonia argyraneura (nerve plant) 110
flamingo flower (*Anthurium*) 117
flax 65
flaxseed 254

Fleur-de-lis 337
Flora (2004) 336
Flora Graeca (Sibthorpe & Bauer) 79
Flora Londonensis (Curtis) 76, 79
floral plants (Phanerogamae) 126
floribundus (free-flowering) 133
Florilegium (Banks) 76
The Florist and Pomologist periodical 304
floristry 35, 72
Flos Solis Maior (common sunflower) 74
Flower Decoration in the House (Jekyll) 72
Flower of Kent (*Malus pumila*) 70
Flower Power (Boston) 83
flower waters 263
flowering plants (Angiosperms) *24, 25*, 40–1, 101, 126
The Flowering Plants and Ferns of Great Britain 124
The Flowering Plants, Grasses, Sedges and Ferns of Great Britain (Pratt) 193
flowers 55, 56, 65, 72, 82–3, *88*, 98, 115, 258
 art forms 311
 cut 172
 edible 167, 204, 212, 214
 infusions 270
 language of 72
 meadow 138
Flowers (Penn) 336
Foeniculum vulgarae (fennel) 45, 138, *147*, 151, 164, 194, 203, 214, 219, 222, 273
Foeniculum vulgare Azoricum Group (Florence fennel) *223*
Food For Free (Mabey) 161
foraging 160, 161
forest gardens 60
forget-me-not family (Boraginaceae) 129
forget-me-not (*Myosotis arvensis*) 98
forget-me-not, wood (*Myrosotis sylvatica*) 137
forked viburnum (*Viburnum furcatum*) 141
fossil plant (*Archaefructus*) 41
Fothergill, Dr John 316, 386
Four Changes (Snyder) 181
Fowler, Alys 381
Fox, Jeremy 184
Fox Talbot, William Henry 80, 333
foxglove (*Digitalis purpurea*) 45, 98, 137
Fragaria vesca (wild strawberry) 97, 158
Fragaria x ananassa (strawberry) 45, 102, *103*, 203, 250
fragrant white water lily (*Nymphaea odorata*) 56, *57*
frangipani (*Plumeria*) 313

frankincense (*Boswellia carteri*) 288, 291
Franklin tree (*Franklinia alatamaha*) 316
Fraxinus excelsior (ash) 97
French green 292
French rose (*Rosa gallica*) 124
Fritscher, Jack 336
fruit 65, 102, 163
Fruit: Edible, Inedible, Incredible (Kesseler) 335
Fuchs, Leonhart Fuchs 79
fuller's earth clay 292
Fungi Kingdom *24, 25, 27*, 32

Galanthus eluvesii var. *monostictus* (greater snowdrop) 134
Galanthus (snowdrop) 98
galbanum (*Ferula galbaniflua*) 281
Galium aparine (goosegrass/ cleavers) 101, 171
Galium verum (lady's bedstraw) 176
Gameren, Sarah van 333
gametes 115
garden asparagus (*Asparagus officinalis*) 138
Garden Club of America 72
Garden Collage magazine 76
Garden of Cosmic Speculation, Dumfries (Scotland) 382
Garden Design magazine 76
The Garden magazine 76
garden as muse 70
Garden Museum, London 364, 386
The Gardeners' Chronicle 323
Gardener's Dictionary (Miller) 75
The Gardener's Dictionary (Miller) 79
gardening for pleasure 68–72
gardening for purpose 64–6
Gardens Illustrated magazine 76
Gardensque style 72
garlic (*Allium sativum*) 42, 94, 204, 206, 209, 230, 234, 273, 277
garlic, hardneck (*Allium sativum* var. *ophioscorodon*) 199
garlic, softneck (*Allium sativum* var. *sativum*) 198
garlic, wild/Ramsons (*Allium ursinum*) 51, 167, 190, 193
Gavin, Paola 187, 234
Gavin, Seana 372
 Fairyville; Homage to Richard Dadd 372
Gaylussacia baccata (huckleberry) 51
geitonogamy 98
The Genera and Species of Orchidaceous Plants (Lindley) 323
'General Sherman' (giant sequoia) (California Sequoia National Park) 38
Genesis project 339
genetics 126, 143

genus 130
Geraniaceae (geranium family) 129
geranium family (Geraniaceae) 129
geranium, rose (*Pelargonium graveolens*) 291, 295
Gerard, John 76
German chamomile (*Matricaria chamomilla*) *272*, 273
Geum 'Totally Tangerine' (avens) *154*, 155
Ghini, Luca 75
giant sequoia (*Secuoidendron gigonteum*) 113
giant water lily (*Nymphaea gigantea*) 49
Giant Waterlily *315*
gin 38
ginger, Japanese 212
ginger, root 191, 193, 212, 226, 230, 251
ginger, shell 4
ginger, western wild (*Asarum caudatum*) 46
ginger (*Zingiber officinale*) 51, 94, 241, 261, 273, *276*, 277, 283, 284
ginkgo (*Ginkgo biloba*) 25, *39*, 101, 175
Ginkgophyta 24
Ginsberg, Allan 181
ginseng (*Pinax ginseng*/ *Eleutherococcus sentilocosus*) 273
Giono, Jean 149
Giverny *71*, 345
Gladstar, Rosemary 274, 277
Glass Flowers: The Ware Collection of Blaschka Glass Models of Plants (Harvard University) 375
Glaucophytes *see* green-blue algae
Global Strategy for Plant Conservation (GSPC) 124
global warming 52
globe artichoke (*Cynara cardunculus* var. *scolymus*) *129*, 189
Glycyrrhiza glabra (licorice) 273
GM crops 143
gnetophytes (*Gnetophyta*) 25, 39
Gnetum (mostly lianas) 39
Goethe, Johann Wolfgang von 95
The Golden Age of Botanical Art (Rix) 319
golden tickseed/calliopsis (*Coreopsis tinctoria*) 176, *177*
goldenrods (*Solidago*) 55
Goldsworthy, Andy 363
goosegrass/cleavers (*Galium aparine*) 101, 171
Gossypium barbadense (Sea Island cotton) 67
Gossypium (cotton) 45, 62, 66, 67, 101, 102
Gotjawal Forest, Jeju Island 347
gourd family (Cucurbitaceae) 129, 212, *216*

grains 212
Grande, John K. 362
grandiflora (large-flowered) 133
grandis (large) 133
grape (*Vitis*) 94, 102
grapefruit (*Citrus x paradisi*) 102, 301, 304
grapeseed 292
grapevine, common (*Vitis vinifera*) 62, 152
grass, Bermuda (*Cynodon dactylon*) 55
grass, big bluestem (*Andropogon gerardi*) 55
grass, blue grama (*Bouteloua gracilis*) 55
grass, buffalo (*Bouteloua dactyloides*) 55
grass, elephant (*Pennisetum purpureum*) 55
grass, Indian (*Sorghastrum nutans*) 55
grasses 52, 55, 97
grassland *54*, 55
Gray, Dr Asa 75, 375
great burnet (*Sanguisorba officinalis*) 158
Great Oxygen Events 22, 27
Great Salt Lake, Utah 360
greater snowdrop (*Galanthus eluvesii* var. *monostictus*) 134
green algae (*Chlorophyta* & *Charophyta*) *24, 28, 325*
green plants (*Viridiplantae*) 25
green-blue algae (Glaucophytes) *24, 25, 27*
Grew, Nehemiah *317*
Grierson, Mary 328
ground ivy leaf 206
ground pine (*Lycopodium*) 34
guelder rose (*Viburnum opulus*) 51, 157
Gulbenkian Science Institute, Portugal 335
gum acacia, Senegal (*Acacia senegal*) 55
gymnosperm *see* non-flowering seed plants
Gyngell, Skye 187, 222

Haarkon *343*
Habitat 67 housing complex (Canada) 383
Haeckel, Ernst *29*, 126, *127*, 311, 314, *319*
 Art Forms in Nature 28
haircap moss (*Polytrichum*) 31
Halla Mountain, Jeju Island 347
Hamatreya (Emerson) 83
Hamilton's Gallery, London 336
Hanekamp, Deborah 287
Hanging Gardens of Babylon *61*
Hanson, Thor 101
hard pine (*Pinus*) 38
hardy plants 141
Harford, Robin 187
Harpagophytum procumbens (devil's claw) 101

Harris, Miria 149, 151, 152, 154, 158, *379*, 386
Harris, Tom 184, 187, 221, 233
hartwort (*Tordylium*) 124
Harvard Houghton Library 388
Harvard Museum of Natural History (HMNH) 375
Harvard University Herbaria 375
Hatshepsut, Queen 66
Hatzius, Lynn 24, 25, *58*, *88*, *120*, *148*, *160*, *186*, *260*, *310*, 373
hawthorn (*Crataegus monogyna*) 281
hazelnut (*Corylus avellana*) 102, 164, 206, 219
health remedies 258
    Herbal Decoction (simmering) 271
    Herbal Infusion (steeping) 270
    Herbal Inhalation/Facial Cream 274
    Herbal Poultice/Compress 274
    Herbal Salve/Ointment 275
    Herbal Syrup 275
    Herbal Tincture 270
    immune support syrup 278
    Infused Oil 271
heather (*Calluna vulgaris*) 97, *140*, 141
heather family (Ericaceae) 129
Hedera helix (ivy) 51, 79, 80, 93, 97, 138, 172
Hedera hibernica (Atlantic ivy) 141
Hedysarum boreale (sweet vetch) 102
Heemskerck, Marten van 60
Heidjens, Simon 360
    Lightweeds 360, *361*
    Tree 360
Heinrich von Ofterdingen (Novalis) 320
heirloom plants 134
Helianthus annuus (sunflower) 45, 98, 102, 130, 176, 311
Helianthus petiolaris (prairie sunflower) 55
Heliconia (false bird-of-paradise) *114*
hellebore, Stern's hybrid (*Helleborus x sterni*) 134
hellebores (*Helleborus*) 11, 102
Helleborus argutifolius 134
Helleborus genus (includes Christmas rose helleborus) 130
Helleborus lividus 134
Helmont, Franciscus Mercurius van 86
Helmont, Jan Baptist van 86, 111
hemlock (*Conium maculatum*) 124, *125*
hemp (*Cannabis sativa*) 97
hen and chick/houseleek (*Sempervivum*) 45, 56
Henrot, Camille 367
    Grosse Fatigue (film) 367

Is it possible to be a revolutionary and like flowers? 367
Henslow, John 91
Hepworth, Barbara 70
Heracleum (hogweed) 124
Heracleum maximum (cow parsnip) *125*
Herb Garden at The Marksman (Ellis) 146
herbaceous plants 138
Herball or General Historie of Plants (Gerard) 76
herbal medicine 65, *170*, 171, *259*, *260*, 261
herbals 76, 79
herbarium 11, 92, 175, 388
The Herbarium Project 11, 175, 386
Herbarium Specimen Painting (Pedder-Smith) *326*
Herbarius (Platonicus) 79
herbs 45, 171, 212, 214, 258, 263, 270, 271, 274–5, 282
herman de vries 362, *363*
    108 pound rosea damascena 363
    from the laguna of venice 363
    to be all ways to be 363
Herschel, John 80, 333
hesperidium 102
Hesperocallis (desert lily) 56
heterospory 36
Hevea brasiliensis (rubber tree) 51, 62
hibiscus (*Hibiscus rosa-sinensis*) 167, 241
hickory nuts (*Carya*) 51, 164
High Line, New York City 69
Hilma af Klint 348
    Paintings for the Temple 348
    Tree of Knowledge 349
Himalayan bath salts 282
Hippocrates 65, 66
hips 102
Historia animalium (*History of Animals*) (Aristotle) 123
Historia Plantarum ('Enquiry into Plants') (Theophrastus) 22, 66, 89
Historia Plantarum (*The History of Plants*) (Ray) 123, 133
HMS *Beagle* 91
HMS *Endeavour* 76, 77, 323
HMS *Investigator* 79
Hockney, David 355
    The Arrival of Spring in Woldgate, East Yorkshire 355
hogweed (*Heracleum*) 124
Hokusai, Katsushika 69, 355
    Big Flower series 68
    The Great Wave off Kanagawa 355
    Peony and Canary 355
    Poppies 355
    Thirty-six Views of Mount Fuji 355
holly, European (*Ilex aquifolium*) 51, 98, 138, 172

hollyhock (*Alcea rosea*) 166, 167
home remedies 300
    All-purpose Cleaner 303
    Aromatic Reed Diffuser 302
    Herbal Cleaning Spray 304
    Lemon Magic 303
    Scented Candle 302
honesty (*Lunaria annua*) 101, 102, *137*, 172
honey 157, 198, 203, 210, 221, 225, 229, 249, 250, 251, 254, 262, 263, 275, 277, 278, 283, 287, 292, 295
honey mesquite (*Prosopis glandulosa*) 211
honeysuckle (*Lonicera*) 214, *215*
Hooke, Robert 58, 91
Hooker, Sir Joseph 75, 76
Hooker, Sir William 75, 133
hornbeam, European (*Carpinus*) 101
hornworts (*Anthocerotophyta*) 24, 25, 30, 33
horse chestnut (*Aesculus hippocastanum*) 97, 98, 179
horseradish (*Armoracia rusticana*) 138, 277
horsetails (*Equisetophyta*) 24, 25, 28, 33, *34*, 36
horticultural plants 35
Horticulture magazine 76
Horticulture Week journal 323
Hortus Botanicus, Amsterdam 69, 343
Hortus Botanicus, Leiden 75
Hortus Cliffortianus 79
Hortus Eystettensis (Besler) 75, 76, 319
hosta (*Hosta*) 97, 138
hot lips (*Psychotria elata*) 99
House of Hackney, London *154*
houseleek *see Sempervivum*
houseplants 35, 38, 42, 45
Howl (Ginsberg) 181
Hoyle, Matt 212
huckleberry (*Gaylussacia baccata*) 51
Hum-Ishu-Ma (Mourning Dove) 256
Humboldt, Alexander von 84
Hyacinthoides non-scripta (British bluebell) 51
hybrid plants 134
Hyde Park, London 69
hydrangea (*Hydrangea*) 97, 138
Hydrangea macrophylla 133
Hymenopus coronatus (orchid mantis) *117*
Hypericum perforatum (St John's wort) 281, 296
Hypnum iponens (sheet moss) 31
hyssop 214

iceberg rose (*Rosa* 'KORbin') 134
Inky Leaves 320, *321*, 415
Ilex aquifolium (European holly) 51, 98, 138, 172

Illicium anisatum (Japanese star anise) 49
Illicium verum (star anise) *49*
Illustrationes florae Novae Hollandiae (Bauer) 79
Illustrations of the Natural Orders of Plants with Groups and Descriptions (Twining) 158, *159*
India grass (*Sorghastrum nutans*) 55
Indian banyan (*Ficus benghalensis*) 93
Indian mango (*Mangifera indica*) 97
Indian timber bamboo (*Bambusa tulda*) 51
indigo, blue wild (*Baptisia australis*) 55
indigo, true (*Indigofera tinctoria*) 176, 177
inflorescences 65
Ingenhousz, Jan 86, 111
inspirational planting 152
installation *see* sculpture and installation
International Botanical Congress, Paris (1867) 124
International Code of Nomenclature (ICN) 124, 126
International Plant Names Index (IPN) 124
International Union for Conservation of Nature (IUCN) 181
interrupted fern (*Osmunda claytoniana*) 35
Ipomoea batatas (sweet potato) 93
Iridaceae (iris family) 129
iris, bearded (*Iris germanica*) 152
iris (*Iris*) 42, 94, 115, *312*
Irish moss 27
Isatis tinctoria (woad) 176, 177
Isle of Man cabbage (*Coincya monensis* subsp. *monensis*) 134
Isoetes lacustris (lake quillwort or Merlin's grass) 34
Italian alder tree (*Alnus cordata*) 108
ivy, Atlantic (*Hedera hibernica*) 141
ivy (*Hedera helix*) 51, 93, 97, 138, 172

Japanese anemone (*Anemone x hybrida* 'September Charm') 155
Japanese gourd (calaba) 212
Japanese maple (*Acer palmatum*) 141, 175
Japanese rose with white flowers (*Rosa rugosa f. alba*) 134
Japanese sago palm (*Cycas revoluta*) 38
Japanese star anise (*Illicium anisatum*) 49
Jardin de Cactus, Lanzarote 70, 372
Jardin des Plantes de Montpelier 75

Jardin des Plantes, Paris 69
Jardin du Roi, Versailles 313
Jardin Majorelle, Marrakech 70, 154
Jarman, Derek 70, 342
jasmine, common (*Jasminum officinale*) 45, 138, 152, 205, 282, 283, *300*
jasmine (Spanish, Royal, Catalan and others) (*Jasminum grandiflorum*) 301
Jefferson, Thomas 70
Jekyll, Gertrude 70, 72, 152, 381
Jencks, Charles Alexander *382*
Jencks, Maggie Keswick 382
Johannsen, Wilhelm 143
'jointfirs' (*Ephedra*) 39
jojoba 271, 284, 292, 293
Joshua tree (*Yucca brevifolia*) 56, *342*
A Journey to Avebury (Jarman) *342*
Juglans cinerea (white walnut) 164
Juglans regia see English walnut
juices *see* drinks and bowls
Juncaceae (sedges) 52, 56
junegrass (*Koeleria macrantha*) 55
Jung, Heinrich 324
Jung Koch and Quentell wall charts *324*
Jungalow style 379
juniper (*Juniperus communis*) 38, 171, 246, 281, 282
Juniperus virginiana (red cedar) 38

Kahlo, Frida 70, 154, *173*, 348
Self-Portrait with Monkey 348
Self-Portrait with Thorn Necklace and Hummingbird 348
Kahn, Fritz 314
Kalanchoe daigremontianum (Mexican hat plant) *114*
kale (*Brassica oleracea* cultivar) 72, 229, 230, 250
Kali tragus (prickly Russian thistle, common tumbleweed) 56
Kandinsky, Wassily 84
kangaroo paw (*Anigozanthos manglesii*) 55
kaolin 292
Karnak, Temple of (Egypt) 66
kava-kava (*Piper methysticum*) 46
Keats, John 70
keck *see* cow parsley
kefir 292
Kelly, Commander Scott 83
Kelly, Ellsworth 351
Tropical Plant, St Martin 350
kelp 27, 299
Kent, William 72, 149
Kerala, India 60
Kesseler, Rob 80, *81*, 83, 308, *334*, 335
Kew *see* Royal Botanic Gardens, Kew
Kew's *State of the World's Plants* report 63, 261, 392
Kiker, Jenny *370*, *371*
Living Pattern 370

Monstera Deliciosa or *Maidenhair Fern* 370
Kingdom Animalia *see* Animal Kingdom
Kingdom Fungi *see* Fungi Kingdom
Kingdom Plantae *see* Plant Kingdom
kiwi 248
Klein, Carol 381
Klint *see* Hilma af Klint
Kngwarreye, Emily Kame 354
Earth's Creation 354
Knight, Nick 336, 376
Koch, Gottlieb von 324
Koeleria macrantha (junegrass) 55
Köhler, Franz Eugen 168
Kordes, Reimer 134
Kusama, Yayoi 357
All the Eternal Love I have for the Pumpkins 357
Infinite Mirrors 356
Kwan, Jeong 184
Kwang Ho Lee 347
Kyoto Garden, Holland Park (London) 372

La Casa Azul, Mexico 348
Labrador tea (*Ledum groenlandicum*) 52
Laburnum anagyroides (common laburnum) 97
laburnum, common (*Laburnum anagyroides*) 97
lady fern (*Athyrium filix-femina*) 35, 51, 175
lady fern with stems, 'Lady in Red' (*Athyrium angustum f. rubellum*) 134
Lady Gaga 130
lady's bedstraw (*Galium verum*) 176
lady's mantle (*Alchemilla vulgaris*) 171
lake quillwort (or Merlin's grass) (*Isoetes lacustris*) 34
Lamiaceae (mint/deadnettle family) 129
land plants (*Embrophytes*) *24*, 28
larch (*larix dedicua*) 38, 101
larix dedicua see larch
Lathyrus latifolius (perennial peavine) 138
Lathyrus odoratus 'Matucana' (sweet pea 'Matucana') 141
Lathyrus odoratus (sweet pea) 10, 45, 98, 101, 137
Lauraceae (laurel family) 129
Laurales 126
laurel family (Lauraceae) 126, 129
Laurent, Anna 324
Laurus nobilis (bay laurel) 46
lavandin (*lavendula x intermedia*) *64*
lavender, English (*Lavandula angustifolia*) 138, *150*, 151, 291, 292, 293, 302, 303, 304

lavender (*Lavendula* spp.) 11, *12*, 45, *150*, 205, 241, 261, 282, 284
lavender, spike (*Lavendula latifolia*) 98
Lavendula spp. (lavender) 11, *12*, 45, 205, 241, 261
Law, Rebecca Louise 308, *364*, 365
The Beauty of Decay I 365
The Flower Garden Display'd 364
Life in Death 365
Still Life 365
Lawler, Charlie 378
Lawless, Julia 284
Le Cuisinier François (Sieur de La Varenne) 190
leatherleaf fern (*Rumora adiantiformis*) 35
leaves 65, 80, 97, 168, 212
Leaves of Grass (Whitman) 47
Ledum groenlandicum (Labrador tea) 52
Lee, Kwang-Ho 308, *346*, 347
leek (*Allium ampeloprasum* cultivar) 90, 229
legumes 65, 102
leiocarpus (with smooth fruits) 133
Leipzig Botanical Garden 75
lemon balm (*Melissa officinalis*) 282, *286*, 287, 296, *297*, 301, 304
lemon (*Citrus limon*) 97, 102, 190, 194, 198, 204, 212, 218, 226, 230, 231, 233, 237, 242, 246, 264, 301, *303*, 304, *305*
lemon, Meyer (*Citrus x limon*) 134
lemon verbena (*Aloysia citrodora*) 241
lemongrass (*Cymbopogon citratus*) 288, 301
Leninism Under Lenin (Liebman) 367
lentils 65, 102
Leo Castelli Gallery, New York 354
Leonardo da Vinci 135
Lepidozamia peroffskyana ('dinosaur food') 38
lettuce 65, 102, 212, 250
Leucanthemum vulgare (oxeye daisy) 55, 172
Levack, Samuel *342*
Lewandowski, Jennifer *342*
L'Héritier de Brutelle, Charles Louis 313
lianas (*Gnetum*) 39
Liatris (blazing star) 55
Liber Abaci (Book of Calculations) (Fibonacci) 311
Liberman, Alexander 336
lichen, reindeer (*Cladonia rangiferina*) 52
lichens 52
licorice (*Glycyrrhiza glabra*) 273

Liebman, marcel 367
Liliaceae (lily family) 86, 129
Liliales 126
Lilium (lily) 42
lily, desert (*Hesperocallis*) 56
lily family (Liliaceae) 86, 129
lily (*Lilium*) 42, 126, 336
lily turf (*Liriope*) 94
lime (*Citrus*) 240, 241, 246
linden (*Tilia x europaea*) 51, *106*, 164, *165*, 281
Lindley, John 49, *322*
Lindley Library, London 326
lingonberry (*Vaccinium vitis-idaea*) 52
Linnaeus, Carl 58, 70, 75, 76, 79, 91, 121, *122*, 123, 124, 126, 127, 129, 130, 143, 313
Liquidambar (seetgum tree) 51
Liriodendron (tulip tree) 46
Liriope (lily turf) 94
Litchi chinensis (lychee) 101
little bluestem (*Schizachyrium scoparium*) 55
liverworts (*Marchantiophyta*) *24*, 25, *30*, 33, 52
Livingstone, David 326
Lodoicea maldivica (coco de mer) 42, 101
Lois de la nomenclature botanique 124
Lonicera (honeysuckle) 214, *215*
Loose Leaf 378
Loose Leaf: Plants Flowers Projects Inspiration (Wona Bae & Lawler) 378
lords-and-ladies (*Arum maculatum*) 98
lotus (*Nelumbo nucifera*) 94, 101, *114*, 340
lotus, white Egyptian (*Nymphaea lotus*) 49
Loudon, J.C. 72
love-in-the-mist (*Nigella damascena*) 11, 137, 152, *332*
Lucerne (*Medicago sativa*) 138
Lunaria annua (honesty) 101, 102, *137*, 172
Lunularia cruciata (crescent-cup liverwort) 30
Lunzer, Alice 135
lupin (*Lupinus* 'Masterpiece') 158
Lupinus arcticus (Arctic lupine) 52
lychee (*Litchi chinensis*) 101
lycophytes 28
Lycopodiophyta (club mosses) *24*, 25, 33, 34, 36
Lycopodium (ground pine) 34

Mabey, Richard 70, 161
Macaroni penguins in Zavodovski Island (Salgado) *338*
mace/nutmeg (*Myrstica fragrans*) 46, 254
McEwen, Rory 79, 323, 328
Old English and Striped Tulip 'Sam Barlow' 308

Machaeranthera tanacetifolia (Tahoka daisy) 55
Mackintosh, Charles Rennie 81
McQueen, Alexander 376
McQuillan, Alan 187, 244
McVicar, Jekka 151
Madagascar palm (*Tahina spectabilis*) 41
madder, common (*Rubia tinctorum*) 176
Magnol, Pierre 129, 133
magnolia family (Magnoliaceae) 129
Magnolia grandiflora (bull bay or southern magnolia) 47, 51, 133
Magnolia liliiflora (formerly *Magnolia discolor*) 196
Magnolia macrophylla (bigleaf magnolia) 133
magnolia (*Magnolia*) 9, 46, 97, 132
Magnolia stellata (star magnolia) 133
Magnoliaceae (magnolia family) 129
Magnoliales 46
Magnoliids 24, 25, 46
mahogany (*Swietenia macrophylla*) 51
maiden grass (*Miscanthus sinensis*) 138
maidenhair fern (*Adiantum*) 35
maize (*Zea mays* subsp. *mays*) 42, 65, 93, 98, 137
Majorelle, Jacques 70
Makoto, Azuma 72, 314, 358, 359
    Exobiotanica (Botanical Space Flight) 358, 359
    In Bloom 359
    Shiki 358
    World Flowers 359
    World Flowers II 359
male fern (*Dryopteris filix-mas*) 35, 138, 155
Malki Museum, Morongo Reservation (California) 211
mallow family (Malvaceae) 129
Malpighi, Marcello 317
Malus domestica 'Braeburn' (Braeburn apple) 134
Malus pumila (apple) 65, 70, 102, 135, 205, 217, 250, 251, 374
Malus sulvestris (crab apple) 163
Malvaceae (mallow family) 129
Mama Medicine's Ritual Bath 16, 286, 287
The Man Who Planted Trees (Giono) 149
Manchantia polymorpha (umbrella liverwort) 30
mandarin (*Citrus reticulata*) 134, 245, 301, 302
Mangifera indica (Indian mango) 97
mango butter 293
mango, Indian (*Mangifera indica*) 97

mangoes 65, 250
mangrove, red (*Rhizophora mangle*) 93, 101
mangrove tree (*Rhizophora*) 56
Manihot esculenta (casava) 93
manioc 65
Manrique, César 70, 372
maple (*Acer*) 45, 51, 101, 102
maple, field (*Acer campestre*) 97
maple, Japanese (*Acer palmatum*) 141, 175
maple, silver (*Acer saccharinum*) 56
maple, sugar (*Acer saccharum*) 168
Mapplethorpe Flora: The Complete Flowers (2016) 336
Mapplethorpe, Robert 80, 336
Marchantiophyta (liverworts) 24, 25, 30, 33, 52
Margaret Mee in Search of Flowers of the Amazon Forest (Mee) 328
marigold, desert (*Baileya*) 56
marigold, pot (*Calendula officinalis*) 171, 282, 286, 287, 291, 292
marjoram 207, 284
The Marksman Public House and Dining Room, Columbia Market (London) 379, 386
Marley, Christopher 370
marsh samphire (*Salicornia europaea*) 168
marshmallow (*Althaea officinalis*) 167, 291
Masai Mara, Kenya 54
materials plants 51
Matisse, Henri 345
    The Parakeet and the Mermaid 345
Matricaria chamomilla (German chamomile) 272, 273
Matthiola incana (Brompton stock) 137
Mayer, Julius Robert 111
meadow flowers 138
meadow rue, Chinese (*Thalictrum delavayi*) 152, 153
meadowgrass (*Poa*) 93
Medicago sativa (Lucerne) 138
Medical Botany: or Illustrations and descriptions of the medicinal plants (Stephenson & Churchill) 184, 278
Medicinal Herbs (Gladstar) 274
medicinal plants 45, 51, 65, 171, 272, 273
Medicinal Plants (Koehler) 66
Medizinal-Pflonzen of Medicinal Plants (Köhler) 168
Mee, Greville 328
Mee, Margaret 79, *328*
Meilland, Francis 124
Melaleuca alternifolia (tea tree) 291, 292, 303, 304
Melissa officinalis (lemon balm) 282, 286, 287, 296, 297, 301, 304
Meller, Gill 225

Mendel, Gregor 58, 143
Mendel's Principles of Heredity: a Defence (Bateson) 142, 143
Mentha genus (garden mint, peppermint) 130, 241, 303, 304
Mentha x piperita (peppermint) 273, 282, 293, 302
Merlin's grass (or lake quillwort) (*Isoetes lacustris*) 34
Mesozoic period 37
mesquite tree *see* honey mesquite
Methodus Plantarum Sexualis (Sexual System of Plants) (Linnaeus) 313
Metropolitan Museum of Art, New York 351
Metrosideros collino (springfire shrub) 77
Metrosideros excelsa (New Zealand Christmas tree) 323
Mexican hat plant (*Kalanchoe daigremontianum*) 114
mexicana (from Mexico) 133
Meyer lemon (*Citrus x limon*) 134
Micrographia (Hooke) 91
The Middle at the Wallace Collection (exhibition, 2017) 352
milkweed (*Asclepias*) 55
Millennium Seed Bank, Kew 83, 101, 335
Miller, Philip 75, 79, 130
millet 65
mimosa (*Acacia dealbata*) 97
Mimosa pudica 90, 118
Mineralogical & Geological Museum, Harvard University 375
mint (*Mentha*) 130, 241, 304
mint/deadnettle family (Lamiaceae) 129, 214
Miscanthus sinensis (maiden grass) 138
Miscanthus sinensis 'Silberfeder' (silver grass) 158
Missouri Botanic Garden 69
mistletoe, European (*Viscum album*) 51
mitochondria 104
Mnium hornum (carpet moss) 51
Mobasser, Ali 379
Modern Photography magazine 354
Mondrian and Colour (exhibition, 2014) 351
Mondrian Flowers (Shapiro) 351
Mondrian, Piet 351
    Chrysanthemum 351
Monet, Claude 58, 69, 70, 71, 152, 345
    The Waterlily Pond: Green Harmony 344
monkey mask plant (*Monstera obliqua*) 96

monkey puzzle tree (*Araucaria araucana*) 38
monk's hood cactus (*Astrophytum ornatum*) 141
Monocots (one seed leaf & one-aperture pollen) 24, 42, 43, 126
Monstera deliciosa (Swiss cheese plant) 42, 43, 51, 134, 141, 345, 378
Monstera obliqua (monkey mask plant) 96
montbretia (*Crocosmia*) 94
Montgomery, Andrew 225
Montreal World's Fair (1967) 383
Morel Tales (Fine) 161
morphine 261
Morris, William 70, 80, 81, 174, 308, 368, 369
    Chrysanthemum 368, 369
    Daisy 369
    Strawberry Thief 81
    Trellis 369
Morus alba (white mulberry) 62
moss, Arctic (*Calliergon giganteum*) 52
moss, carpet (*Mnium hornum*) 51
moss, knight's plume (or ostrich-moss) (*Ptilium crista-castrensis*) 52
moss, peat (*Ptilium ceista-castrensis*) 52
moss, Spanish (*Tilandsia usneaides*) 57
mosses (*Bryophyta*) 24, 25, 31, 33, 52
moth repellant 38
mother die *see* cow parsley
MR Studio London 388
mugwort 282
Muir, John 53, 117, 181
    My First Summer in the Sierra 35
mulberry, white (*Morus alba*) 62
multicaulis (with many stems) 133
mung bean (*Vigna radiata*) 93
Muray, Nickolas 348
Murayama, Macoto 79, 308, 330, 331
Musa acuminata (cultivated banana) 42, 102, 138, 249, 254
Musa (wild bananas/plantain) 51, 139
Musaceae (banana family) 129
Museum of Comparative Zoology, Harvard University 375
Museum van Loon, Amsterdam 308
mushrooms, reishi 278
Musk rose (*Rosa moschata*) 124
mustard seeds 204
Myosotis arvensis (forget-me-not) 98
Myrosotis sylvatica (wood forget-me-not) 137

myrrh (*Commiphora myrrha*) 65, 281
Myrstica fragrans (nutmeg/mace) 46, 254
myrtle, common (*Myrtus communis*) 151
Myrtus communis (common myrtle) 151
Mysteries of the Unseen World (Schwartzberg) 343
Mystical England (Levack & Lewandowski) *342*

nanus (dwarf) 133
Narcissus (daffodil) 42, *92*, 94, 115, 138, 172
Nardostachys jatamansi (spikenard) 281
NASA 137
Nasella pulchra (purple needlegrass) 55
nasturtium (*Tropaeolum*) 45, *185*, 191, 205
National Endowment for Science, Technology and the Arts (NESTA) 335
natural dyes 176
Natural History of Carolina, Florida and the Bahama Islands (Catesby) 76, 79
Natural History Museum, London 336
Naturalis Historia (Pliny the Elder) 65, 66, 187
The Nature Conservancy 83
nature, living with 180–1
The Nature-Printer (Östlund) 373
nature's larder 184, 186–7
nectar 115
nectarine/peach (*prunus persica*) 45, 65, 203
Nelumbo nucifera (lotus) 94
Neolithic period 65, 66
Nepenthaceae (tropical pitcher plants) 51
Nepenthes northiana (pitcher plant) *119*
Nephrolepis exaltata (Boston fern) 35
neroli (*Citrus aurantium*) 291
nerve plant (*Fittonia argyraneura*) *110*
nettle (*Urtica dioica*) 97, 168, 189, 190, 194, 273, *294*, 295
New American Garden 382
New York Botanical Garden 75, 329, 348
New York Central Park 58, 69, 381
The New York Times 329
New Zealand Christmas tree (*Metrosideros excelsa*) *323*
News About Flowers (Benjamin) 333
Niagara Reservation 381
Nicolson, Harold 381
Nicotiana tabacum (tobacco plant) 62

Nigella damascena (love-in-the-mist) 11, 137, 152
Nigella damascena Spinnerkotpf 332
nightshade, black (*Solanum nigrum*) 130
NILS-UDO 308, *362*, 363
  Birch Tree Planting 363
  Clemson Clay Nest 363
  Crack in Lava, Flower Petals Called Tongues of Fire 362
nocturna (nocturnal) 133
nomenclature *see* binomial nomenclature
non-flowering seed plants (gymnosperms) 24, 25, 36, 37, 101
non-green algae 27
non-vascular land plants (Bryophytes) 24, 28, 29
North American redbud (*Cercis canadensis* 'Forest Pansy') 152
North, Marianne 47, 79, 118, *119*, *322*
Norway spruce (*Picea abies*) 38
Novalis 320
Nozedar, Adele 187, 226
nutmeg/mace (*Myrstica fragrans*) 46, 254
nuts 65, 102, 158, 164, 206, 271
Nympaea alba var. *Rosea 48*
Nymphaea alba (white water lily) 49
Nymphaea caerulea (blue Egyptian water lily) 49
Nymphaea candida (dwarf water lily) *114*
Nymphaea gigantea (giant water lily) 49, *315*
Nymphaea lotus (white Egyptian lotus) 49
Nymphaea mexicana (yellow water lily) 49
Nymphaea odorata (fragrant white water lily) 56, 57
Nymphaeales 49

oak, English (*Quercus robur*) 51, 97
oak (*Quercus*) 45, 62, 93, 101
oak, white (*Quercus alba*) 51, 97
oatmeal 264, 292
oats (*Avena sativa*) 249, 254
Ocimum basilicum (sweet basil) 141
Oddsberg, Josefina 157
odoratus (fragrant) 133
Oehme, Wolfgang *382*
Oenothera biennis (evening primrose) 171
officianlis (with herbal uses) 133
oil palm (*Elaeis*) 62
oils 65
O'Keefe, Georgia 348, 370
  Black Iris VI 348
  Flower of Life II 348
  Jimson Weed 348

olive (*Olea europaea*) 94, *95*, 102, 152
Olmsted, Frederick Law 69, *381*
Olneya tesota (desert ironwood) 56
Olsson, Nina 187, 209
Olympic National Park, Washington State 289
On the Causes of Plants (Theophrastus) 89
On the Origin of Species (Darwin) 70, 76, 86, 91, 116, 123
one seed leaf & one-aperture pollen (Monocots) 24
onion (*Allium cepa*) 42, 65, 94, 209, 234, 273, 277
onion, wild (*Allium canadense*) 130
opium poppy (*Papaver somniferum*) 62, 65, 261
Opuntia ficus-indica (prickly pear) 45, 56, 94, *95*, 167
Opuntia santa-rita (Santa Rita prickly pear) *57*
orange, bitter (*Citrus aurantium*) *64*, 243
orange, blood (*citrus x sinensis*) 220, 229, 241
orange, sweet (*Citrus sinensis*) *64*, 101, 237, 242, *245*, 246, 304
oranges 65, 191, 221, 292, 304
orchid mantis (*Hymenopus coronatus*) 117
Orchidaceae family 133
orchids (Orchidaceae) 40, 42, 51, 93, 101, *133*, 311, *319*, 336
oregano (*Origanum vulgare*) 138
organ pipe cactus (*Stenocereus thurberi*) 56
orientalis (eastern) 133
ornamental gardens 60
Orto Botanico di Pisa 58, 69
Oryza sativa (Asian rice) 42, 56, 62, 102, 137
Osmunda claytoniana (interrupted fern) *35*
Östlund, Pia *373*
Ottolenghi, Yotam 184
Oxalidaceae (wood sorrel family) 129
Oxalis acetosella (wood sorrel) 97
oxeye daisy (*Leucanthemum vulgare*) 55, 172

Paeonia (peony) 45, 80, 94, 102, 138
painting and drawing 344–55
Paleolithic period 65
palm, acai (*Euterpe precatoria*) 51
palm, betel (*Areca catechu*) *136*
palm, date (*Phoenix dactylifera*) *239*
palm, oil (*Elaeis*) 62
palm, parlour (*Chamaedorea elegans*) 141
palm, walking (*Socratea exorrhiza*) 51

palmate leaflet 97
Palmeral, House of Hackney *154*
Palo santo (*Bursera graveolens*) 281
pampas grass (*Cortaderia selloana*) 55
panicle 98
Panicum virgatum (switch grass) *139*, 157
Pantone 155
Papaver rhoeas (common, corn or field poppy) 11, 45, 101, 137, 157, 172
Papaver somniferum (opium poppy) 62, 65, 261
Papaveraceae (poppy family) 129, *318*
paper birch (*Betula papyrifera*) 51
Paperwork and the Will of Capital (Simon) *342*
Pará rubber tree (*Hevea brasiliensis*) 62
Parkinson, Sydney 76, 77, *323*
parlour palm (*Chamaedorea elegans*) 141
parsley, edible (*Petroselinum crispum*) 124, 129, 137, 190, 207, 209
parsnip (*Pastinaca sativa*) 93, *224*, 225, 229
passion flower (*Passiflora laurifolia*) 44, *99*
Pastinaca sativa (parsnip) 93, *224*, 225, 229
patchouli (*Pogostemon cablin*) 291
Paulley, James 314
paw paw, common (*Asima triloba*) 46
Pawnee National Grassland, Colorado 54
Paxton, Sir Joseph 49, 142
Paxton's Flower Garden (Paxton) 49
pea (*Pisum sativum*) 65, 102, *142*, 143, 189, 194, *195*
peace lily (*Spathiphyllum wallisii*) 42, 111, 151
peace rose (*Rosa* 'Madame A. Meilland') 124
peach/nectarine (*prunus persica*) 45, 65, 203
pear cactus (*Opuntia ficus-indica*) 56
pear (*Pyrus*) 97, 217, 219
peat moss (*Sphagnum*) 31
peavine, perennial (*Lathyrus latifolius*) 138
pecan (*Carya illinoiensis*) 51, 164, 206
Pedder-Smith, Rachel *326*, *327*
pedicel 94
peduncle 94
Peking spurge (candelabra tree, poinsetta) (*Euphorbia ingens*) 55, 130
Pelargonium graveolens (rose geranium) 291, 295
Penn, Irving 80, 336

INDEX 409

Pennisetum purpureum (elephant grass) 55
peony (*Paeonia*) 45, 80, 94, 102, 138, 336
pepo 102
pepper *see* black pepper
peppermint (*Mentha x piperita*) 273, 282, 293, 302, 303
peregrinus (exotic) 133
perennial 138
perennial peavine (*Lathyrus latifolius*) 138
perfectus (complete) 133
The Perfumier and the Stinkhorn (Mabey) 161
periwinkle, rosy (*Catharanthus roseus*) 51
Persea americana (avocado) 46, 65, 118, *248*, 249, 250
persimmon (*Diospyros kaki*) 97
petiolated 97
Petroselinum crispum (edible parsley) 124, 129, 137
Phaeoceros laevis (smooth hornwort) 30
Phaeophyta (brown algae) *24*, 27
Phalaenopsis genus (moth and cultivated orchids) 130
Phanerogamae (floral plants) 126
Phaseolus coccineus (runner bean) 101
Phillips, Roger 80
Philosophia Botanica (Linnaeus) 129
phloem 107
phlox family (Polemoniaceae) 129
Phlox (phlox) 138
Phoenix dactylifera (date) 229
Photographs of British Algae: Cyanotype Impressions (Atkins) 80, 333
photography and film 80, 322–43
photosynthesis *24*, 27, 111
Phragmites australis (common reed) 179
Physeterostemon 329
physic gardens 75
Picea abies (Norway spruce) 38
Picea (spruce) 52
Picturesque style 72
pignut (*Carya glabra*) 164
Pinax ginseng/Eleutherococcus sentilocosus (ginseng) 273
Pinax theatri botanici (Illustrated Exposition of Plants) (Bauhin) 123
Pincushion flower seed (*Scabiosa cretica*) *81*
pine (incl. pine nut, pine nut) (*Pinus*) 52, 93, 94, 97, 164, 179, 190, 209, 238
pine (*Pinus sylvestris*) 288, 301
pineapple (*Ananas comosus*) *42*, 250
pinnate leaflet 97
Pinophyta see conifers
Pinus (hard pine) 38

Pinus longaeva (bristlecone pine) 138
Pinus (pine incl. pine nut, pine cone) 52, 93, 94, 97, 164, *178*, 179, 206
Piper methysticum (kava-kava) 46
Piper nigrum (black pepper) 46, *47*, 143, 190, 194, 198, 203, 204, 209, 210, 225, 233, 234, 241, *285*
Piperales 46
pistachio 127
Pisum sativum (pea) 65, 102, *142*, 143, 194, *195*
pitcher plant (*Nepenthes northiana*) 119
pitcher plant, tropical (*Nepenthaceae*) 51
Pitton de Tournefort, Joseph 126, 127
plane tree 45
plant cell *88*
Plant Kingdom (Kingdom Plantae) 22, *24*, 25, 27, 80, 83, 86, 91, 121, 123, 326, 342
The Plant magazine 76
Plant Power (exhibition, 1992) 336
Plantago foliis ovato-lanceolatis pubescentibus, spica cylindrica, scapo tereti (now known as *Plantago media*) 124
Plantago lanceolata (ribwort plantain) 171, 291
Plantago major (greater plantain) 291
plantain, greater (*Plantago major*) 291
plantain (*Musa*) 51, 65
plantain, ribwort (*Plantago lanceolata*) 171, 291
The Planthunter magazine 76
Plants: From Roots to Riches (Willis) 314
plants
  adaptation 116
  art forms 311
  early writers on 22
  growth and decay 116
  healing powers 260–1
  history and development 22
  naming of 124
  ornamental 42, 45, 51
  and people 58, 59
  society-shaping 62
  tender 141
  what they know 118
Plants of the World: An Illustrated Encyclopedia of Vascular Plants (Kew Royal Botanic Gardens) 80
Plato 311
Platonicus, Apuleius 79
Platycerium see staghorn fern
Pleasure Garden magazine 76
Pliny the Elder 65, 66, 130, 168, 187
Poa (bluegrass/meadowgrass) 93

Pogostemon cablin (patchouli) 291
poinsetta (Peking spurge, candelabra tree) (*Euphorbia ingens*) 55, 130
Polemoniaceae (phlox family) 129
Pollan, Michael, *In Defense of Food* 184
Pollard, Mike 388
Pollen: The Hidden Sexuality of Flowers (Kesseler) 335
pollen, pollination 36, 41, 42, 98, *108*, *109*, 115, 157, *334*, 335
Polystichum munitum (western sword fern) 35
Polytrichum (haircap moss) *31*
pomegranate (*Punica granatum*) 241, 242
pomes 102
Pop Art 360
poppy, Californian (*Eschscholzia california*) *174*, 175
poppy, common (or corn or field) (*Papaver rhoeas*) 11, 45, 101, 137, 157, 172
poppy family (Papaveraceae) 129, *318*
poppy, opium (*Papaver somniferum*) 62
poppy seeds 102
Portland rose (*rosa* 'Comte de Chambord') 151
Postmodernism 382
pot marigold (*Calendula officinalis*) 171, 282, *286*, 287, 291, 292
potato family (Solanaceae) 129, 130
potato (*Solanum tuberosum*) 45, 65, 94, 102, 115, 130, 189, *208*, 209
Potter, Beatrix 76
Power, Nancy Goslee 381
The Practice of the Wild (Snyder) 181
prairie dropseed (*Sporobolus heterolepis*) 55
prairie sunflower (*Helianthus petiolaris*) 55
prairie-clover (*Dalea purpurea*) 55
Pratt, Anne 193
pressed botanicals 174
prickly pear cactus (*Opuntia ficus-indica*) 45, 56, 94, *95*, 167
prickly pear, Santa Rita (*Opuntia santa-rita*) 57
prickly Russian thistle (*Kali tragus*) 56
Priestley, Joseph 86, 111
primrose, evening (*Oenothera biennis*) 171
primrose family (Primulaceae) 129, 205
Primula veris (cowslip) 94
Primulaceae (primrose family) 129
The Private Life of Plants (TV) 80

Prodromus historiae generalis plantarum (Magnol) 129
progymnosperms 36
propolis 299
Prosopis glandulosa (honey mesquite) *211*
prostrate rosemary (*Rosmarinus officinalis* 'Prostratus Group') 158
Proteales 45, *322*
Protected Planet 181
prunes 237
Prunus armeniaca (apricot) 249, 251
Prunus avium (wild cherry) 163
Prunus cultivars (cherry) 203
Prunus domestica subsp. *insititia* (damson) 217
Prunus dulcis syn. *Prunus amygdalus* (almond) 194, 206, 219, 237, 249, 251, 253
prunus persica (peach/nectarine) 45, 65, 203
Prunus spinosa (sloe) 163
Psychotria elata (hot lips) 99
Ptak, Claire 184, 187, 201
Pteridium (bracken) 35
Pteridophyta (ferns) *24*, 25, 28, 33, 35, 36, 88
Ptilium crista-castrensis (knight's plume or ostrich-moss) 52
Ptilota plumosa 333
pumpkin (*Cucurbita maxima*) 65, 94, 206, *216*, 217, 237
Punica granatum (pomegranate) 241
Purbrick, Lentil and Matt 187, 237
purple coneflower (*Echinacea purpurea*) *54*, 55, 138, *139*, 157, 172, 273
purple needlegrass (*Nasella pulchra*) 55
purpletop verbena (*Verbena bonariensis*) 138
purpuratus (purple) 133
Pursell, Dr JJ 278
pussy willow (*Salix*) 72, 97, *178*, 179
Putorana Plateau, Siberia 52, *53*
Puya rainmondii (Queen of the Andes) 41
Pyropia 27
Pyrus (pear) 97, 217
Pythagoras 311

Queen of the Andes (*Puya rainmondii*) 41
Queen Anne's lace *see* wild carrot
Quentell, Dr Friedrich 324
Quercus (acorn) 164
Quercus alba (white oak) 51, 97
Quercus (oak) 45, 62, 93, 130
Quercus robur see English oak
quillworts 33, 34
quince (*Cydonia oblonga*) 217, 233, *317*, *334*

quinine (*Chincona officinalis*) 51, 62, 261
Quinn, Marc 360
  The Engine of Evolution 360
  Garden 360
  Nurseries of El Dorado 360
  Out of Body 360

raceme 98
racemosus (with flowers in racemes) 133
radicchio (red chicory) *232*, 233
radish (*Raphanus raphanistrum* subsp. *sativus*) 93, 189, 230
Rafflesia arnoldii (corpse lily) 51
rain forest *50*, 51, 80, *289*
Rake's Progress magazine 76
ramsons/wild garlic (*Allium ursinum*) 51, 167, *192*, 193
rank order 126
Ranunculaceae (buttercup family) 45, 126, 129, 311
Ranunculates 45, 126
Ranunculus genus (buttercup, spearwort, water crowfoot) 130
rapeseed 65
Raphanus raphanistrum subsp. *sativus* (radish) 93, 189
raspberry (*Rubus*) 102, 204, 250
Ratibia 54
rattan palm (rattan genera) 51, 302
Ray, John 70, 123, 133
Rebecca Louise Law: Life in Death (2017) 364
recipes
  Blackberry Jam 218
  Blood Orange Marmalade 231
  bouquet garni 190
  Bunch of Herbs Sheep's Milk Ice Cream 214–15
  Cranberry Jelly 231
  Crystallised Flowers 205
  Elderberry Flu Remedy 226
  Fig Leaf Ice Cream 201
  Garlic Scape Mimosa 198
  Hand Roll 212–13
  Hazelnut Butter 217
  Herbed Potatoes 209
  Kale Crisps 230
  Kimchi 230
  Magnolia Petals Lightly Pickled in Elderflower Vinegar 197
  Mango Coconut Raw Oat Porridge 254, *255*
  Native 'Power Food' (Energy Bar) 238
  Orange, Fennel Seed and Almond cake 221
  Pea and Nettle Gnudi with Braised Fennel Almond Crema *16*, 194, *195*
  Pickled Cucumber 205
  Poached Pears 217
  Poor Man's Capers 191
  Raw Cabbage, Fennel and Pecorino Salad 222
  Rawow Chocolate Butter 253
  Red Chicory, Pickled Quince, Goat's Curd and Walnut Salad 233
  Rhubarb Compote 191
  Roast Parsnips with Blackberries, Honey Chicory and Rye Flakes *224*, 225
  Rosehip Syrup 218
  Skillet Mesquite Cornbread 210
  Stuffed Courgette Flowers 204
  Summer Fruit Coulis 204
  Sweet and Spicy Turnips *236*, 237
  Swiss Chard with Cannelini Beans and Chilli 234
  Wild Garlic Kimchi 193
  Wild Green Pesto 206–7
  Wild Pesto 190
red algae (*Rhodophyta*) 24, 27, 213
red cedar (*Juniperus virginiana*) 38
red chicory (radicchio) *232*, 233
Red House, Bexleyheath 70
red mangrove (*Rhizophora mangle*) 93, 101
redcurrants 204
Redouté, Pierre-Joseph 78, 79, 197, 308, *312*, 313
redwood (*Sequoia sempervirens*) 38, 93
Redzepi, René 184
reed, common (*Phragmites australis*) 179
reindeer lichen (*Cladonia rangiferina*) 52
reishi mushrooms 278
Renaissance 80, 91
Rensten, John 184, 187
reproduction 115
Repton, Humphry 72
Reseda luteola (weld) 176
respiration 112
Resting in the Garden (Bonnard) 70
rhassoul 292
Rheum rhabarbarum (rhubarb) 97, 102, 138, 189
Rhinanthus minor (yellow rattle) 137
rhizome 94
Rhizophora mangle (red mangrove) 93, 101
Rhizophora (mangrove tree) 56
rhododendron (*Rhododendron*) 51
Rhodophyta see red algae
RHS *Botany for Gardeners* 121
RHS *Latin for Gardeners* 121
rhubarb (*Rheum rhabarbarum*) 97, 102, 138, 189
Ribes nigrum (blackcurrant) 203
ribwort plantain (*Plantago lanceolata*) 171
Riccia fluitans (aquatic liverwort) 30
rice 65, 102

rice, Asian (*Oryza sativa*) 42, 56, 62, 102, 137
Richards, Albert G. 336
Ricinus communis (castor-oil plant) 101
Rix, Martyn 312, 319
Robinson, William 70, 72
Rocha, Simona 72
rocket/arugula (*Eruca sativa*) 207
Rollinia deliciosa (biriba/ wild-sugar apple) 46
Roman chamomile (*Chamaemelum nobile*) 273
Root, Abiah 388
root ginger 191, 193, 212, 230, 251
roots 65, 93, 115, 271
roots, edible 168
Rosa canina (dog rose) 97, 102, *120*, 311
Rosa caucasea 322
Rosa fedtschenkoana (Russian rose) 124
Rosa gallica (French rose) 78, 124
Rosa genus (wild/ornamental roses) 130
Rosa 'KORbin' (iceberg rose) 134
Rosa 'Madame A. Meilland' (Peace rose) 124
Rosa moschata (Musk rose) 124
Rosa rugosa f. alba (Japanese rose with white flowers) 134
Rosa rugosa/*Rosa canina* (rosehip) 163, 218, 278, 295
Rosa x damascena (Damask rose) 121, 124, 291
Rosaceae (rose family) 129, *329*
Rosarum monographia: or a botanical history of Roses (Lindley) 323
rose, Damask (*Rosa x damascena*) 121, 124, 291
rose, dog (*Rosa canina*) 97, 102, *120*, 311
rose family (Rosaceae) 129
rose, French (*Rosa gallica*) 78, 124
rose geranium (*Pelargonium graveolens*) 291, 295
rose, guelder (*Viburnum opulus*) 51, 157
rose, iceberg (*Rosa* 'KORbin') 134
rose, Japanese with white flowers (*Rosa rugosa f. alba*) 134
rose, Musk (*Rosa moschata*) 124
rose otto (*Rosa damascena*) 291
rose, Peace (*Rosa* 'Madame A. Meilland') 124
rose, Portland (*rosa* 'Comte de Chambord') 151
rose (*Rosa*) 45, 72, 98, 124, 205, 214, 241, 282, 283, *286*, 287, *290*, 293, 336
rose, Russian (*Rosa fedtschenkoana*) 124
rosehip (*Rosa rugosa*/*Rosa canina*) 163, 218, 278, 295

rosemary, prostrate (*Rosmarinus officinalis* 'Prostratus Group') 158
rosemary (*Rosimarinus officinalis*) 97, 138, *139*, 141, 225, 246, *247*, 282, 284, 288, 301, 302, 304
rosewood (*Antiba rosaeodora*) 295
Rosie Sanders' Flowers (Sanders) 328
Rossetti, Dante Gabriel 369
rosy periwinkle (*Catharanthus roseus*) 51
'rotten flesh' corpse flower (*Amorphophallus titanum*) 75
rough-stalked feather moss (*Brachythecium rutabulum*) 31
Rousseau, Jean-Jacques 76
Royal Botanic Garden, Edinburgh 75
Royal Botanic Garden, Sydney 75
Royal Botanic Gardens, Kew 58, 69, 75, 76, 79, 86, 121, 313, 314, 326, 335
Royal Horticultural Society (RHS) 72, 134, 141
rubber tree (*Hevea brasiliensis*) 51, 62
Rubia tinctorum (common madder) 176
Rubus chamaemorus (cloudberry) 52
Rubus fruticosus (blackberry) *162*, 163, 217
Rubus (raspberry) 102, 204, 250
Rudbeckia 54
Rudbeckia fulgida var. *sullivantii* 'Goldsturm' (black-eyed susan) 155
rudis (wild) 133
Rumes obtusifolius (dock leaf) 179
Rumex acetosa (sorrel) 203
Rumora adiantiformis (leatherleaf fern) 35
runner bean (*Phaseolus coccineus*) 101
Ruscus aculeatus (butcher's broom) 94
Ruskin, John 369
Russell, Ben 366
Russian rose (*Rosa fedtschenkoana*) 124
Ruysch, Frederik 348
Ruysch, Rachel 308, 348, 352
  Flowers in a Glass Vase with a Tulip 348
  Still-Life with Flowers 73
rye, Canada wild (*Elymus canadensis*) 55
rye flakes 225
Ryhanen, Sarah 72

Saccharum see sugar cane
Sachs, Julius von 111
Sackville-West, Vita 72, *380*, 381
  the *Garden* 381
  The Land 381

Some Flowers 381
Safdie, Moshe *383*
saffron (*Crocus sativus*) 281
sage, clary (*Salvia sclarea*) 138
sage, desert (*Salvia eremostachya*) 56
sage (*Salvia officinalis*) 205, 281, 282, 304
sage, white 287
St John's wort (*Hypericum perforatum*) 281, 296
St Petersburg Botanical Garden 75
Salgado, Sebastião *338, 339*
Salicornia europaea (marsh samphire) 168
Salix pulchra (diamond-leaf willow) 52
Salix (pussy willow) 72, 97, *178, 179*
salt cedar 60
salts 263
Salvia eremostachya (desert sage) 56
Salvia genus (common/sacred white sage, chia) 130
Salvia hispanica (chia) 238, 249, 251
Salvia officinalis (sage) 205, 281, 282, 304
Salvia sclarea (clary sage) 138, 291
Sambucus nigra (elderflower) 167, 197, 214, 226, 227, 241, 242, 273, 278, *279*
samphire, marsh (*Salicornia europaea*) 168
sandalwood (*Santalum album*) 301, 302
Sanders, Rosie 79, 308, *328*
sangueneus (blood-red) 133
Sanguisorba officinalis (great burnet) 158
Sansevieria trifasciata (snake plant) 42
Santa Rita prickly pear (*Opuntia santa-rita*) *57*
Santalum album (sandalwood) 301, 302
São Paulo Instituto de Botânica 328
Sargassum 27
sassafras (*Sassafras*) 36
Sassen, Viviane 80, 308, 340, *341*
Mud and Lotus (exhibition, 2017) 340, *341*
Saussure, Nicolas-Théodore de 111
Savage Beauty (exhibition, 2015) *376, 377*
savannah *54*, 55
Saxifraga caespitosa (tufted saxifrage) 52
saxifrage (Saxifragaceae) 129
saxifrage, tufted (*Saxifraga caespitosa*) 52
Scabiosa cretica (Pincushion flower seed) *81*

scallions / spring onions (*Allium cepa* var. *cepa*) 189
Schefflera arboricola (dwarf umbrella tree) 97
Schizachyrium scoparium (little bluestem) 55
Schwartzberg, Louie 80, *343*
scizocarp 102
Scott, Katie 79, 83, *127*, 308, *314, 315*
sculpture and installation 356–67
sea beet (*Beta vulgaris* subsp. *maritima*) 168
Sea Island cotton (*Gossypium barbadense*) 67
sea lettuce *28*, 193
seaweed 212, *213*, 299
Seba, Albertus 314
The Secret Garden - 100 Floral Radiographs (Richards) 336
Secuoidendron gigonteum (giant sequoia) *113*
sedges (Juncaceae) 52, 56
sedum (*Sedum*) 138
seed ferns 28
seed producing plants (Spermatophytes) 24, 36
seeds 65, 101, 164, 206, 212, 271
Seeds: Time Capsules of Life (Kesseler) 335
Selaginella see spike moss
Semff, Michael 351
sempervirens (evergreen) 133
Sempervivum (hen and chick/house leek) 45, 56, *57*
Senebier, Jean 111
Senegal gum acacia (*Acacia senegal*) 55
sequoia, giant (*Secuoidendron gigonteum*) *38*, *113*
Sequoia sempervirens see redwood
sesame seeds 212
sessile 97
shagbark (*Carya ovata*) 164
shallot (*Allium cepa* var. *aggregatum*) 42, *131*, 137
shallot, ornamental (*Allium hollandicum*) *131*
Shapiro, David 351
shea butter 293, 296
shea tree (*Vitellaria paradoxa*) 267
The Sheaf 345
sheet moss (*Hypnum iponens*) 31
shellbark (*Carya lacinios*) 164
Shepherd, Jess *320*, *321*
Sherwood, Dr Shirley 312
Shipman, Ellen Biddle 381
shiso leaf 212
shoots 93
shrubs 51, 56
Siberian dogwood (*Cornus alba*) 141
sideoats grama (*Bouteloua curtipendula*) 55
Sidney Jannis Gallery, Manhattan 351
Sierra Club 83, 181

Sieur de La Varenne, François Pierre 190
silkworms 62
Silurian period 28
silver birch (*Betula pendula*) 151
silver grass (*Miscanthus sinensis* 'Silberfeder') 158
silver maple (*Acer saccharinum*) 56
silver-leaved box (*Eucalyptus pruinosa*) 316
Simblet, Sarah 121
Simon, Taryn *342*
simple fruit 102
simplex (unbranched) 133
Simpson, Niki *329*
Simpson, Tim 333
Singapore Botanical Gardens 75
single celled organisms (Archaea & Bacteria) 24, 27
Sissinghurst Castle, Kent 381
Six Gallery reading (1955) 181
sloe (*Prunus spinosa*) 163, 243
Smithson, Robert 360
Spiral Jetty 360
smoke bush, European (*Cotinus coggyria*) 94, 155
smooth crocus (*Crocus laevigatus*) 141
smooth hornwort (*Phaeoceros laevis*) 30
smoothies *see* drinks and bowls
Smyrnium olusatrum (Alexanders) 168
snake plant (*Sansevieria trifasciata*) 42
snapdragons (*Antirrhinum*) 179, 311
snowdrop (*Galanthus*) 98, *99*
snowdrop, greater (*Galanthus eluvesii* var. *monostictus*) 134
Snyder, Gary 181
soaptree yucca (*Yucca elata*) 56
Socratea exorrhiza (walking palm) 51
Solanaceae (potato family) 129
Solanum genus (potato, tomato, aubergine/eggplant) 130
Solanum lycopersicum (tomato) 65, 93, 130, 134, *202*, 203
Solanum melongena (aubergines or eggplants) 65, 104, 130
Solanum nigrum (black nightshade) 130
Solanum tuberosum (potato) 45, 65, 94, 102, 115, 130, 189, *208*, 209
Solidago (goldenrods) 55
Sorghastrum nutans (Indian grass) 55
sorrel (*Rumex acetosa*) 203, 206, 207
soursop (*Annona muricata*) 46
south giant horsetail (*Equisetum giganteum*) 34
Sowerby, James 79, 184
spadix 98
Spanish moss (*Tilandsia usneaides*) *57*

spathe 117
Spathiphyllum wallisii (peace lily) 42, 141, 151
spear-leaved orache (*Atriplex prostrata*) 97
species 133
Species Plantarum (*The Species of Plants*) (Linnaeus) 75, 123
A Specimen of the Botany of New Holland 79
Spermatophytes *see* seed-producing plants
Sphagnum (peat moss) 31
spider plant (*Chlorophytum comosum*) 42, 93
spike 98
spike lavender (*Lavendula latifolia*) 98
spike mosses (*Selaginella*) 33, *34*
spikenard (*Nardostachys jatamansi*) 281
spinach (*Spinacia oleracea*) 65, 249, 250
spiral aloe (*Aloe polyphylla*) 311
spores 32, 101
Sporobolus heterolepis (prairie dropseed) 55
spring crocus (*Crocus vernus*) 51
spring greens (cultivar of *Brassica oleracea*) 189
spring onions 230
Spring plants/recipes 188–201
springfire shrub (*Metrosideros collino*) 77
spruce (*Picea*) 52, 97
Spry, Constance 72
squash (*Cucurbita pepo* var. *pepo*) 65, 94
staghorn fern (*Platycerium*) 33, 35
stalks 97, 168
star anise (*Illicium verum*) *49*, 226, 237
star magnolia (*Magnolia stellata*) 133
Starkey, Flora 72, *376*, *377*
State of the World's Plants report (2017) 63, 261, 392
Stearn, William T. 312
Stellaria media (chickweed) 137, 168
stellaris (star-like) 133
stems 65, 94
Stenocereus thurberi (organ pipe cactus) 56
Stephenson, John 278
Stern's hybrid hellebore (*Helleborus x sterni*) 134
Stevens, Margaret 312
stinging nettle *see* nettle (*Urtica dioica*)
stomata *108*
Story of Flowers (Botanicum Animation) 314
Strand, Paul 348
stranglers (*Ficus*) 51
strawberry (*Fragaria x ananassa*) 45, 102, *103*, 203, 250

strawberry tree (*Arbutus unedo*) 157
strawberry, wild (*Fragaria vesca*) 97
Streptophyta see ancestral green algae
Stuppy, Dr Wolfgang 335
style planting 155
subspecies 134
succulents 56
sugar 264
sugar cane (*Saccharum*) 42, 62
sugar maple (*Acer saccharum*) 168
sultanas 237
summer plants/recipes 202–15
sunflower, common (*Flos Solis Maior*) 74
sunflower (*Helianthus*) 45, 98, 102
sunflower (*Helianthus annuus*) 45, 98, 102, 130, 176, 311
sunflower, prairie (*Helianthus petiolaris*) 55
sunflower seeds 206
Sweden, James van 382
sweet almond oil 302
sweet basil (*Ocimum basilicum*) 141
sweet chestnut (*Castanea sativa*) 164, *165*, 217
sweet grass 282
sweet orange (*Citrus sinensis*) 64, 101, 237, 242, *245*, 246, 304
sweet pea (*Lathyrus odoratus*) 10, 45, 98, 101, 137, *142*, 311, *331*
sweet pea 'Matucana' (*Lathyrus odoratus* 'Matucana') 141
sweet potato (*Ipomoea batatas*) 93
sweet vetch (*Hedysarum boreale*) 102
sweet violet (*Violet odorata*) 101, 167
sweetcorn (*Zea mays*) 217
sweetgum tree (*Liquidambar*) 51
Swietenia macrophylla (mahogany) 51
Swiss chard (*Beta vulgaris* subsp. *vulgaris*) 93, 217, 234, *235*, 250
Swiss cheese plant (*Monstera deliciosa*) 42, *43*, 51, 134, 141, *345*
switch grass (*Panicum virgatum*) *139*, 157
sycamore (*Acer pseudoplatanus*) 179
sylvestris (found wild) 133
Symphytum officinale (comfrey) 291
Systema Naturae (Ehret) 313
Systema Naturae (*The System of Nature*) (Linnaeus) 75, 123, 126
Syzygium aromaticum (clove) 226, 243, 301

Tahina spectabilis (Madagascar palm) 41
Tahoka daisy (*Machaeranthera tanacetifolia*) 55
taiga 52, *53*
Tamarix 60
Tanacetum parthenium (feverfew) 171
Taraxacum officinale (dandelion) 93, 168, *169*
taro 65
Tasker, Dain L. 336, *337*
Tasmanian pepper (*Tasmannia lanceolata*) 46
Tate Modern, London 357
Taxodium distichum (bald cypress tree) 57
taxonomy 120, 121, 126, 129
Taxus baccata (European yew) 38, 101
tea (*Camellia sinensis*) 45, 62, *63*, 130
tea, Chinese (*Camellia sinensis* var. *sinensis*) 134
tea, Labrador (*Ledum groenlandicum*) 52
tea tree (*Melaleuca alternifolia*) 291, 292, 303, 304
temperate grassland 55
The Temple of Flora (Thornton) *4*, 309
The Ten Biggest 348
tequila 56
thale-cress (*Arabidopsis thaliana*) 137
Thalictrum delavayi (Chinese meadow rue) 152, *153*
Theobroma cacao (cocoa been) 51, 130, *252*, 253
Theophrastus 22, 58, 62, 66, 89, 91, 121, 123, 168
This is the American Earth (Sierra Club) 339
thistle, prickly Russian (*Kali tragus*) 56
Thompson, Anne Blackwell 389
Thoreau, Henry David 161, 165, 181, 182
Thornton, Robert J. *4*, *122*, 309
thyme (*Thymus vulgaris*) 151, 190, 191, 205, 225, 245, *265*, 291, 304
Tilandsia usneaides (Spanish moss) 57
Tilia x europaea (linden or common lime) 51, *106*, 164, *165*, 281
Tillandsia (air plant) 42
tinctorius (used for dyeing) 133
titan arum (*Amorphophallus titanum*) *19*, 75
tobacco plant (*Nicotiana tabacum*) 62
tomato (*Solanum lycopersicum*) 65, 93, 130, 134, *202*, 203
Tordylium (hartwort) 124
Trachaeophytes (Vascular Land Plants) 24
trade 45, 66
transpiration 112
Travers, Kitty 187, 214

tree fern species, larger (*Dicksonia/Cyathea*) 35
tree houseleek 57
trees 51, 56, *88*
Trifolium repens (clover) 97, 179
Trigonella foenum-graecum (fenugreek) 101
Triticum aestivum (most common wheat) 42, 137
The Triumph of Seeds (Hanson) 101
Tropaeolum (nasturtium) 45
tropical pitcher plants (*Nepenthaceae*) 51
true indigo (*Indigofera tinctoria*) 176
Tscherman von Seysenegg, Erich 143
tuber 94, 102
tufted hair grass (*Deschampsia cespitosa*) 158
tufted saxifrage (*Saxifraga caespitosa*) 52
tulip tree (*Liriodendron*) 46
tulip (*Tulipa*) 42, 62, 77, 87, 94, 98, 155, *309*, 336
Tulipa genus (lady, eyed, garden tulips) 130
Tulip mania 58, 72, *73*, 155
tumbleweed, common (*Kali tragus*) 56
tundra 52
Turk, Gavin 308
Turkish Tulips 308
Turkish Tulips (exhibition, 2017) 308
turmeric (*Curcuma longa*) 51, 254, *2/3*
Turner Contemporary, Margate 351
Turner, Tiffanie 375
  Cremon Mum 375
  The Fine Art of Paper Flowers 375
turnip (*Brassica rapa* subsp. *rapa*) 229, *236*, 237
Twining, Elizabeth 158
Typha minima (bulrush/cattail) 158

Ulothrix zonata (green algae) 325
Ultramarine 341
umbrella liverwort (*Manchantia polymorpha*) 30
umbrella thorn (*Vachellia tortilis*) 55
umbrella tree, dwarf (*Schefflera arboricola*) 97
understorey 51
undulatus (wavy) 133
United Nations 124
United States Botanic Garden 75
United States Department of Agriculture (USDA) 141
United States National Herbarium, Washington 121
University of Oxford Botanic Garden 69, 75

University of Pisa 75
Uppsala University, Sweden 69, 70, 75
Urtica dioica (nettle) 97, 168, 189, 190, 194, 273, 295

Vaccinium corymbosum (blueberry) 151, 238, 249, 250, 251
Vaccinium myrtillus (bilberry/whortleberry) 163
Vaccinium oxycoccus (cranberry) 52, 163
Vaccinium vitis-idaea (ligonberry) 52
Vachellia tortilis (umbrella thorn) 55
vacuole 104
The Vacuum at Kunsthalle Winterthur (exhibition, 2010) 352
valerian (*Valeriana officinalis*) 138, 311
Van Niel, Cornelius 111
Vanek, Bernadett 16, 187, 194
vanilla orchid (*Vanilla planifolia*) 51
vanilla pod 219, 283
variation 134
variegatus (variegated) 133
Vasco da Gama 66
vascular land plants (*Trachaeophytes*) 24, 25, 32–6
vascular spore producers (*Pteridophytes*) 33
  diploid generation (sporophyte) 33
  haploid generation (gametophyte) 33
Vedas 65
vegetable glycerin 264
vegetables 65
Venus flytrap (*Dyonaea muscipula*) 118
Verbena bonariensis (Argentinian vervain) 157
Verbena bonariensis (purpletop verbena) 138, 214
Vernonnet 70
Versailles, France 69, 313
Vertical Forest (Bosco Verticale) 383
vervain, Argentinian (*Verbena bonariensis*) 157
vervain (*Viberna officinalis*) 281
vetiver (*Vetiveria zizanioides*) 281, 288
Viberna officinalis (vervain) 281
viburnum, forked (*Viburnum furcatum*) 141
Viburnum opulus (guelder rose) 51, 157
viburnum (*Viburnum*) 97
Vicia faba (broad bean) 203
Victoria and Albert Museum, London 376
Victoria amazonica (Amazon water lily) 51
Victoria Park, London 69

Vigna radiata (mung bean) 93
vine 51
viola heartsease (*Viola tricolor*) 45
viola (*Viola*) 94, 97, 205
violet family (Violaceae) 129
violet, sweet (*Violet odorata*) 101, 167, 205, 206
virens (green) 133
viridescens (becoming green) 133
Viridiplantae (green plants) 25
Viscum album (European mistletoe) 51
vitamin E 295
Vitellaria paradoxa (shea tree) *267*
Vitis (grape) 94
Vitis vinifera (common grapevine) 62, 152
Vogue magazine 336, 376
Vries, Marie de 143
vulgaris (common) 133

Walker, Kim 296
walking palm (*Socratea exorrhiza*) 51
walnut, English (*Juglans regia*) 51, 97, 190, 233
walnut, white (*Juglans cinerea*) 164
Ware, Mary Lee 375
Warhol, Andy 354
Warm Springs Reservation, Oregon 181
Wat Mahathat, Thailand *82*, 83
water lily, Amazon (*Victoria amazonica*) 51, 70
water lily, blue Egyptian (*Nymphaea caerulea*) 49
water lily, dwarf (*Nymphaea candida*) 114
water lily, fragrant white (*Nymphaea odorata*) 56, *57*
water lily, giant (*Nymphaea gigantea*) 49, *315*
water lily, white (*Nymphaea alba*) 48, 49
water lily, yellow (*Nymphaea mexicana*) 49
water lotus 45
watercress 207
watermelon (*Citrullus lanatus* var. *lanatus*) 137
Waters, Alice 184
waxes 263
Weddell, H. 278
Weenix, Jan, *Flowers on a Fountain with a Peacock* 352
Weiwei, Ai *see* Ai Weiwei
weld (*Reseda luteola*) 176
wellbeing 280
   Breath in theForest Salt Bath and Balm 288
   Inner Radiance Ritual Bath 287
   Love Potion 283
   Moon Bath 282
   RitualTea 283
   SmudgeStick 282
   Warming Body oil 284

Welwitschia mirabilis (gnetophyte) 39, *323*
West Ham Park, London 386
western sword fern (*Polystichum munitum*) 35
western wild ginger (*Asarum caudatum*) 46
wetlands 56, *159*
What a Plant Knows: A Field Guide to the Senses (Chamovitz) 118
wheat, common (*Triticum aestivum*) 42, 137
wheat (*Triticum*) 62, 65, 98, 102
wheatgerm 288, 292
white Egyptian lotus (*Nymphaea lotus*) 49
white mulberry (*Morus alba*) 62
white oak (*Quercus alba*) 51, 97
white walnut (*Juglans cinerea*) 164
white water lily, fragrant (*Nymphaea odorata*) 56, *57*
white water lily (*Nymphaea alba*) 49
Whitman, Walt 47
whortleberry/bilberry (*Vaccinium myrtillus*) 163
wild bananas (*Musa*) 51
wild burdock (*Arctium*) 168
wild carrot (*Daucus carota*) 98, 124, 175
wild cherry (*Prunus avium*) 163
wild chervil *see* cow parsley
wild cinnamon (*Canella winterana*) 46
Wild Flowers of Britain (Phillips) 80
Wild Food (Phillips) 80
Wild Fruits (Thoreau) 161
The Wild Garden (Robinson) 70
wild garlic (*Allium ursinum*) 51, 167
wild onion (*Allium canadense*) 130
wild strawberry (*Fragaria vesca*) 97, 158, 163
wild-beaked parsley *see* cow parsley
wild-sugar apple (*Rollinia deliciosa*) 46
Wilhelm, G.T. 177
Williams, Tennessee 384
Willis, Kathy 63, 86, 314
willow, desert (*Chilopsis*) 56
willow, diamond-leaf (*Salix pulchra*) 52
willow (*Salix*) 97, 111
Wings of Life (Disneynature film) 343
winter plants/recipes 228–38
winter's bark (*Drimys winteri*) 46
wisteria, Chinese (*Wisteria sinensis*) 138, 152
witch hazel 264, 295
Witt, Sarah 187, 210
woad (*Isatis tinctoria*) 176, *177*
Woese, Carl 126
Wollheim, Bruno 355

The Wondergarden of Nature (Blossfeldt) 333
wood anemone (*Anemone nemorosa*) 51
wood forget-me-not (*Myrosotis sylvatica*) 137
wood, polishing 34
wood sorrel family (Oxalidaceae) 129
wood sorrel (*Oxalis acetosella*) 97
Woodward, John 86, 111
woody liana (*Austrobaileya scandens*) 49
woody plants 138
Woolsthorpe Manor, Lincolnshire 70
Wordsworth, William 70, 316

xenogamy 98
xylem 107

Yamasaki, Rika *388*
yarrow (*Achillea millefolium*) 138, 171, 273
Yayoi Kusama *see* Kusama, Yayoi
yellow pear tomato *see* tomato
yellow rattle (*Rhinanthus minor*) 137
yellow water lily (*Nymphaea mexicana*) 49
Yellowstone National Park 69
yew, European (*Taxus baccata*) 38, 101
ylang ylang (*Cananga odorata*) 46, 301
Yosemite National Park, California 181, 339
Yucca brevifolia (Joshua tree) 56, 342
Yucca elata (soaptree yucca) 56

Zamia pungens (cycad) 38
Zantedeschia (calla lily) *43*, 93
Zavatto, Amy 187
Zea mays subsp. *mays* (maize) 42, 65, 93, 98, 137
Zea mays (sweetcorn) 217
Zen Buddhism 181
Zingiber officinale (ginger) 51, 94, 241, 261, 273, *276*, 283, 284
zinnia, common (*Zinnia elegans*) 72, 83, 137

PICTURE CREDITS

All reasonable efforts have been made by the author and publishers to trace the copyright owners of the material quoted in this book and of any images reproduced in this book. In the event that the author or publishers are notified of any mistakes or omissions by copyright owners after publication, the author and publishers will endeavour to rectify the position accordingly for any subsequent printing.

Key: t: top, b: below, l: left, r: right.

123RF: 56, 82;

Alamy: 29, 30l, 30r, 34l, 38r, 39b, 40, 42, 43t, 43bl, 54tl, 54tr, 77tl, 77b, 78, 99tl, 99tr, 103, 107l, 108l, 127tr, 130, 135b, 137, 139tr, 139b, 142, 165l, 180, 188, 192, 211, 215, 216, 227, 235, 240, 290, 316l, 317l, 318, 323r, 325, 382r;

Amy Zavatto: Photography by Claire Lloyd Davies ©HarperCollinsPublishers (2015) 247;

Andrew Montgomery: ©Quadrille Publishing, Ltd (2016) 224;

Anne Blackwell Thompson (www.blackwellbotanicals.com): 389l;

Anne Ten Donkelaar (www.anneten.nl): 366l;

Arcangel: 43;

Archive.org: /Missouri Botanical Garden, *New Illustration of the Sexual System of Carolus von Linnaeus and the Temple of Flora, or Garden of Nature* by Robert Thornton (1807) 4, 122, 309; *Plantae Selectae Quarum Imagines ad Exemplaria Naturalia Londini, in Hortis Curiosorum Nutritamanu Artificiosa Doctaque* by Georg Dionysius Ehret, Christoph Jacob Trew and Johann Jacob Haid (1750) 31r; / *The Anatomy of Plants: With an idea of a philosophical history of plants: and several other lectures read before the Royal Society* by Nehemiah Grew (1682) 317r; / *Rosarum Monographia; or, A Botanical History of Roses* by John Lindley (1820) 322r;

Azuma Macoto: ©AMKK. Photography by Shunsuke Shiinoki 358, 359;

Ben Russell (www.benrussell.com): 366r;

Bernadett Vanek: 16t, 195;

Biodiversity Heritage Library: Missouri Botanical Garden /*L' Illustration horticole* Vol 38: t. 120 (1891) 128; /*Curtis's Botanical Magazine* Vol 15-16 39t; / University Library, University of Illinois Urbana-Champaign 45, 108, /Natural History Museum Library, London/*Musa paradisiaca*, Illustration by Jean Theodore Descourtilz, Paris 1834 139tl;

Bobbi Angell (www.bobbiangell.com): 329l;

Bridgeman Images: Magnolia Grandiflora, 2003 (w/c on paper), Barron, Jenny (Contemporary Artist), / Private Collection /©Jenny Barron 47t; /Private Collection /©Look and Learn 48; / Granger 64; /Private Collection /Photo ©Christie's Images /©Ellsworth Kelly Foundation 350;

©The Trustees of the British Museum: 44;

Brown Brothers Continental Nurseries Catalogue 1909: 135;

©The Cecil Beaton Studio Archive at Sotheby's: 380;

Charlie Lawler of Looseleaf (www.looseleafstore.com.au): 378;

Claire Lloyd Davies: ©HarperCollinsPublishers (2015) 247;

The Corning Museum of Glass, photo by: 374;

© Emily Kame Kngwarreye/Copyright Agency. Licensed by DACS 2018 354;

Flora Starkey: 377; /Photography by Yolanda Chiaramello 376;

Freer Gallery of Art and Arthur M. Sackler Gallery, Smithsonian Institution, Washington, D.C.: Gift of Charles Lang Freer, F1916.95, 259;

Getty Images: 12, 15t, 15b, 19, 37, 38l, 50, 53, 54b, 57tl, 57tr, 57b, 61, 63t, 68, 71, 73b, 73t, 74, 77tr, 81t, 87, 90, 92, 95t, 95b, 99b, 100tl, 100tr, 100b, 106, 109, 110, 112, 113, 114tl, 114tr, 114b, 116, 117t, 117b, 125, 127bl, 129, 131, 132, 140, 150, 154tr, 156, 162, 165b, 170, 173t, 173b, 177tl, 177tr, 177b, 178tr, 178b, 185, 199, 202, 223, 252, 265, 266, 269, 272, 276, 279, 280, 289, 294, 297, 298, 300, 305, 339, 344, © Succession H. Matisse/ DACS 2018/ Walter Carone 345, 355, 357, 368, 381, 382l, 383l, 383r;

Haarkon (www.haarkon.co.uk): 343;

Hagemann & Partner Bildungsmedien: 324;

The Hilma af Klint Foundation (www.hilmaafklint.se), courtesy of: Photo Albin Dahlström/Moderna Museet, Stockholm, Sweden: 349;

Houghton Library: MS Am 1118.11, © President and Fellows of Harvard College 388l;

©House of Hackney: 'Palmeral print', 154b;

©Inky Leaves Publishing: Jess Shepherd, 320, 321;

i-stock: 32, 34r, 228, 245, 248, 262;

Jenny Kiker: 370, 371;

Joseph Bellows Gallery, courtesy of: 337;

Katie Scott: ©Katie Scott 315t; ©Big Picture Press, 2016: 127tl, 314, 315b;

Kwang-Ho Lee: Images courtesy of Johyun Gallery: 346, 347;

Kyle Books: ©Nina Olsson, 2017 208;

©Lindsey Carr (www.littlerobot.org.uk): 159;

Lucy Augé (www.lucyauge.co.uk): 369;

©Lynn Hatzius (www.lynnhatzius.com): 24, 58, 88, 120, 148, 160, 186, 260, 310, 373l;

Macoto Murayama: 330; Image courtesy of Frantic Gallery, 331;

Mama Medicine: Photography ©Ashley Glynn (2018) 16bl, 286;

The Metropolitan Museum of Art: / Warner Communications Inc. Purchase Fund, 1978. Accession number 1978.602.2 332;

Miria Harris (www.miriaharris.com): 153, 154tl; ©Miria and Tom Harris, stylist: Miria and Tom Harris 213, 220, 232; ©Miria Harris and Ali Mobasser, 2013, photography by Ali Mobasser 379;

MR Studio London (www.mrstudiolondon.co.uk): 388r;

NASA: Landsat imagery courtesy of NASA Goddard Space Flight Center and U.S. Geological Survey: 26;

©The Trustees of the Natural History Museum, London: 316r, 328l;

New York Public Library Digital Collections: /*Phalaenopsis grandiflora aurea* illustrated by Joseph Mansell (1888–1894) 133; /*Traité des Arbres et Arbustes que L'on Cultive en France en Pleine Terre* illustrated by Pierre Joseph Redouté (1801–1819) 'Magnolia discolor = Magnolier bicolore' 196, 'Cluster of Date Palms' 239; /*Pomona Italiana, Vol I* by Giorgio Gallesio (1817–1839) 200; /*Hortus Romanus Juxta Systems Tournefortianum Paulo* by Giorgio Bonelli (1772–1793) 207, /*A Curious Herbal* by Elizabeth Blackwell (1737–1739) 'Black Pepper' 285; /*Les Liliacées* by Pierre Joseph Redouté (1805–1816) 'Iris pumila floribus caeruleis' 312, /*Photographs of British Algae* by Anna Atkins (1853) 333;

Nicola Cunningham (www.stemapothecarystore.com): 389r;

Niki Simpson (www.nikisimpson.co.uk): 329r;

NILS-UDO (www.nils-udo.com): 362t;

Nina Olsson: ©Nina Olsson, 2017, 208;

Phoebe Cummings (www.phoebecummings.com): Photography by Sylvain Deleu 385;

Pia Ostlund: 373r;

plantcurator.com: /*Medizinal-Pflanzen (Medicinal Plants)*, Vol I by Franz Eugen Köhler (1887), 35, 166, 169, 322; Vol II, 63, 67, 136; /*Kunst-Formen der Natur (Art Forms of Nature)* by Ernst Haeckel (1899–1904) 319;

Plants on Pink (www.plantsonpink.com): Photography by Lotte van Baalen 378l;

©Quadrille Publishing Ltd: Photography by Andrew Montgomery, 224

Rachel Dein (www.racheldein.com): 384;

Rachel Pedder-Smith: 326, 327;

Rebecca Louise Law: Photo by Fabio Affuso 364; Photo by Katherine Mager and Broadway Gallery 365t; Photo by Charles Emerson, 365br; Photo by Rachel Warne, 365bl;

©Rob Kesseler: 81b, 334t, 334br; /from *Pollen the Hidden Sexuality of Flowers* by Rob Kesseler and Madeline Harley (Papadakis Publisher, 2014) 334bl;

Rosie Sanders (www.rosiesanders.com): 328r;

©The Board of Trustees of the Royal Botanic Gardens, Kew: 47b, 119, 322l, 323l;

Samuel Levack and Jennifer Lewandowski: (www.levacklewandowski.com): 342l;

Sangtae Kim, Penn State: 23;

Science Photo Library: 28, 104;

Seana Gavin (www.seanagavin.com): 372;

©Sebastiao Salgado/Amazonas/nbpictures: 338;

Shantanu Starick: ©Shantanu Starick, 2016, 236;

Shutterstock: 20, 21, 31, 46, 49, 55, 84, 85, 96, 107r, 141, 144, 145, 178tl, 182, 183, 256, 257, 306, 307;

Simon Heijdens (www.simonheijdens.com): 361;

Solomon R. Guggenheim Museum, New York 61.1589. ©2018 Mondrian/Holtzman Trust: 351;

©Sonya Patel Ellis: 6, 10, 147, 174, 255, 387, /Stylist: Sonya Patel Ellis: 255 / Photography by Simon Pask 9; Photography by Carmel King 386

studio herman de vries Joana Schwender 2015, private collection, Germany: 362b, 363;

©Taryn Simon Projects (www.tarynsimon.com): *Memorandum of Understanding between the Royal Government of Cambodia and the Government of Australia Relating to the Settlement of Refugees in Cambodia*

Ministry of Interior, Phnom Penh, Cambodia, 26 September, 2014.

Paperwork and the Will of Capital, 2015. Framed text and archival inkjet prints, 215.9 x 186.1 x 7cm (85 x 73.25 x 2.75 inches), 342r;

Tiffanie Turner (www.papelsf.com): Photo ©Scott Chernis 375;

©Tom Ellis: 352, 353b; Photo by Rob Murray 353t;

Tom Harris: ©Miria & Tom Harris, stylist: Miria & Tom Harris 213, 220, 232;

Viviane Sassen: 16br, 341;

Will Cruickshank (www.willcruickshank.net): 367;

©YAYOI KUSAMA: 356.